Clifford Algebras and Zeons

Geometry to Combinatorics and Beyond

Clifford Algebras and Zeons

Geometry to Combinatorics and Beyond

George Stacey Staples
Southern Illinois University Edwardsville, USA

World Scientific

NEW JERSEY • LONDON • SINGAPORE • BEIJING • SHANGHAI • HONG KONG • TAIPEI • CHENNAI • TOKYO

Published by

World Scientific Publishing Co. Pte. Ltd.

5 Toh Tuck Link, Singapore 596224

USA office: 27 Warren Street, Suite 401-402, Hackensack, NJ 07601

UK office: 57 Shelton Street, Covent Garden, London WC2H 9HE

Library of Congress Control Number: 2019027697

British Library Cataloguing-in-Publication Data
A catalogue record for this book is available from the British Library.

CLIFFORD ALGEBRAS AND ZEONS
Geometry to Combinatorics and Beyond

ISBN 978-981-120-257-5

For any available supplementary material, please visit
https://www.worldscientific.com/worldscibooks/10.1142/11340#t=suppl

For Nancy, Josh, Arsenal, Sparkle, Hancock, and Pecan.

Preface

Clifford algebras have many well-known applications in physics, engineering, and computer graphics. Zeon algebras are subalgebras of Clifford algebras whose combinatorial properties lend them to graph-theoretic applications such as enumerating minimal cost paths in dynamic networks. The goal of this book is to provide foundational knowledge of zeon algebras, their properties, and their potential applications in an increasingly technological world. The material is organized in a careful progression from geometry to combinatorics and beyond.

Designed as a graduate- or advanced undergraduate-level mathematics textbook, it is also suitable for self-study by researchers interested in new approaches to existing combinatorial problems and applications (e.g., wireless networks, Boolean satisfiability, coding theory, etc.).

Acknowledgments

Since development of the CliffMath package for *Mathematica* began in 2008, I have received comments, suggestions, and feedback from a number of people. I am particularly appreciative of Arturas Acus and Marco Budinich for their insights and suggestions. I have also benefited from conversations with Rafal Ablamowicz and Stephen Sangwine.

I would also like to acknowledge my master's students: Cody Cassiday, Amanda Davis, Lisa Dollar, Erin Haake, Glenn Harris, Theresa Lindell, Tiffany Stellhorn, Alex Weygandt, and David Wylie. Beyond our joint papers and understanding gained from our collaborations, the necessity of being able to computationally verify our guesses and conjectures has also contributed to the development and improvement of CliffMath code.

Contents

Introduction

I usually teach this material as a graduate-level "Topics in Algebra" course. The course touches on several areas of mathematics, allowing flexibility in prerequisites. One semester of algebraic structures and one semester of senior-level linear algebra is the minimum. Ideally, this would be supplemented with a semester of real analysis and some background in combinatorics. Some useful references are itemized by category in Table 0.1.

Table 0.1 Useful Background References

Subject Area	References
Clifford (geometric) algebras	[1], [9], [67], [80]
Abstract algebra	[46], [84], [109]
Linear algebra	[47]
Analysis and operator theory	[5], [48]
Combinatorics and graph theory	[25], [107], [112], [113]

While Part II consists mostly of material known for many years, the presentation is my own. Parts III and IV are less well known. The earliest material along these lines was published in my doctoral dissertation in 2004. Material accumulated quickly, thanks in part to the wealth of applications known to René Schott, resulting in our 2012 book *Operator Calculus on Graphs* [91].

Whereas Parts II and III represent "Geometry to Combinatorics," Part IV represents the "Beyond," where the interrelationships of geometry, combinatorics, and operator theory are drawn together to provide a quantum probabilistic interpretation of earlier ideas.

Throughout the book, examples have been computed using *Mathematica* with the CliffMath [102] package. The current version, **CliffMath2018**, and documentation can be found through the book's web page `https://www.worldscientific.com/worldscibooks/10.1142/11340` or via the *Research* link at my web page: `http://www.siue.edu/~sstaple`.

Table 0.2 Summary of Notation

Notation	Meaning
$\mathcal{C}\ell_Q(V)$	Clifford algebra of quadratic form Q of V.
$\mathcal{C}\ell_{p,q,r}$	Clifford algebra of signature (p, q, r).
$\mathcal{C}\ell_{p,q}$	Clifford algebra of signature $(p, q, 0)$.
$\mathcal{C}\ell_n$	Euclidean Clifford algebra of $\mathbb{R}^n$; signature $(n, 0, 0)$.
$\mathbf{v}_i$, $\mathbf{v}_{\{i\}}$	Vector: lowercase, bold, single index.
$\mathbf{v}_I$	Multi-index notation for basis blades: $\mathbf{v}_I := \prod_{\ell\in I}\mathbf{v}_\ell$
v_I, v_i	Scalar coefficients in canonical expansions.
$\sharp w$	(Maximal) grade of element w
$\natural u$	Minimal grade of element u
$\langle u\rangle_\ell$	Grade-ℓ part of u
$v \rfloor u$, $v \lfloor u$	Geometric left and right contraction, respectively.
$v \wedge u$	Exterior product.
$\mathbf{x}\vert\mathfrak{w}$	$\mathbf{x}$ "divides" $\mathfrak{w}$; i.e., $\mathfrak{w} = \pm\mathbf{x}\mathfrak{v}$ for decomposable $\mathfrak{v}$, invertible $\mathbf{x}$.
$\tilde{u}$	Reversion of $u \in \mathcal{C}\ell_Q(V)$: $\tilde{u} = \sum_{k=0}^{\dim V}(-1)^{\frac{n(n-1)}{2}}\langle u\rangle_k$
$\hat{u}$	Grade involution: $\hat{u} = \sum_{k=0}^{\dim V}(-1)^k\langle u\rangle_k$
$\overline{u}$	Clifford conjugate: $\overline{u} = \sum_{k=0}^{\dim V}(-1)^{\frac{n(n+1)}{2}}\langle u\rangle_k$
	Sets
$\mathbb{R}^*$	Invertible real numbers, $\mathbb{R}^* := \mathbb{R}\setminus\{0\}$
$\mathbb{N}_0$	Nonnegative integers $\{0, 1, 2, \ldots\}$
$[n]$	The n-set: $[n] = \{1, \ldots, n\}$
$I\triangle J$	Set-symmetric difference $I\triangle J = (I\cup J)\setminus(I\cap J)$
$\vert X\vert$ or $\sharp X$	Cardinality of set X
	Numbers
B_n	The nth Bell number.
B_n^*	The nth Bessel number.
C_n	The nth Catalan number.
$\left\{ {n \atop k} \right\}$	Stirling number of the second kind.
$S_1(n; k)$	Stirling number of the first kind.
	Zeon functions
$\mathfrak{exp}\, u$	Zeon exponential of u.
$\mathfrak{log}\, u$	Zeon logarithm of u.
$\mathfrak{cos}\, u$	Zeon cosine of u.
$\mathfrak{sin}\, u$	Zeon sine of u.
$\mathfrak{tan}\, u$	Zeon tangent of u.
$\mathfrak{cosh}\, u$	Zeon hyperbolic cosine of u.
$\mathfrak{sinh}\, u$	Zeon hyperbolic sine of u.
$\mathfrak{tanh}\, u$	Zeon hyperbolic tangent of u.
	Miscellaneous
$\mathcal{S}_n$	Symmetric group of order $n!$
$\mathbf{u}\,.\,\mathbf{w}$	Concatenation of finite sequences $\mathbf{u}$ and $\mathbf{w}$

PART I
The Essentials

Chapter 1

Algebra

This part of the book is intended as a very brief review of some concepts and terminology that will appear in subsequent chapters. The background material appearing here is fairly standard and can be found in any number of textbooks. A typical linear algebra reference is the text by Friedberg, et al. [47]. Abstract algebra references include the texts by Rotman [84] and Fraleigh [46]. References on Clifford algebras are numerous and not quite as standard.

1.1 Algebraic Structures

Definition 1.1.1. A *group* consists of a set G together with a binary operation $\cdot$ (typically called multiplication) such that the following conditions are satisfied:

(1) G is closed under the operation.
(2) The operation is associative.
(3) There exists an element $e \in G$ such that $eg = g$ for all $g \in G$.
(4) For each element $g \in G$, there exists an element $g' \in G$ such that $g'g = e$. This element will be denoted g^{-1}.

If $g_1 g_2 = g_2 g_1$ holds for all $g_1, g_2 \in G$, the group is said to be *abelian*. In that case, the operation is typically written as addition. The additive identity is written as 0.

Definition 1.1.2. A subgroup H of a group G is said to be *normal* in G if every left coset of H is also a right coset. That is, $gH = Hg$ for every $g \in G$. Equivalently, $gHg^{-1} = H$ when H is a normal subgroup of G. Notation: We write $H < G$ when H is a subgroup of G, and $H \triangleleft G$ when

H is a normal subgroup of G.

Definition 1.1.3. A *ring* R is an additive abelian group with a second binary operation $\cdot$ (typically called multiplication) satisfying:

(1) R is closed under multiplication.
(2) For all $x, y, z \in R$, $x(y+z) = xy + xz$ and $(x+y)z = xz + yz$.

Examples of rings include $\mathbb{Z}$, $n\mathbb{Z}$, $\mathbb{Z}_k$, and rings of polynomials such as $\mathbb{R}[x]$, $\mathbb{C}[x]$, and $\mathbb{Z}[x]$.

Definition 1.1.4. Given a group $(G, \cdot)$ and a commutative ring R with nonzero unity 1, the group ring of G over R is the ring consisting of the set

$$RG = \{rg : r \in R, g \in G\}$$

with the operations of addition and multiplication satisfying:

(1) For all $r_1, r_2 \in R$ and $g \in G$, $r_1 g + r_2 g = (r_1 + r_2)g$.
(2) For all $r_1, r_2 \in R$ and $g_1, g_2 \in G$, $r_1 g_1\, r_2 g_2 = (r_1 r_2)(g_1 g_2)$.
(3) $1\, g = g$.

Example 1.1.5. Consider the multiplicative group $\mathfrak{Q} = \{\pm 1, \pm\mathbf{i}, \pm\mathbf{j}, \pm\mathbf{k}\}$ of unit quaternions, satisfying

$$\begin{aligned} \mathbf{ij} &= \mathbf{k} = -\mathbf{ji}, \\ \mathbf{jk} &= \mathbf{i} = -\mathbf{kj}, \\ \mathbf{ki} &= \mathbf{j} = -\mathbf{ik}, \\ \mathbf{i}^2 = \mathbf{j}^2 &= \mathbf{k}^2 = -1. \end{aligned}$$

The group ring of quaternions over $\mathbb{Z}$ is

$$\mathbb{Z}\mathfrak{Q} = \{m_0 + m_1\mathbf{i} + m_2\mathbf{j} + m_3\mathbf{k} : m_0, m_1, m_2, m_3 \in \mathbb{Z}\}.$$

In the group ring of quaternions over $\mathbb{Z}$,

$$(3\mathbf{i} + 7\mathbf{j})\mathbf{k} + 5\mathbf{j} = -3\mathbf{j} + 7\mathbf{i} + 5\mathbf{j} = 2\mathbf{j} + 7\mathbf{i}.$$

Remark 1.1.6. When $\Bbbk$ is a field, $\Bbbk G$ is the *group algebra of G over $\Bbbk$*. We'll return to the group algebra idea later.

Definition 1.1.7. The unitary group $U(n)$ is the group of $n \times n$ unitary matrices. These are $n \times n$ square matrices with complex entries satisfying $\Psi^\dagger\Psi = \Psi\Psi^\dagger = I$, where $\Psi^\dagger$ denotes the conjugate transpose of Ψ.

Example 1.1.8. The unitary group $U(1)$ is the multiplicative group of unit complex numbers. In particular,

$$U(1) := \{z \in \mathbb{C} : z\overline{z} = 1\}.$$

Vector Spaces

Definition 1.1.9. A *vector space* V over a field $\Bbbk$ consists of a set on which two operations (*addition* and *scalar multiplication*) are defined so that for each pair $\mathbf{x}, \mathbf{y} \in V$ there exists a unique element $\mathbf{x} + \mathbf{y} \in V$, and for each $\alpha \in \Bbbk$ and each $\mathbf{x} \in V$, there exists a unique element $\alpha\mathbf{x} \in V$ such that the following conditions hold:

(1) For all $\mathbf{x}, \mathbf{y} \in V$, $\mathbf{x} + \mathbf{y} = \mathbf{y} + \mathbf{x}$.
(2) For all $\mathbf{x}, \mathbf{y}, \mathbf{z} \in V$, $\mathbf{x} + (\mathbf{y} + \mathbf{z}) = (\mathbf{x} + \mathbf{y}) + \mathbf{z}$.
(3) There exists an element $\mathbf{0} \in V$ such that $\mathbf{x} + \mathbf{0} = \mathbf{x}$ for all $\mathbf{x} \in V$.
(4) For each $\mathbf{x} \in V$, there exists $\mathbf{y} \in V$ such that $\mathbf{x} + \mathbf{y} = \mathbf{0}$.
(5) For all $\mathbf{x} \in V$, $1\mathbf{x} = \mathbf{x}$.
(6) For all $a, b \in \Bbbk$ and all $\mathbf{x} \in V$, $(ab)\mathbf{x} = a(b\mathbf{x})$.
(7) For all $a, b \in \Bbbk$ and all $\mathbf{x} \in V$, $(a + b)\mathbf{x} = a\mathbf{x} + b\mathbf{x}$.
(8) For all $a \in \Bbbk$ and all $\mathbf{x}, \mathbf{y} \in V$, $a(\mathbf{x} + \mathbf{y}) = a\mathbf{x} + a\mathbf{y}$.

A collection of vectors $\{\mathbf{x}_1, \ldots, \mathbf{x}_n\}$ is said to be *linearly independent* if and only if

$$\alpha_1 \mathbf{x}_1 + \cdots + \alpha_n \mathbf{x}_n = \mathbf{0} \Rightarrow \alpha_1 = \alpha_2 = \cdots = \alpha_n = 0.$$

The *span* of a set $\{\mathbf{x}_1, \ldots, \mathbf{x}_n\}$ is the set of all linear combinations of elements of the set; i.e.,

$$\text{span}(\{\mathbf{x}_1, \ldots, \mathbf{x}_n\}) = \{\alpha_1 \mathbf{x}_1 + \cdots + \alpha_n \mathbf{x}_n : \alpha_i \in \Bbbk, \ \forall 1 \le i \le n\}.$$

A vector space V is said to be *spanned by* $\{\mathbf{x}_1, \ldots, \mathbf{x}_n\}$ if $V \subseteq \text{span}(\{\mathbf{x}_1, \ldots, \mathbf{x}_n\})$.

If V is spanned by a linearly independent set of vectors $\mathcal{B} = \{\mathbf{x}_1, \ldots, \mathbf{x}_n\}$, then $\mathcal{B}$ is said to be a *basis* for V. The *dimension* of V is the number of vectors in a basis of V.

Examples of vector spaces include the following:

(1) $\mathbb{R}^n$: n-dimensional Euclidean space
(2) $\mathbb{C}^n$: n-dimensional complex space
(3) $\mathbb{R}[x]$: polynomials with real coefficients
(4) $\mathcal{C}[0, 1]$: continuous functions defined on the unit interval
(5) $\mathcal{C}$: continuous functions
(6) $M_{m \times n}(\mathbb{R})$ collection of $m \times n$ matrices with real entries
(7) $M_n(\mathbb{R})$ collection of order-n square matrices with real entries

Algebras

With vector space concepts in hand, we now formalize the idea of an *algebra.*

Definition 1.1.10 (Rotman [84]). If R is a ring that is also a vector space over a field $\Bbbk$, then R is called a $\Bbbk$-*algebra* if

$$(\alpha u)v = \alpha(uv) = u(\alpha v)$$

for all $\alpha \in \Bbbk$ and all $u, v \in R$.

Quadratic Forms

An n-ary *quadratic form* over a field $\Bbbk$ is a map $Q : V \to \Bbbk$ from a finite dimensional $\Bbbk$-vector space V to $\Bbbk$ such that $Q(av) = a^2Q(v)$ for all $a \in \Bbbk$ and $v \in V$ and such that $Q(u+v) - Q(u) - Q(v)$ is bilinear.

Equivalently, an n-ary quadratic form is a homogeneous quadratic polynomial in n variables. A symmetric $n \times n$ matrix $A = (a_{ij})$ determines a quadratic form Q_A in n variables via

$$Q_A(x_1, \ldots, x_n) = \sum_{1 \le i,j \le n} a_{ij}x_ix_j = \mathbf{x}^\top A\mathbf{x},$$

where $\mathbf{x}^\top = (x_1, \ldots, x_n)$ is a row vector in $\Bbbk$. One can similarly construct the symmetric matrix for a given quadratic form.

Any real quadratic form Q has an orthogonal diagonalization of the form

$$Q(x_1, \ldots, x_n) = \sum_{j=1}^{n} \lambda_j \check{x_j}^2$$

such that the corresponding matrix is diagonal. Moreover, there always exists a change of variables such that the coefficients λ_i are 0, 1, and -1. The *signature* of the quadratic form is the triple (p, q, r), where p is the number of ones, q is the number of -1s, and r is the number of zeros. When all λ_i have the same sign, the quadratic form is said to be *definite.* When $\lambda_i = 1$ for all i, Q is said to be *positive definite*, while $\lambda_i = -1$ for all i makes Q *negative definite.* When all λ_i are nonzero, the form is said to be *nondegenerate.* A real vector space with an indefinite nondegenerate quadratic form of index $(p, q, 0)$ (or simply (p, q)) is often denoted as $\mathbb{R}^{p,q}$, particularly in the physical theory of spacetime.

1.2 Linear Operators on Inner Product Spaces

Definitions in this section are standard and can be found in any number of linear algebra textbooks. The formalism here closely follows the text of Friedberg, Insel, and Spence [47]. Other useful references include Friedman's text on modern analysis [48] and the work of Akhiezer and Glazman [5].

Definition 1.2.1. An *inner product* on V is a mapping $V \times V \to \Bbbk$ satisfying the following:

(1) $\langle \mathbf{x}, \mathbf{y} \rangle = \overline{\langle \mathbf{y}, \mathbf{x} \rangle} \in \Bbbk$,
(2) $\langle \alpha\mathbf{x}, \mathbf{y} \rangle = \langle \mathbf{x}, \overline{\alpha}\mathbf{y} \rangle = \alpha \langle \mathbf{x}, \mathbf{y} \rangle$,
(3) $\langle \mathbf{x} + \mathbf{y}, \mathbf{z} \rangle = \langle \mathbf{x}, \mathbf{z} \rangle + \langle \mathbf{y}, \mathbf{z} \rangle$, and
(4) $\langle \mathbf{x}, \mathbf{y} + \mathbf{z} \rangle = \langle \mathbf{x}, \mathbf{y} \rangle + \langle \mathbf{x}, \mathbf{z} \rangle$.

A vector space with inner product is called an *inner product space.*

Two vectors $\mathbf{x}, \mathbf{y} \in V$ are said to be *orthogonal* if and only if $\langle \mathbf{x}, \mathbf{y} \rangle = 0$.

Example 1.2.2 ($\mathbb{R}^n$: n-dimensional Euclidean space)**.** Let $\mathbf{e}_1, \ldots, \mathbf{e}_n$ denote the standard orthonormal basis of $\mathbb{R}^n$. That is, $\mathbf{e}_i$ is the column vector of all zeros except for the ith entry, which is 1.

Note that $\mathbf{x} \in \mathbb{R}^n$ can be written in terms of the standard basis according to

$$\mathbf{x} = \begin{pmatrix} x_1 \\ \vdots \\ x_n \end{pmatrix} = \sum_{i=1}^{n} x_i \, \mathbf{e}_i .$$

The *Euclidean inner product* is defined by

$$\langle \mathbf{x}, \mathbf{y} \rangle = \sum_{i=1}^{n} x_i \, y_i .$$

For ease of notation, let the Euclidean inner product $\langle \mathbf{x}, \mathbf{y} \rangle$ be denoted by $\mathbf{x} \cdot \mathbf{y}$ for all $\mathbf{x}, \mathbf{y} \in \mathbb{R}^n$.

Example 1.2.3. Given $f, g \in \mathcal{C}[0, 1]$

$$\langle f, g \rangle := \int_0^1 f\overline{g} \, dx$$

is an example of an inner product.

Definition 1.2.4. Let V be a vector space over $\Bbbk$, where $\Bbbk$ is either $\mathbb{R}$ or $\mathbb{C}$. A *norm* $\|\cdot\|$ on V is a real-valued function satisfying the following conditions for all $\mathbf{x}, \mathbf{y} \in V$ and $\alpha \in \Bbbk$:

- $\|\mathbf{x}\| \geq 0$, and $\|\mathbf{x}\| = 0 \Leftrightarrow \mathbf{x} = \mathbf{0}$;
- $\|\alpha\,\mathbf{x}\| = |\alpha|\,\|\mathbf{x}\|$;
- $\|\mathbf{x} + \mathbf{y}\| \leq \|\mathbf{x}\| + \|\mathbf{y}\|$.

In $\mathbb{R}$, absolute value defines a norm. In $\mathbb{C}$, the modulus is a norm.

Definition 1.2.5. The *Euclidean norm* of a vector $\mathbf{x} \in \mathbb{R}^n$, denoted $\|\mathbf{x}\|$, is defined by

$$\|\mathbf{x}\| = (\mathbf{x} \cdot \mathbf{x})^{1/2}.$$

Linear Operators

This section is a very brief review of linear operator theory intended to provide context for our continued discussion. More interesting operator ideas will appear in later sections.

Definition 1.2.6. Given a vector space V over a field $\Bbbk$, a *linear operator* on V is a function $T : V \to V$ such that for all $\mathbf{x}, \mathbf{y} \in V$ and all $\alpha \in \Bbbk$, the following hold:

$$\begin{aligned} T(\mathbf{x} + \mathbf{y}) &= T(\mathbf{x}) + T(\mathbf{y}), \\ T(\alpha\mathbf{x}) &= \alpha T(\mathbf{x}). \end{aligned}$$

For convenience, the space of all linear operators on V will be denoted $\mathcal{L}(V)$.

Definition 1.2.7. Let $T \in \mathcal{L}(V)$. The *kernel* (or *null space*) of T is the set

$$\ker(T) := \{\mathbf{x} \in V : T(\mathbf{x}) = \mathbf{0}\}.$$

The *image* (or *range*) of T is the collection

$$\mathrm{Im}(T) := T(V) = \{T(\mathbf{x}) : \mathbf{x} \in V\}.$$

The dimension of $\ker(T)$ is called the *nullity* of T, while the dimension of $\mathrm{Im}(T)$ is called the *rank* of T. One can easily verify that T is one-to-one if and only if the nullity of T is zero.

Theorem 1.2.8 (Rank+Nullity). *Let V be a vector space of dimension n, and let T be a linear operator on V. Then,*

$$\mathrm{rank}(T) + \mathrm{nullity}(T) = n.$$

Differentiation is a linear operator on the space of polynomial functions $\mathbb{R}[x]$. Integration, on the other hand, is a *linear transformation* from $\mathbb{R}[x]$ to $\mathbb{R}$; i.e., it is a linear mapping from one vector space to another.

When V is finite-dimensional, linear operators on V can be represented using square matrices of order n. This matrix representation depends on the ordered basis being used for V.

Definition 1.2.9. Let $\beta = \{\mathbf{u}_1, \ldots, \mathbf{u}_n\}$ denote an ordered orthonormal basis for V, and let T be a linear operator on V. The *regular representation of T with respect to β* is the $n \times n$ matrix $[T]_\beta = (t_{ij})$ whose entries are defined by $t_{ij} := \langle \mathbf{u}_i, T(\mathbf{u}_j) \rangle$.

It is not difficult to verify that $T(\mathbf{u}_i) = [T_\beta]\mathbf{u}_i$ for each $i = 1, \ldots, n$. By linearity of T, $T(\mathbf{x}) = [T]_\beta(\mathbf{x})$ for all $\mathbf{x} \in V$.

Orthogonal Projections

Let $\mathbf{u} \in \mathbb{R}^n$ be a fixed unit vector. Then, any vector $\mathbf{x} \in \mathbb{R}^n$ is coplanar with $\mathbf{u}$ and has canonical decomposition into parallel and perpendicular components relative to $\mathbf{u}$. That is,

$$\mathbf{x} = \mathbf{x}_u + \mathbf{x}_{u_\perp},$$

where $\mathbf{x}_u = (\mathbf{x} \cdot \mathbf{u})\mathbf{u}$. This allows us to write

$$\mathbf{x}_{u_\perp} = \mathbf{x} - \mathbf{x}_u = \left(\sqrt{\|\mathbf{x}\|^2 - (\mathbf{x} \cdot \mathbf{u})^2}\right) \mathbf{u}_\perp,$$

for unit vector $\mathbf{u}_\perp$.

Note that the collection of vectors orthogonal to $\mathbf{u}$ generate a hyperplane.

It is well known that given an orthonormal basis $\{\mathbf{e}_i\}$ for $\mathbb{R}^n$, any vector $\mathbf{x} \in \mathbb{R}^n$ has the canonical expansion

$$\mathbf{x} = \sum_{i=1}^{n} \langle \mathbf{x}, \mathbf{e}_i \rangle \, \mathbf{e}_i.$$

The coefficients $x_i = \langle \mathbf{x}, \mathbf{e}_i \rangle$ of this expansion are referred to as *Fourier coefficients.*

Definition 1.2.10. Given an inner product space V and unit vector $\mathbf{u} \in V$, the *orthogonal projection onto* $\mathbf{u}$ is the operator $\pi_\mathbf{u}$ defined on V by

$$\pi_\mathbf{u}(\mathbf{v}) = \langle \mathbf{u}, \mathbf{v} \rangle \mathbf{u}.$$

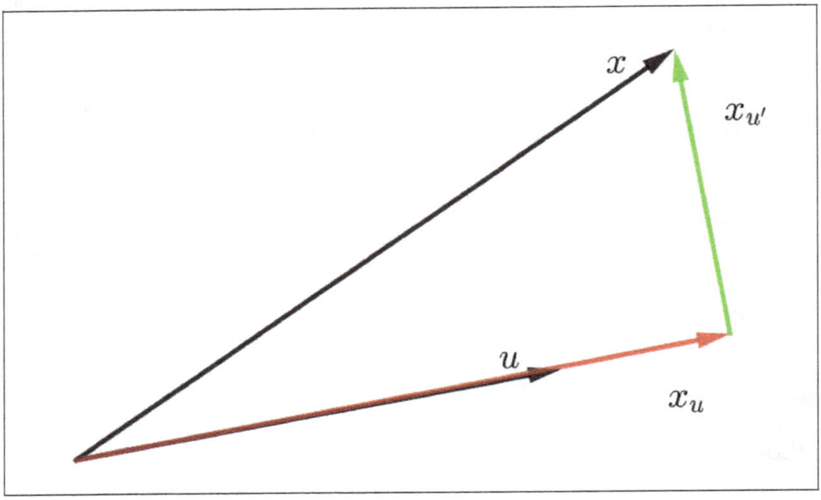

Figure 1.1 Decomposition of a vector $\mathbf{x} \in \mathbb{R}^2$ into components parallel and orthogonal to a fixed unit vector $\mathbf{u}$.

Outer Products, Projections, and Idempotents

Recall that in an n-dimensional vector space V, any translation of an $(n-1)$-dimensional subspace is referred to as a *hyperplane*.

When $\mathbb{R}^n$ is equipped with the standard basis, the canonical projection operators π_i project onto the coordinate axes. The operator $I-\pi_i$ then gives orthogonal projection onto the unique hyperplane containing the origin and orthogonal to the ith coordinate axis.

The inner product $\mathbf{x} \cdot \mathbf{y}$ can be written as the matrix product $\mathbf{x}^\dagger \mathbf{y}$. Exchanging the order of multiplication $\mathbf{x}\mathbf{y}^\dagger$ gives the *outer product* of $\mathbf{x}$ and $\mathbf{y}$. While the inner product is scalar-valued, the outer product is an $n \times n$ matrix.

Gram-Schmidt Orthogonalization

The following application of orthogonal projections is useful. Suppose $S = \{\mathbf{v}_1, \ldots, \mathbf{v}_k\}$ is a linearly independent collection in an inner product space V. In many instances, it is possible to create a collection S' of pairwise orthogonal vectors that generate the same subspace of V. The process is known as *Gram-Schmidt orthogonalization*.

(1) First, set $\mathbf{u}_1 := \mathbf{v}_1$.
(2) For each $\ell = 2, \ldots, k$, set

$$\mathbf{u}_\ell := \mathbf{v}_\ell - \sum_{j=1}^{\ell-1} \frac{\langle \mathbf{v}_\ell, \mathbf{u}_j \rangle}{\|\mathbf{u}_j\|^2} \mathbf{u}_j.$$

(3) The collection $S' = \{\mathbf{u}_1, \ldots, \mathbf{u}_k\}$ now satisfies $\text{span}(S) = \text{span}(S')$. If unit vectors are required, normalize the collection S' by dividing each element by its norm.

1.3 Group and Semigroup Representations

Group Actions

An action of a group is a formal way of interpreting the manner in which the elements of the group correspond to transformations of some space in a way that preserves the structure of that space. Historically, the first group action studied was the action of the Galois group on the roots of a polynomial. Actions of groups on vector spaces are called representations of the group.

A group G is said to *act on a set* X if there exists a map $G \times X \to X$ (written here as $*$) such that the following conditions hold for all $x \in X$:

- $e * x = x$ where e is the identity of the group G.
- For all $g, h \in G$, $g * (h * x) = (gh) * x$.

In this case, G is called a transformation group, X is called a G-set, and $*$ is called the *group action*.

If, for each pair $x, y \in X$, there exists $g \in G$ such that $y = g * x$, then G is said to act *transitively* on X.

For a given x, the set $\{g * x : g \in G\}$, where the group action moves x, is called the *group orbit* of x. The subgroup of G fixing x is called the *isotropy group* of x.

The group $\mathbb{Z}_2 = \{0, 1\}$ acts on the real numbers by multiplication by $(-1)^n$. The identity leaves everything fixed, while 1 sends x to $-x$. For $x \neq 0$, the orbit of x is $\{x, -x\}$, and the isotropy subgroup is trivial: $\{0\}$. The only group fixed point of this action is $x = 0$.

In a group representation, a group acts by invertible linear transformations of a vector space V. In fact, a representation is a group homomorphism from G to $GL(V)$, the general linear group of V.

Given a group G, it is not difficult to see that G *acts on itself by conjugation*. In particular, defining $h * g = hgh^{-1}$, one can verify the required axioms of a group action.

Group Representations

The collection of invertible linear transformations on $\mathbb{R}^n$ constitutes a group called the *general linear group*, which we denote by $GL_n(\mathbb{R})$. Utilizing the regular representations of these linear transformations, this group is typically defined to be the $n \times n$ matrices with real entries and nonzero determinants. The *special linear group* $SL_n(\mathbb{R})$ is the subgroup of $GL_n(\mathbb{R})$ corresponding to matrices of determinant 1.

As we have already seen, the orthogonal group $O(n)$ corresponds to $n \times n$ matrices with real entries, determinant ± 1, and satisfying $A \in O(n) \Rightarrow A^\dagger A = I$. The *special orthogonal group* $SO(n)$ is the subgroup of $O(n)$ corresponding to matrices of determinant 1.

In fact, the orthogonal group $O(n)$ consists of those linear transformations of $\mathbb{R}^n$ that are rotations, reflections, or improper reflections (reversals). The special orthogonal group consists of rotations only.

Related to these groups are the unitary and special unitary groups. Recall first that the unitary group $U(n)$ is the group of $n \times n$ matrices with complex entries satisfying $\Psi^\dagger \Psi = \Psi \Psi^\dagger = I$, where $\Psi^\dagger$ denotes the conjugate transpose of Ψ. The *special unitary group* $SU(n)$ is the subgroup of $U(n)$ whose elements have determinant 1.

The definitions and basic results appearing here can be found in texts on representation theory. Useful references include the texts by Sengupta [92] and Serre [93].

Definition 1.3.1. Given a group G, a *linear representation* of G is a homomorphism $\rho : G \to GL_n(\mathbb{k})$.

Recalling that $GL_n(\mathbb{k})$ can be regarded as the group of invertible linear operators acting on an n-dimensional vector space V over the field $\mathbb{k}$, the space V is called the *representation space*. The *degree* of the representation is the dimension of the representation space.

Representations always exist. For example, one can always define the *trivial representation* $\vartheta : G \to \mathbb{k}$ by

$$\vartheta(g) = 1, \quad (g \in G).$$

Note that when a group is finitely generated, a representation is uniquely specified by its action on the group's generators. This follows from the fact that representations are homomorphisms. Consequently, the identity of the group must always be represented by the identity operator of the representation space.

Example 1.3.2. Consider the group of unit quaternions: $\mathfrak{Q} = \{\pm 1, \pm \mathbf{i}, \pm \mathbf{j}, \pm \mathbf{k}\}$. Observe that $\mathfrak{Q}$ is generated by $\mathbf{i}$ and $\mathbf{j}$. One representation over $\mathbb{R}$ (i.e., a real representation) is given by setting

$$\rho(\mathbf{i}) = \rho(\mathbf{j}) = -1.$$

From this, one immediately sees $\rho(\mathbf{k}) = \rho(\mathbf{ij}) = \rho(\mathbf{i})\rho(\mathbf{j}) = (-1)(-1) = 1$. Further, $\rho(-1) = \rho(\mathbf{i}^2) = \rho(\mathbf{i})^2 = (-1)^2 = 1$.

Given a group G, the *regular representation* R of G is the representation on the vector space over $\Bbbk$ with basis G, where for every $g \in G$,

$$R(g)(h) = gh \ \ (h \in G).$$

Example 1.3.3. A degree-8 represention of $\mathfrak{Q}$ is specified as follows: First, let $\{\mathbf{e}_i : 1 \leq i \leq 8\}$ be the standard basis for $\mathbb{R}^8$ and associate $1 \sim \mathbf{e}_1$, $-1 \sim \mathbf{e}_2$, $\mathbf{i} \sim \mathbf{e}_3$, ..., $-\mathbf{k} \sim \mathbf{e}_8$. Then $R : \mathfrak{Q} \to GL_8(\mathbb{R})$ is determined by

$$R(\mathbf{i}) = \begin{pmatrix} 0&0&0&1&0&0&0&0 \\ 0&0&1&0&0&0&0&0 \\ 1&0&0&0&0&0&0&0 \\ 0&1&0&0&0&0&0&0 \\ 0&0&0&0&0&0&0&1 \\ 0&0&0&0&0&0&1&0 \\ 0&0&0&0&1&0&0&0 \\ 0&0&0&0&0&1&0&0 \end{pmatrix}, \quad R(\mathbf{j}) = \begin{pmatrix} 0&0&0&0&0&1&0&0 \\ 0&0&0&0&1&0&0&0 \\ 0&0&0&0&0&0&1&0 \\ 0&0&0&0&0&0&0&1 \\ 1&0&0&0&0&0&0&0 \\ 0&1&0&0&0&0&0&0 \\ 0&0&0&1&0&0&0&0 \\ 0&0&1&0&0&0&0&0 \end{pmatrix}.$$

Before considering complex representation spaces, an important set of matrices from $U(2)$ will be recalled. The *Pauli matrices* are defined as follows: $\sigma_0 = \begin{pmatrix} 1&0 \\ 0&1 \end{pmatrix}$, $\sigma_{\mathrm{x}} = \begin{pmatrix} 0&1 \\ 1&0 \end{pmatrix}$, $\sigma_{\mathrm{y}} = \begin{pmatrix} 0&-i \\ i&0 \end{pmatrix}$, $\sigma_{\mathrm{z}} = \begin{pmatrix} 1&0 \\ 0&-1 \end{pmatrix}$.

By an exercise, one can show that the non-identity Pauli matrices σ_{x}, σ_{y}, and σ_{z} are pairwise-anticommutative and satisfy ${\sigma_{\mathrm{x}}}^2 = {\sigma_{\mathrm{y}}}^2 = {\sigma_{\mathrm{z}}}^2 = \sigma_0$. This suggests we are close to another representation of $\mathfrak{Q}$; i.e., the non-identity Pauli matrices "behave" like the unit quaternions $\mathbf{i}$, $\mathbf{j}$, and $\mathbf{k}$. Regarding the Pauli matrices as 2×2 matrices with complex entires, we can simply multiply each matrix by the imaginary unit i.

Example 1.3.4. A complex degree-2 representation of $\mathfrak{Q}$ is given as follows: define $\tau(\mathbf{i}) = i\sigma_{\mathrm{x}} = \begin{pmatrix} 0&i \\ i&0 \end{pmatrix}$ and $\tau(\mathbf{j}) = \bar{i}\sigma_{\mathrm{y}} = \begin{pmatrix} 0&1 \\ -1&0 \end{pmatrix}$. It follows that $\tau(1) = \sigma_0$ and $\tau(\mathbf{k}) = i\sigma_z = \begin{pmatrix} i & 0 \\ 0 & -i \end{pmatrix}$. Note that these matrices are elements of $SU(2)$.

Definition 1.3.5. Let ρ be a representation of a group G. The *character* of ρ is the map $\chi : G \to \Bbbk$ defined by

$$\chi(g) = \operatorname{tr}(\rho(g)).$$

Three properties of the character follow immediately from the definition.

Lemma 1.3.6. *For every $g, h \in G$,*

i. $\chi(hgh^{-1}) = \chi(g)$,
ii. $\chi(g^{-1}) = \overline{\chi(g)}$, *and*
iii. $\chi(e)$ *is the degree of the representation.*

Proof. Let A, B be square matrices of order n, and observe that

$$\operatorname{tr}(AB) = \sum_{i=1}^{n}\sum_{j=1}^{n} a_{ij}b_{ji} = \sum_{i=1}^{n}\sum_{j=1}^{n} b_{ji}a_{ij} = \sum_{j=1}^{n}\sum_{i=1}^{n} b_{ji}a_{ij} = \operatorname{tr}(BA).$$

Hence, assuming B is invertible,

$$\begin{aligned}\operatorname{tr}(BAB^{-1}) &= \operatorname{tr}((BA)B^{-1})\\ &= \operatorname{tr}(B^{-1}(BA))\\ &= \operatorname{tr}(A).\end{aligned}$$

Then, for $g, h \in G$,

$$\begin{aligned}\chi(hgh^{-1}) &= \operatorname{tr}(\rho(hgh^{-1}))\\ &= \operatorname{tr}(\rho(h)\rho(g)\rho(h^{-1}))\\ &= \operatorname{tr}(\rho(h)\rho(g)\rho(h)^{-1})\\ &= \operatorname{tr}(\rho(g))\\ &= \chi(g).\end{aligned}$$

Next, notice we can find a basis β_g such that $\rho(g)$ is a diagonal matrix with each diagonal entry being a root of unity [92]. It follows that

$$\rho(g)^{-1} = \overline{\rho(g)}.$$

Then

$$\begin{aligned}\chi(g^{-1}) &= \operatorname{tr}(\rho(g^{-1}))\\ &= \operatorname{tr}(\rho(g)^{-1})\\ &= \operatorname{tr}(\overline{\rho(g)})\\ &= \overline{\operatorname{tr}(\rho(g))}\\ &= \overline{\chi(g)}.\end{aligned}$$

Finally, if ρ is a degree-n representation, $\rho(e) = \mathbb{I} \Rightarrow \chi(e) = n$. $\square$

Considering again the group $\mathfrak{Q}$, if φ is the character of τ, then one can see

$$\varphi(g) = \begin{cases} \pm 2 & g = \pm 1, \\ 0 & \text{otherwise.} \end{cases}$$

Since ρ is a degree-1 representation, the character will be the same as the representation itself.

Finally, if χ_R is the character for the regular representation R,

$$\chi_R(g) = \begin{cases} 8 & g = 1, \\ 0 & \text{otherwise.} \end{cases}$$

Given an n-dimensional vector space V over a field $\mathbb{F}$, we know that if a basis is fixed, $GL(V) \simeq GL(n, \mathbb{F})$ where $GL(V)$ is the set of linear operators from V to itself. Using this, we can view a representation as a homomorphism from the group to $GL(V)$. We will call V the *representation space*. Given a representation ρ and a space W, we say W is *G-invariant* if $\rho(g)W \subseteq W$ for every $g \in G$. If the only invariant spaces are $\{0\}$ and V, we say the representation is *irreducible*. Equivalently, we say ρ is an irreducible representation. Using again our example of Q, we can quickly verify that ρ and τ are irreducible and that R is not irreducible (in which case we call R *reducible*). Given two representations $\rho : G \to GL(V)$ and $r : G \to GL(W)$ of a group G, we say ρ is *isomorphic* to r if there exists an invertible mapping $f : V \to W$ such that

$$f \circ \rho = r \circ f.$$

The next result can be found in [93].

Lemma 1.3.7 (Schur). *Let $\rho : G \to GL(V)$ and $r : G \to GL(W)$ be two irreducible representations of G, and let f be a linear mapping of V onto W such that $r(s) \circ f = f \circ \rho(s)$ for all $s \in G$. Then if ρ and r are not isomorphic, we have $f = 0$. If $V = W$ and $\rho = r$, then f is a scalar multiple of the identity.*

Proof. If $f = 0$ the result is trivial. If $f \neq 0$, then for $x \in \ker(f)$ we have $f \circ \rho(x) = r(x) \circ f = 0$. Thus $\rho(x) \in \ker(f)$, and $\ker(f)$ is G-invariant. Since V was irreducible, $\ker(f)$ is either V or $\{0\}$; however, we have already excluded the former by assuming $f \neq 0$. A similar argument shows the image of f must equal W. Thus, f is an isomorphism of V onto W.

If, in addition, $V = W$ and $r = \rho$, let λ be an eigenvalue of f. Define

$$f' = f - \lambda\mathbb{I}.$$

Since λ is an eigenvalue of f, it follows $\ker f' \neq \{\mathbf{0}\}$; however, $r \circ f' = f' \circ \rho$. The first part of the proof says $f' = 0$, or $f = \lambda\mathbb{I}$. □

We can also define an inner product on the characters of representations, which will prove to be useful later.

Definition 1.3.8. For a group G, given two characters χ and ξ we define $\langle\chi|\xi\rangle$ as

$$\langle\chi|\xi\rangle = \frac{1}{|G|}\sum_{g\in G}\chi(g)\xi(g^{-1}).$$

A representation ρ with character χ is irreducible if and only if $\langle\chi|\chi\rangle = 1$. With this inner product one could show that the irreducible characters of a group G form an orthonormal basis in the space of class functions [93]. We conclude this section with a number of results from [93] that will be incredibly useful later on.

Lemma 1.3.9. *Let V be a representation space of G with corresponding character φ. Suppose that V can be written $V = W_1 \oplus W_2 \oplus \cdots \oplus W_k$, where each W_i is an irreducible representation space. If W is an irreducible representation space with corresponding character χ, the number of W_i isomorphic to W is $\langle\varphi|\chi\rangle$.*

Proof. Let χ_i be the character of each W_i. One can see

$$\varphi = \chi_1 + \cdots + \chi_k.$$

Then

$$\langle\varphi|\chi\rangle = \langle\chi_1|\chi\rangle + \cdots + \langle\chi_k|\chi\rangle.$$

However each $\langle\chi_i|\chi\rangle$ is equal to 1 if W_i is isomorphic to W and 0 otherwise. The result follows. □

Theorem 1.3.10. *The number of irreducible representations of a group G is equal to the number of conjugacy classes of G.*

Proof. Let $C_1, \ldots, C_k$ be the distinct conjugacy classes of G. If f is a class function on G, then f is constant on each C_i. Thus each class function is determined by the values it assumes on each C_i. These values can be chosen arbitrarily which shows the dimension of the space of class functions is equal to k. However, since we have previously stated that the dimension of the space of class functions has dimension equal to the number of irreducible representations, this completes the proof. □

Theorem 1.3.11. *If m is the number of irreducible representations and n_i is the degree of ρ_i for $i = 1, \ldots, m$ then*

$$|G| = \sum_{i=1}^{m} {n_i}^2.$$

Proof. If ξ is the character of the regular representation, notice

$$\xi(s) = \begin{cases} |G| & s = e, \\ 0 & \text{otherwise.} \end{cases}$$

If χ_i is an irreducible character of degree n_i, it is contained in the regular representation $\langle \xi | \chi_i \rangle$ times. Expanding this out we get

$$\begin{aligned} \langle \xi | \chi_i \rangle &= \frac{1}{|G|} \sum_{g \in G} \xi(g^{-1}) \chi_i(g) \\ &= \frac{1}{|G|} |G| n_i \\ &= n_i. \end{aligned}$$

Now for every $s \in G$,

$$\xi(s) = \sum_{i=1}^{m} n_i \chi_i(s).$$

Taking $s = 1$ yields the desired result. □

Theorem 1.3.12. *If G is an Abelian group, then every irreducible representation is of degree* 1.

Proof. Since G is Abelian, the number of conjugacy classes of G is $|G|$. By the previous theorem, if $n_1, \ldots, n_{|G|}$ are the degrees of the irreducible representations, we have

$$|G| = {n_1}^2 + \cdots + {n_{|G|}}^2.$$

Thus, each $n_i = 1$. □

Semigroup Representations

Essential definitions and notational conventions for semigroup representation theory follow the formalism of Izhakian, Rhodes, and Steinberg [58]. As previously noted, all semigroup representation spaces are complex.

Given a semigroup S, two elements $a, b \in S$ are said to be $\mathfrak{J}$-*equivalent* (written $a \mathfrak{J} b$) if $SaS = SbS$. The set of all things $\mathfrak{J}$-equivalent to $a \in S$ forms a $\mathfrak{J}$-class. The $\mathfrak{J}$-classes partition the semigroup S. A $\mathfrak{J}$-class is said to be *regular* if it contains an idempotent.

For every idempotent e of a semigroup S, we call G_e the *maximal subgroup of* S *at* e where $G_e = \{\text{invertible elements of } eSe\}$. Two idempotent elements $e, f \in S$ are said to be *isomorphic* if there exists an $x \in eSf$ and $x^* \in fSe$ such that $xx^* = e$ and $x^*x = f$.

The regular $\mathfrak{J}$-classes will play a large role in determining the number of irreducible representations of a semigroup S. Before we give the exact number to expect, we need a few more results. The following useful lemma can be found in [23].

Lemma 1.3.13. *If* $e, f \in S$ *are isomorphic idempotents, then* $G_e \simeq G_f$. *Moreover,* e *and* f *are isomorphic if and only if* $e \mathfrak{J} f$.

For a semigroup S we define a representation to be a homomorphism to the set of endomorphisms of $\mathbb{C}^n$, which can be realized as $n \times n$ matrices with entries in $\mathbb{C}$. In other words, a representation ρ of S is a homomorphism

$$\rho : S \to \text{End}(\mathbb{C}^n).$$

The familiar terminology of a faithful representation, trivial representation and character of a representation follows. The idea of an irreducible semigroup representation is again the same, except we require that the representation is not constantly 0.

The next theorem, based on results of Clifford-Suchkewitch [23] and Munn (as found in Rhodes and Zalcstein [82]), will be useful for determining the number of irreducible representations.

Theorem 1.3.14. *Let* $G_1, \ldots, G_m$ *be a choice of exactly one maximal subgroup from each regular* $\mathfrak{J}$*-class of* S. *Then, letting* k_i *denote the number of conjugacy classes of* G_i, *the number of irreducible representations of* S *is* $\sum_{i=1}^{m} k_i$.

Exercises

Exercise 1.1: Let G be a group. Show that the identity and inverses are two-sided.

Exercise 1.2: Let $\mathbf{x}, \mathbf{y} \in \mathbb{R}^2$ and show that $\mathbf{x} \cdot \mathbf{y} = \|x\|\,\|y\| \cos\theta$, where θ is the angle between vectors $\mathbf{x}$ and $\mathbf{y}$.

Exercise 1.3: Let $\mathbf{x}, \mathbf{y} \in \mathbb{R}^3$ and show that $\|\mathbf{x} \times \mathbf{y}\| = \|x\|\,\|y\| \sin\theta$, where θ is the angle between vectors $\mathbf{x}$ and $\mathbf{y}$.

Exercise 1.4: Let $\mathbf{x}, \mathbf{y}, \mathbf{z} \in \mathbb{R}^3$ and show that $|\mathbf{x} \cdot \mathbf{y} \times \mathbf{z}|$ is the volume of the parallelepiped determined by $\mathbf{x}$, $\mathbf{y}$, and $\mathbf{z}$

Exercise 1.5: Show that $\mathcal{L}(V)$ is a vector space.

Exercise 1.6: Show that $\ker(T)$ and $\mathrm{Im}(T)$ are subspaces of V.

Exercise 1.7: Let $\mathbf{u} = (1, -2, 1) \in \mathbb{R}^3$ and write the vector $\mathbf{x} = (-3, 0, 2)$ as a sum of vectors, one parallel and one perpendicular to $\mathbf{u}$.

Exercise 1.8: Let $\mathbf{u}, \mathbf{v} \in \mathbb{R}^n$ be nonparallel unit vectors. Give a direct proof that the operator on $\mathbb{R}^n$ defined by

$$\mathbb{I} - 2\mathbf{v}\mathbf{v}^\dagger - 2\mathbf{u}\mathbf{u}^\dagger + 4\langle \mathbf{v}, \mathbf{u} \rangle \mathbf{v}\mathbf{u}^\dagger$$

is an element of $\mathrm{SO}(n)$.

Exercise 1.9: Show that $SO(n)$ is a normal subgroup of $O(n)$.

Exercise 1.10: Compute the remaining six matrices of the regular representation of the group of unit quaternions, $\mathfrak{Q}$.

Exercise 1.11: Show that the non-identity Pauli matrices σ_{x}, σ_{y}, and σ_{z} are pairwise-anticommutative and satisfy ${\sigma_{\mathrm{x}}}^2 = {\sigma_{\mathrm{y}}}^2 = {\sigma_{\mathrm{z}}}^2 = \sigma_0$.

Chapter 2

Combinatorics

We begin with a brief review of combinatorics associated with polynomials.

The *multinomial theorem* states that

$$(x_1 + \cdots + x_m)^n = \sum_{\substack{0 \le k_1, \ldots, k_m \\ k_1 + \cdots + k_m = n}} \binom{n}{k_1, \ldots, k_m} x_1{}_1^k \cdots x_m{}^{k_m}$$

where the *multinomial coefficient*

$$\binom{n}{k_1, \ldots, k_m} := \frac{n!}{k_1! \cdots k_m!}$$

represents the number of ways n objects may be chosen from a collection of m objects with repetition.

When $m = 2$, the *binomial theorem* arises as a special case:

$$(x_1 + x_2)^n = \sum_{k=0}^{n} \binom{n}{k} x_1{}^k x_2{}^{n-k}.$$

Here the *binomial coefficient* $\binom{n}{k}$ is determined by

$$\begin{aligned} \binom{n}{k} &:= \frac{n!}{k!(n-k)!} \\ &= \binom{n}{n-k} \end{aligned}$$

and represents the number of ways k objects may be chosen from a set of n objects.

Lemma 2.0.1 (Pascal's recurrence). *Let n and k be positive integers. The binomial coefficients satisfy*

$$\binom{n}{0} = \binom{n}{n} = 1.$$

Moreover, when $0 < k < n$*, the binomial coefficients satisfy* Pascal's recurrence*:*

$$\binom{n}{k} = \binom{n-1}{k} + \binom{n-1}{k-1}.$$

Proof. The recurrence is established via combinatorial arguments by supposing $n-1$ objects are "red" and one object is "blue." From this collection, k objects will be chosen. Counting the combinations is then done in two ways, interpreted as the two sides of the equation. Details of the proof are left as an exercise. □

Lemma 2.0.2 (Vandermonde's identity)**.** *For nonnegative integers* $r, m,$ *and* n*, the binomial coefficients satisfy the following identity:*

$$\binom{m+n}{r} = \sum_{k=0}^{r} \binom{m}{k} + \binom{n}{r-k}.$$

As in the proof of Lemma 2.0.1, the proof of Lemma 2.0.2 again proceeds by counting one thing in two ways. In this case, it is helpful to visualize a collection of $n+m$ objects (n "red" and m "blue") from which r objects will be chosen.

Lemma 2.0.3 (Generalized Vandermonde's identity)**.** *For nonnegative integers* $m, n_1, n_2, \ldots, n_p$*, the following identity holds:*

$$\binom{n_1 + \cdots + n_p}{m} = \sum_{k_1+\cdots+k_p=m} \binom{n_1}{k_1} \cdots \binom{n_p}{k_p}.$$

2.1 Partitions and Counting Numbers

The partition notation here follows that of Rota and Wallstrom [83]. Given a finite set b, let $\Pi(b)$ denote the set of all partitions of b. In particular, an element $\sigma \in \Pi(b)$ is a collection of nonempty disjoint subsets, called *blocks*, whose union is b. Denote by $|\sigma|$ the number of blocks contained in σ.

The set $\Pi(b)$ is partially ordered by defining $\sigma \leq \pi$ if and only if every block of σ is a subset of some block in π. Accordingly, two partitions of particular interest are defined by

$$\hat{1}_b := b, \text{ and}$$

$$\hat{0}_b := \{b_1\} \cup \cdots \cup \{b_{|b|}\}.$$

When the set being partitioned is clear, one writes simply $\hat{1}$ or $\hat{0}$.

The meet of two partitions $\sigma \wedge \pi$ is defined as the partition whose blocks are the nonempty pairwise intersections of some block of σ with some block of π. The *join* of two partitions, denoted $\sigma \vee \pi$, is the smallest partition containing both σ and π. Note that $\Pi(b)$ is a *lattice*.

A *segment* $[\sigma, \pi]$ of the lattice $\Pi(b)$ is defined by

$$[\sigma, \pi] := \{\rho \in \Pi(b) : \sigma \leq \rho \leq \pi\}.$$

For positive integer n, the number of ways of partitioning an n-set into k equivalence classes is $S(n,k) = \left\{ {n \atop k} \right\}$, a *Stirling number of the second kind.* For fixed $n > 0$, summing $S(n,k)$ over k from 1 to n gives the total number of ways of partitioning the n-set into equivalence classes, defined as the nth *Bell number*, B_n. In other words, the nth Bell number is the sum

$$B_n = \sum_{k=0}^{n} \left\{ {n \atop k} \right\}.$$

Values for $1 \leq n \leq 8$ appear in Table 2.1.

Table 2.1 Stirling numbers of the second kind and Bell numbers.

$n \backslash k$	1	2	3	4	5	6	7	8	B_n
1	1	0	0	0	0	0	0	0	1
2	1	1	0	0	0	0	0	0	2
3	1	3	1	0	0	0	0	0	5
4	1	7	6	1	0	0	0	0	15
5	1	15	25	10	1	0	0	0	52
6	1	31	90	65	15	1	0	0	203
7	1	63	301	350	140	21	1	0	877
8	1	127	966	1701	1050	266	28	1	4140

A partition π of the n-set is said to be *crossing* if for some pair of blocks $\gamma, \delta \in \pi$ there exist $i, k \in \gamma$ and $j, \ell \in \delta$ such that $i < j < k < \ell$. For disjoint sets $\gamma, \delta \in 2^{[n]}$, the notation $\gamma \pitchfork \delta$ is defined to indicate that γ and δ form a *crossing* partition of $\gamma \cup \delta$.

For each $n \in \mathbb{N}$, let C_n denote the number of non-crossing partitions of the n-set. The numbers C_n are called the *Catalan numbers.* These numbers have numerous combinatorial interpretations: for example, C_n is the number of full binary trees having $n+1$ leaves [95]. For $n \geq 0$, the Catalan numbers have the following closed formula:

$$C_n = \binom{2n}{n} - \binom{2n}{n+1} = \frac{1}{n+1}\binom{2n}{n}.$$

The partition π is said to be *overlapping* if there exist blocks $\gamma, \delta \in \pi$ such that $\min\gamma < \min\delta < \max\gamma < \max\delta$. For disjoint sets $\gamma, \delta \in 2^{[n]}$, the notation $\gamma \between \delta$ is defined to indicate that γ and δ form an *overlapping* partition of $\gamma \cup \delta$.

Given blocks γ and δ, define the notation $\gamma \pitchfork \delta$ to indicate that γ and δ are non-overlapping. A partition π of the n-set is said to be *non-overlapping* if $\gamma \pitchfork \delta$ whenever $\gamma \neq \delta \in \pi$. Associating with each block δ the closed interval $[\min(\delta), \max(\delta)] \subset \mathbb{R}$, it is evident that two blocks overlap if and only if their associated intervals are disjoint or one contains the other.

Example 2.1.1. The blocks $\{1, 3, 5\}$ and $\{2, 4\}$ form a non-overlapping partition of the 5-set, while blocks $\{1, 2, 4\}$ and $\{3, 5\}$ constitute an overlapping partition.

Let B_n^* denote the number of non-overlapping partitions of the n-set. For $n > 0$, the numbers B_n^* are called the *Bessel numbers* [44]. The first six Bessel numbers are $\{1, 2, 5, 14, 43, 143\}$.

Useful Combinatorial Identities

The *Stirling numbers of the second kind*, written $\left\{ n \atop k \right\}$, count the number of ways a set of n labelled objects can be partitioned into k nonempty subsets.

By definition, $\left\{ n \atop 0 \right\} = 0$ for all positive integers n. Similarly, $\left\{ n \atop k \right\}$ is defined to be zero whenever $k > n$. A well-known closed formula for $\left\{ m \atop k \right\}$ is the following [110]:

$$\left\{ m \atop k \right\} = \frac{1}{k!} \sum_{j=0}^{k} (-1)^j \binom{k}{j} (k-j)^m. \tag{2.1}$$

The *Stirling numbers of the first kind* are defined by the following property: $S_1(j, \ell)$ is the number of permutations of j elements that contain exactly ℓ cycles.

Lemma 2.1.2. *For $n, m \in \mathbb{N}$,*

$$\sum_{k=1}^{n} \frac{m!}{(m-k)!} \left\{ n \atop k \right\} = m^n. \tag{2.2}$$

Proof. The sum on the left-hand side of (2.2) is equivalent to the number of ways n objects can be placed into m labeled boxes. For each $k = 1, \ldots, n$, the n objects are first partitioned into k subsets, and k boxes are chosen to contain these subsets. The remaining $m - k$ boxes are left empty. Of the k

chosen boxes, the distinct subsets of objects can be assigned in any order. For fixed k, this results in

$$k!\binom{m}{k}\left\{ {n \atop k} \right\} = \frac{m!}{(m-k)!}\left\{ {n \atop k} \right\}$$

possible assignments. Summing over k gives the claimed interpretation.

The interpretation of the right-hand side of (2.2) is more straightforward. Each of the n objects is assigned (with repetition) to one of the m boxes. Each selection is independent and can be made in m ways. Hence, the result. □

The next lemma will be useful for establishing upper bounds on coefficients appearing in products of invertible zeons.

Lemma 2.1.3. *Let $2^{[n]m}$ denote the m-fold Cartesian product of the power set of $[n]$, defined as the collection of ordered m-tuples of subsets of $[n]$. Given a fixed set $I \in 2^{[n]}$, the number of independent ordered m-tuples whose union is I is given by*

$$\left|\left\{ (I_1|\cdots|I_m) \in 2^{[n]m} : \bigcup_{j=1}^{m} I_j = I \right\}\right| = m^{|I|}. \tag{2.3}$$

Proof. The left-hand side counts the number of ways $|I|$ objects can be distributed among m labeled boxes. The choices are independent; hence for each object in the collection I, one of m boxes is chosen. By the multiplication principle of counting, the number of ways this can be done is $\underbrace{m\cdots m}_{|I|\text{times}} = m^{|I|}$. □

2.2 Graph Theory

The principal motivation for pursuing problems in graph theory is the abundance of real-world applications, notably in computer science, where new methods of tackling computationally difficult problems are needed. Graphs provide natural models for wireless networks, traffic sensors, the World Wide Web, etc. Moreover, these graphs evolve in real time. Random walks on graphs are of interest as models of internet searches, data transmission, and even error propagation.

The terminology appearing here can be found in any number of graph theory texts. The reader is referred to [112] for graph theory beyond the essential notation and terminology found here.

A *graph* $G = (V, E)$ is a collection of vertices V and a set E of unordered pairs of vertices called *edges*. A *directed graph* is a graph whose edges are *ordered pairs* of vertices. Two vertices $v_i, v_j \in V$ are *adjacent* if there exists an edge $e = (v_i, v_j) \in E$.

Given an existing edge $e = (v_i, v_j) \in E$, the edge e is said to be *incident* with the vertices v_i and v_j. The number of edges incident with a vertex is referred to as the *degree* of the vertex. A graph is said to be *regular* if all its vertices are of equal degree. A graph is *finite* if V and E are finite sets, that is, if $|V|$ and $|E|$ are finite numbers.

A *loop* in a graph is an edge of the form (v, v). A graph is said to be *simple* if it contains no loops and no unordered pair of vertices appears more than once in E.

An *independent set* in a graph G is a set of pairwise nonadjacent vertices. A *clique* in a graph G is a set of pairwise adjacent vertices. A *matching* of G is a subset $E_1 \subset E$ of the edges of G having the property that no pair of edges in E_1 shares a common vertex. The largest possible matching on a graph with n vertices consists of $n/2$ edges, and such a matching is called a *perfect matching*.

A *k-walk* $\{v_0, \ldots, v_k\}$ in a graph G is a sequence of vertices in G with *initial vertex* v_0 and *terminal vertex* v_k such that there exists an edge $(v_j, v_{j+1}) \in E$ for each $0 \leq j \leq k-1$. A k-walk contains k edges. A *self-avoiding walk* or *path* is a walk in which no vertex appears more than once. A *closed k-walk* is a k-walk whose initial vertex is also its terminal vertex. A *k-cycle* is a self-avoiding closed k-walk with the exception $v_0 = v_k$. An *Euler circuit* is a closed walk encompassing every edge in E exactly once.

For convenience, two-cycles (which have a repeated edge) will be allowed. The term *proper cycle* will refer to any cycle of length three or greater.

A *Hamiltonian cycle* is an n-cycle in a graph on n vertices; i.e., it contains V. Given a graph G, the *circumference* and *girth* of G are defined as the lengths of the longest and shortest cycles in G, respectively.

Given a graph $G = (V, E)$, a *subgraph* of G is a graph $G' = (V', E')$ such that $V' \subseteq V$ and $E' \subseteq E$. Note that G' must be a graph; i.e., vertices appearing within ordered pairs in E' must be elements of V'. A *cycle cover* of a graph G is a set of subgraphs $\{C_1, \ldots, C_k\}$ of G such that (i) each subgraph is a cycle, and (ii) each vertex of G is contained in exactly one of the subgraphs C_j, $(1 \leq j \leq k)$.

A graph G is said to be *connected* if for every pair of vertices v_i, v_j in

G, there exists a k-walk on G with initial vertex v_i and terminal vertex v_j for some positive integer k.

A *tree* is a connected graph that contains no cycles. A *spanning tree* in a graph is a subgraph that is a tree and contains all of the graph's vertices.

A *connected component* of a graph G is a connected subgraph G' of maximal size. In other words, $V(G') \subseteq V(G)$, $E(G') \subseteq E(G)$, and there is no connected subgraph G'' with the property $V(G') \subsetneq V(G'')$.

The following four basic graph-theoretic results will be useful in later chapters. The first deals with trees.

Lemma 2.2.1. *Let G be a connected graph on $n \geq 2$ vertices. Then G is a tree if and only if G contains $n - 1$ edges.*

Proof. Proof is by induction on the number of vertices n. When $n = 2$, the graph G contains one edge and is a tree by definition. Assuming the lemma is true for some positive integer $n \geq 2$, let G be a connected graph on n vertices, and let the graph H be constructed by appending one vertex v to G. In other words, $V(H) = V(G) \cup \{v\}$. In order to make H connected, one edge must be appended, joining v to some existing vertex u of G. Now H is a connected graph on $n + 1$ vertices and is a tree since v is incident with only one edge.

It remains to be seen that appending two edges incident with v prevents H from being a tree. Suppose a second edge incident with v is appended to H. This edge is incident with some vertex $w \neq u$ of G. Since G is connected, there exists a walk in G having initial vertex u and terminal vertex w. Appending vertex v and its two incident edges to G yields a cycle in H. Thus, H cannot be a tree.

Hence, at most one edge can be appended to G in constructing H. The $(n + 1)$-vertex connected graph H consists of n edges, and the proof is complete. □

The next standard result deals with connected 2-regular graphs.

Lemma 2.2.2. *Let G be a connected graph on $n \geq 3$ vertices. Then G is a cycle if and only if G is regular of degree 2.*

Proof. Proof is by induction on the number of vertices n. Note that when $n = 3$, the only connected graph on three vertices of degree 2 is the 3-cycle. Assume the lemma is true for some $n \geq 3$, and let G be a connected graph on n vertices containing n edges.

Let H be a connected graph constructed from G by appending one vertex v and an edge incident with v. The edge incident with v must also be incident with a vertex u of G, which is now of degree 3. To correct this, one edge incident with u and another vertex w must be removed, lowering the degree of w to 1. In order to make H regular of degree 2, a new edge incident with v and w is appended. This makes H a cycle on $n+1$ vertices. □

The next result makes clear the relationship between the number of edges in a graph and the degrees of its vertices. The name comes from counting handshakes among a group of people.

Lemma 2.2.3 (Handshaking Lemma). *If G is any graph of e edges, then*

$$\sum_{v \in V_G} \deg(v) = 2e.$$

Proof. Since each edge is incident with exactly two vertices, summing degrees over all vertices counts each edge exactly twice. □

When working with a finite graph G on n vertices, one often utilizes the *adjacency matrix* A associated with G. If the vertices are labeled $\{1, \ldots, n\}$, one defines A by

$$A_{ij} = \begin{cases} 1 & \text{if } v_i, v_j \text{ are adjacent} \\ 0 & \text{otherwise.} \end{cases}$$

A simple but useful result of this definition, which can also be generalized to directed graphs, is the following.

Proposition 2.2.4. *Let G be a graph on n vertices with associated adjacency matrix A. Then for any positive integer k, the (i,j)th entry of A^k is the number of k-walks $i \to j$. In particular, the entries along the main diagonal of A^k are the numbers of closed k-walks in G.*

Proof. Proof is by induction on k. Details are left as an exercise. □

Hypercubes

Hypercubes are regular polytopes frequently arising in computer science and combinatorics. They are intricately connected to Gray codes, and their graphs arise as Hasse diagrams of finite Boolean algebras. Hamilton cycles in hypercubes correspond to cyclic Gray codes and have received significant attention in light of the "Middle Levels Conjecture" originally proposed by I. Havel [111].

On the classical stochastic side, random walks on hypercubes are useful in modeling tree-structured parallel computations [63]. On the quantum side, hypercubes often appear in quantum random walks [12, 71]. Random walks on Clifford algebras have also been studied as random walks on directed hypercubes [86].

By considering certain generalizations of hypercubes, combinatorial properties can be obtained for tackling a variety of problems in graph theory and combinatorics [88, 96, 98]. By defining combinatorial raising and lowering operators on the associated semigroup algebras, an operator calculus (OC) on graphs is obtained, making graph-theoretic problems accessible via the tools of algebraic (quantum) probability [85, 87, 89, 91].

Given any finite set S of cardinality $|S|$, the *power set* of S, denoted 2^S, is the set comprised of all subsets of S. The cardinality of 2^S is easily shown to be $2^{|S|}$. Similarly given an orthonormal basis B for an inner product space V, the number of *canonical subspaces of* V, i.e., subspaces of V generated by subsets of B is $2^{|B|}$.

The n-dimensional cube, or *hypercube* $\mathcal{Q}_n$, is the graph whose vertices are in one-to-one correspondence with the n-tuples of zeros and ones and whose edges are the pairs of n-tuples that differ in exactly one position. This graph has natural applications in computer science, symbolic dynamics, and coding theory. The structure of the hypercube allows one to construct a random walk on the hypercube by "flipping" a randomly selected digit from 0 to 1 or vice versa.

Let $b \in \{0,1\}^n$ be a block, or *word*, of length n; that is, let b be a sequence of n zeros and ones. The *Hamming weight* of b, denoted $w_{\mathrm{H}}(b)$, is defined as the number of ones in the sequence. The *binary sum* of two such words is the sequence resulting from addition modulo-two of the two sequences.

The *Hamming distance* between two binary words is defined as the weight of their binary sum. In particular, given two binary strings $a = (a_1 a_2 \cdots a_n)$ and $b = (b_1 b_2 \cdots b_n)$, the Hamming distance between a and b, denoted $d_{\mathrm{H}}(a, b)$, is defined as the number of positions at which the strings differ:

$$d_{\mathrm{H}}(a, b) = |\{i : 1 \leq i \leq n, a_i \neq b_i\}|.$$

With the Hamming distance defined, the formal definition of the *hypercube* $\mathcal{Q}_n$ can be given.

Definition 2.2.5. The *n-dimensional hypercube*, $\mathcal{Q}_n$, is the graph whose vertices are the 2^n n-tuples from $\{0,1\}$ and whose edges are defined by the rule

$$\{v_1, v_2\} \in E(\mathcal{Q}_n) \Leftrightarrow w_{\mathrm{H}}(v_1 \oplus v_2) = 1.$$

Here $v_1 \oplus v_2$ is bitwise addition modulo-two, and w_{H} is the Hamming weight. In other words, two vertices of the hypercube are adjacent if and only if their Hamming distance is 1.

Fixing the set $B = \{e_1, \ldots, e_n\}$, the power set of B is in one-to-one correspondence with the vertices of $\mathcal{Q}_n$ via the binary subset representation

$$(a_1 a_2 \cdots a_n) \leftrightarrow e_I \Leftrightarrow a_i = \begin{cases} 1 & i \in I, \\ 0 & \text{otherwise.} \end{cases}$$

Of particular interest are some variations on the traditional hypercube defined above. First, the *looped hypercube* is the pseudograph obtained from the traditional hypercube $\mathcal{Q}_n$ by appending a loop at each vertex. In particular, ${\mathcal{Q}_n}^\circ = (V_\circ, E_\circ)$, where $V = V(\mathcal{Q}_n)$ and $E = E(\mathcal{Q}_n) \cup \{(v,v) : v \in V(\mathcal{Q}_n)\}$.

Definition 2.2.6. Let V denote the vertex set of the n-dimensional hypercube, $\mathcal{Q}_n$. Let αV denote the set obtained from V by appending the symbol α to each vertex in V. The Hamming weight of α is taken to be zero. A *signed hypercube* is a (possibly directed) graph G on vertex set $V \cup \alpha V$ such that

$$(u, w) \in E(G) \Rightarrow w_{\mathrm{H}}(u \oplus w) = 1.$$

Exercises

Exercise 2.1: Use the binomial theorem to expand $(x+1)^6$.

Exercise 2.2: Prove that the binomial coefficients satisfy *Pascal's recurrence* by showing that

$$\binom{n}{k} = \binom{n-1}{k} + \binom{n-1}{k-1}$$

for $0 < k < n$.

Exercise 2.3: Prove Vandermonde's identity. For nonnegative integers r, m, n, show that the binomial coefficients satisfy

$$\binom{m+n}{r} = \sum_{k=0}^{r} \binom{m}{k} + \binom{n}{r-k}.$$

Exercise 2.4: Prove Proposition 2.2.4.

PART II
Geometric Algebra

Chapter 3

Geometry of the Complex Plane

William Kingdon Clifford (1845-1879) introduced what is now termed "geometric algebra" by building on the work of Hermann Grassmann (1809-1877), who developed the exterior algebra. Clifford's geometric algebra is in fact a special case of a Clifford algebra.

A Clifford algebra is an algebra generated by a vector space with a quadratic form, and is a unital associative algebra. As $\mathbb{k}$-algebras, they generalize the real numbers, complex numbers, quaternions and other hypercomplex number systems. The theory of Clifford algebras is intimately connected with the theory of quadratic forms and orthogonal transformations.

This part of the book is intended to be a somewhat "traditional" introduction to Clifford algebras from a geometric point of view. The first few chapters will be spent on concepts of geometric algebra and Euclidean Clifford algebras. The first example of geometric algebra encountered by students is typically the field of complex numbers: $\mathbb{C} = \{a + bi : a, b \in \mathbb{R}, i^2 = -1\}$.

As a vector space over the real numbers, $\mathbb{C}$ is isomorphic to $\mathbb{R}^2$ via the linear transformation $a + bi \mapsto (a, b)$. Equipping $\mathbb{R}^2$ with (commutative) vector multiplication defined by

$$(a, b) * (c, d) := (ac - bd, ad + bc), \tag{3.1}$$

one can show that the mapping $a + bi \mapsto (a, b)$ is a *field isomorphism* from $\mathbb{C}$ to $\mathbb{R}^2$.

Given complex number $z = a + bi$, the real numbers a and b are respectively called the *real part* and *imaginary part* of z. The real and imaginary parts are respectively denoted by $\Re z$ and $\Im z$.

Writing $z = \Re z + i\Im z$, the *complex conjugate* of z is the complex number

$$\overline{z} = \Re z - i\Im z.$$

Geometrically, complex conjugation is correctly interpreted as *reflection across the real axis.* In other words, the transformation $z \mapsto \overline{z}$ corresponds to the plane transformation $(a,b) \mapsto (a,-b)$.

Writing $z = a + bi \in \mathbb{C}$, $z\overline{z} = a^2 + b^2$. The *modulus* of z is then given by the principal square root of $z\overline{z}$ and is denoted as follows:

$$|z| = \sqrt{z\overline{z}} = \sqrt{a^2 + b^2}.$$

Regarding z as a vector of $\mathbb{R}^2$ with the multiplication defined in (3.1), one sees that the modulus coincides with the inner product norm of $\mathbb{R}^2$; i.e., $|a + bi| = \|(a,b)\|$.

Observe that when $z \in \mathbb{C}$ is nonzero, the following holds:

$$z\frac{\overline{z}}{|z|^2} = \frac{z\overline{z}}{|z|^2} = 1.$$

It thereby follows that the (unique) multiplicative inverse of any nonzero complex number z is given by $z^{-1} = \dfrac{\overline{z}}{|z|^2}$.

Example 3.0.1. Letting $z = 2+4i \in \mathbb{C}$, the inverse of z is $z^{-1} = \overline{z}/(z\overline{z}) = (1-4i)/(20)$, as verified by

$$\begin{aligned}(2+4i)\frac{2-4i}{20} &= \frac{(2+4i)(2-4i)}{20} \\ &= (2^2+4^2)/20 \\ &= 1.\end{aligned}$$

Euler's Formula

A complex number expressed as $x + iy$ is said to be written in *rectangular form.* In light of the vector space isomorphism $\mathbb{C} \cong \mathbb{R}^2$, it is not difficult to convert the rectangular form to the *polar form*:

$$x + iy \mapsto r(\cos\theta + i\sin\theta),$$

where $r = \sqrt{x^2+y^2}$, $x = r\cos\theta$, and $y = r\sin\theta$. Here, θ is an angle measured from the positive real axis to the vector $(x,y) \in \mathbb{R}^2$.

The product of two complex numbers is computed as follows: writing $z_1 = r_1e^{i\theta_1}$ and $z_2 = r_2e^{i\theta_2}$, the product z_1z_2, the product z_1z_2 is given by

$$\begin{aligned}z_1z_2 &= r_1(\cos\theta_1 + i\sin\theta_1)r_2(\cos\theta_2 + i\sin\theta_2) \\ &= r_1r_2(\cos(\theta_1+\theta_2) + i\sin(\theta_1+\theta_2)).\end{aligned}$$

In other words, the modulus of the product is the product of the factors' moduli, and the argument of the product is the sum of the factors' arguments. Geometrically, this means that for fixed $u \in \mathbb{C}$, the mapping $z \mapsto zu$ is a *rotation* of z by the argument of u, coupled with a *dilation* ($|u| > 1$) or a *contraction* ($|u| < 1$) of z by $|u|$.

A more convenient representation of $z = x + iy$ is obtained by first recalling the Maclaurin series expansion of the exponential function, namely:

$$\exp(u) = \sum_{k=0}^{\infty} \frac{u^k}{k!}.$$

Setting $u = i\theta$ for real number θ, one obtains *Euler's formula*:

$$\begin{aligned} e^{i\theta} &= \sum_{k=0}^{\infty} \frac{(i\theta)^k}{k!} \\ &= \sum_{k \text{ even}} \frac{(i\theta)^k}{k!} + \sum_{k \text{ odd}} \frac{(i\theta)^k}{k!} \\ &= \sum_{k \text{ even}} \frac{(-1)^{k/2}\theta^k}{k!} + i \sum_{k \text{ odd}} \frac{(-1)^{\frac{k-1}{2}}\theta^k}{k!} \\ &= \cos\theta + i\sin\theta. \end{aligned}$$

The polar form $z = r(\cos\theta + i\sin\theta)$ now simplifies to the *exponential form of a complex number*, $z = re^{i\theta}$. The exponential form leads to some nice results with natural geometric implications.

Letting $z_1 = r_1e^{i\theta_1}$ and $z_2 = r_2e^{i\theta_2}$, the product z_1z_2 is quickly seen to be

$$z_1z_2 = r_1e^{i\theta_1}r_2e^{i\theta_2} = r_1r_2e^{i(\theta_1+\theta_2)}.$$

Extending to powers of z, one obtains the following theorem of De Moivre.

Theorem 3.0.2 (De Moivre's Formula). *Let $z \in \mathbb{C}$ be arbitrary and let $n \in \mathbb{N}$. Writing $z = re^{i\theta}$, the nth power of z is given by*

$$\begin{aligned} z^n &= r^n e^{in\theta} \\ &= r^n(\cos n\theta + i\sin n\theta). \end{aligned}$$

Working backward from De Moivre's Formula, one finds that for any positive integer k, every nonzero complex number has precisely k distinct k roots. These roots are evenly placed around the circle of radius $r^{1/k}$ centered at the origin in the complex plane.

Proposition 3.0.3. *Suppose $z = re^{i\theta}$, and let $k \in \mathbb{N}$. It follows that z has k distinct complex k^{th} roots, given by*

$$\omega_\ell = r^{1/k} e^{i\frac{\theta+2\ell\pi}{k}},$$

where ℓ runs from 0 to $k-1$.

In order to discuss rotations and reflections in $\mathbb{R}^2$, it is useful to recall the definition of the orthogonal group $O(2)$.

Definition 3.0.4. The *orthogonal group* $O(2)$ is the multiplicative group of 2×2 matrices having real entries satisfying the following conditions: $A \in O(2)$ if and only if $A^\dagger A = \mathbb{I}$ and $\det A = \pm 1$. In the case $\det A = -1$, A represents a *reflection*, while $\det A = 1$ indicates that A represents a rotation. The *special orthogonal group* $SO(2)$ is the subgroup of $O(2)$ defined by

$$SO(2) = \{A \in O(2) : \det A = 1\}.$$

A common tool for computing planar rotations is formalized in the following proposition.

Proposition 3.0.5. *Let $\theta \in \mathbb{R}$ and set $A = \begin{pmatrix} \cos\theta & -\sin\theta \\ \sin\theta & \cos\theta \end{pmatrix}$. Given arbitrary vector $\mathbf{x} \in \mathbb{R}^2$, $\mathbf{x}' = A\mathbf{x}$ is the vector obtained by rotating $\mathbf{x}$ about the origin by angle θ.*

Proof. Once can prove this proposition by showing that the mapping $(x_1, x_2) \mapsto (x_1\cos\theta - x_2\sin\theta, x_1\sin\theta + x_2\cos\theta)$ represents the stated rotation. In particular, when $\mathbf{x}$ has complex representation $\mathbf{x} = x_1 + ix_2 = re^{i\alpha}$, the result follows from

$$\begin{aligned}
\mathbf{x}' &= re^{i(\alpha+\theta)} \\
&= e^{i\theta} re^{i\alpha} \\
&= e^{i\theta}(x_1 + ix_2) \\
&= (\cos\theta + i\sin\theta)(x_1 + ix_2) \\
&= x_1\cos\theta - x_2\sin\theta + i(x_1\sin\theta + x_2\cos\theta).
\end{aligned}$$

The $\mathbb{R}^2$ representation of $\mathbf{x}'$ is therefore

$$\begin{pmatrix} x_1\cos\theta - x_2\sin\theta \\ x_1\sin\theta + x_2\cos\theta \end{pmatrix} = \begin{pmatrix} \cos\theta & -\sin\theta \\ \sin\theta & \cos\theta \end{pmatrix} \begin{pmatrix} x_1 \\ x_2 \end{pmatrix}.$$

□

Theorem 3.0.6. *Suppose* **u** *and* **v** *are unit vectors in the plane, separated by angle* α *measured from* **u** *to* **v**, *and let* ℓ_u, ℓ_v *denote the lines (i.e. subspaces) spanned by* **u** *and* **v**, *respectively. Let* **x** *be an arbitrary vector in the plane. Let* $\mathbf{x}'$ *be obtained by reflecting* **x** *across* ℓ_u, *and let* $\mathbf{x}''$ *be obtained from* $\mathbf{x}'$ *by reflection across* ℓ_v. *Then,* $\mathbf{x}''$ *is obtained from* **x** *by rotation through angle* 2α.

Proof. Let $u, v \in \mathbb{C}$ be the unit complex numbers corresponding to unit vectors **u** and **v** under the canonical vector space isomorphism $\mathbb{C} \cong \mathbb{R}^2$. In exponential form, one sees that $v\overline{u} = e^{i\alpha}$.

Recalling that the mapping $z \mapsto u\overline{z}u$ is the reflection of z across ℓ_u, the composition of the reflections across ℓ_u and ℓ_v is given by

$$\begin{aligned} z \mapsto u\overline{z}u \mapsto v\overline{u\overline{z}u} \\ &= (v\overline{u})z(\overline{u}v) \\ &= e^{i2\alpha}z. \end{aligned}$$

Identifying $z \in \mathbb{C}$ with vector $\mathbf{x} \in \mathbb{R}^2$ and writing $z = re^{i\theta}$, the rotation is clear from $e^{i2\alpha}z = re^{i(\theta+2\alpha)}$. □

Definition 3.0.7. The composition of a rotation with a reflection is called an *improper reflection* or a *reversal.*

Exercises

Exercise 3.1: Prove that $\mathbb{R}^2$ equipped with vector multiplication defined in (3.1) is field isomorphic to $\mathbb{C}$.

Exercise 3.2: Compute using the binomial theorem: $(1+i)^5$.

Exercise 3.3: Compute using De Moivre's Formula: $(1-i\sqrt{3})^6$.

Exercise 3.4: Let $z_0 \in \mathbb{C}$ be fixed. Define a function $f : \mathbb{C} \to \mathbb{C}$ such that for any $z \in \mathbb{C}$, $w = f(z)$ is obtained by rotating z about z_0 by angle $\pi/6$.

Exercise 3.5: Find all complex solutions to $z^8 - 256 = 0$. Express answers in exponential form and sketch them on a circle of appropriate radius.

Exercise 3.6: Let $u \in \mathbb{C}$ be a fixed unit complex number. Show that the mapping $z \mapsto u\bar{z}u$ corresponds to reflection of z across the line ℓ_u passing through u and the origin.

Exercise 3.7: Let $u \in \mathbb{C}$ be a fixed unit complex number. Show that the mapping $z \mapsto u\bar{z}u$ corresponds to reflection of z across the line ℓ_u passing through u and the origin.

Exercise 3.8: Let $u \in \mathbb{C}$ be a fixed unit complex number. Show that the mapping $z \mapsto -u\bar{z}u$ corresponds to reflection of z across the line $u_\perp$ orthogonal to the line ℓ_u as described in the previous exercise.

Exercise 3.9: Give a detailed geometric description of the transformation $f(z) = \left(-\frac{1}{2} + \frac{\sqrt{3}}{2}i\right)\bar{z}$.

Exercise 3.10: Find $w \in \mathbb{C}$ such that $2i = R_w{}^{\pi/6}(0)$. [Here, $R_w{}^{\theta}(z)$ denotes rotation of z by angle θ about fixed point w.]

Exercise 3.11: Compute using De Moivre's Formula: $(1 - i\sqrt{3})^6$.

Exercise 3.12: Establish the triangle inequality for complex numbers. Given $z, w \in \mathbb{C}$, prove that

$$|z + w| \leq |z| + |w|.$$

Exercise 3.13: Show that for positive integer n, the complex number $e^{i2\pi/n}$ generates a multiplicative subgroup of $U(1)$ having order n.

Exercise 3.14: Construct a set $X \in \mathbb{C}$ such that $U(1)$ acts transitively on X.

Exercise 3.15: Construct a finite subgroup $G < U(1)$ and a finite set $X \subset \mathbb{C}$ such that G acts transitively on X.

Exercise 3.16: Determine a transformation $f : \mathbb{C} \to \mathbb{C}$ such that the region on the right is the *image under* f of the region on the left. There can be multiple correct answers.

a.

b.

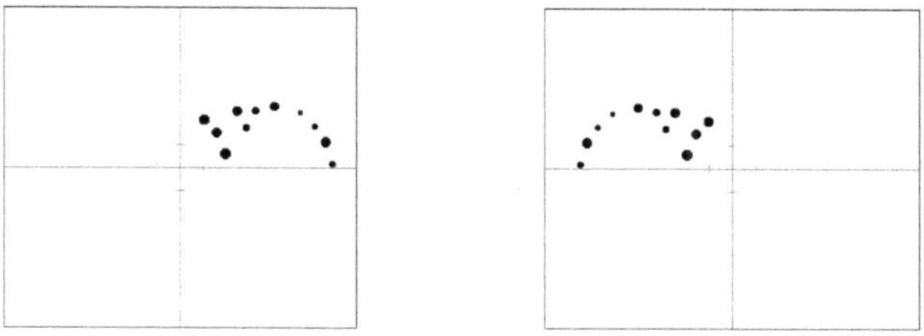

c.

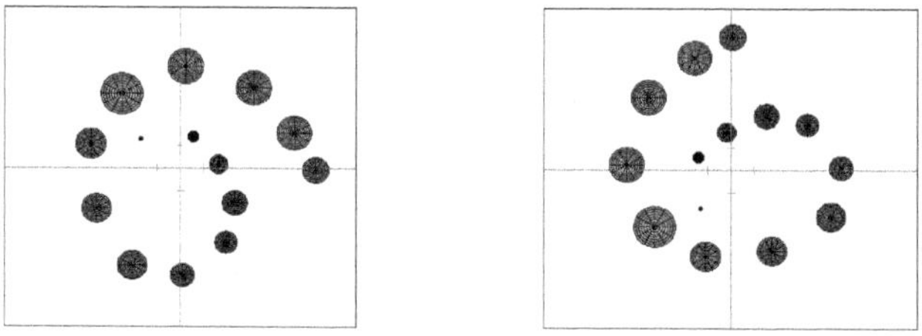

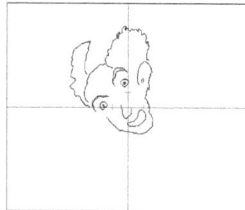

Chapter 4

Quaternions

Leonhard Euler (1707-1783) arithmetically related two-dimensional geometric rotation in the complex plane with the formula $e^{i\theta} = \cos\theta + i\sin\theta$ via the mapping $z \mapsto e^{i\theta}z$. William Rowan Hamilton (1805 - 1865) sought a similar mapping in three dimensions. His initial attempts involved triplets consisting of one real and two imaginary parts, but he was unable to find a successful way to multiply the triplets and thus could not form a division ring.

In 1843, while walking across Brougham Bridge with his wife, Hamilton had an inspirational thought. He carved what we now know as the fundamental rules of quaternion multiplication into the bridge itself:

$$\mathbf{i}^2 = \mathbf{j}^2 = \mathbf{k}^2 = \mathbf{ijk} = -1.$$

At this point, Hamilton had a non-Abelian multiplicative group of unit quaternions, denoted here by $\mathfrak{Q}$. The Cayley table of $\mathfrak{Q}$ is seen in Table 4.1.

Table 4.1 Multiplication in $\mathfrak{Q}$

	1	**i**	**j**	**k**	-1	-**i**	−**j**	−**k**
1	1	**i**	**j**	**k**	-1	-**i**	−**j**	−**k**
i	**i**	-1	**k**	-**j**	-**i**	1	-**k**	**j**
j	**j**	-**k**	-1	**i**	-**j**	**k**	1	-**i**
k	**k**	**j**	-**i**	-1	-**k**	-**j**	**i**	1
-1	-1	-**i**	-**j**	-**k**	1	**i**	**j**	**k**
-**i**	-**i**	1	-**k**	**j**	**i**	-1	**k**	-**j**
-**j**	-**j**	**k**	1	-**i**	**j**	-**k**	-1	**i**
-**k**	-**k**	-**j**	**i**	1	**k**	**j**	-**i**	-1

Remark 4.0.1. Writing $X = \begin{pmatrix} 1 & \mathbf{i} & \mathbf{j} & \mathbf{k} \\ \mathbf{i} & -1 & \mathbf{k} & -\mathbf{j} \\ \mathbf{j} & -\mathbf{k} & -1 & \mathbf{i} \\ \mathbf{k} & \mathbf{j} & -\mathbf{i} & -1 \end{pmatrix}$, the Cayley table of $\mathfrak{Q}$ is of the form $(\sigma_0 - \sigma_{\mathrm{x}}) \otimes X$, where $\otimes$ denotes the Kronecker product of matrices.

The group of unit quaternions extends naturally to an algebra of quaternions.

Definition 4.0.2. The *algebra of quaternions* is the group algebra of $\mathfrak{Q}$ over $\mathbb{R}$:

$$\mathbb{H} := \mathbb{R}\mathfrak{Q} = \{x_0 + x_1\mathbf{i} + x_2\mathbf{j} + x_3\mathbf{k} : x_0, x_1, x_2, x_3 \in \mathbb{R}\}.$$

An element $\mathbf{q} = x_0 + x_1\mathbf{i} + x_2\mathbf{j} + x_3\mathbf{k} \in \mathbb{H}$ is called a *pure quaternion* if $x_0 = 0$.

The quaternions allowed Hamilton to achieve his goal of arithmetically representing an object's orientation and its rotation in three dimensions. Pure quaternions are naturally in one-to-one correspondence with vectors in $\mathbb{R}^3$. In particular, a vector $\mathbf{x} = (x_1, x_2, x_3) \in \mathbb{R}^3$ can be represented by a *pure quaternion* as follows:

$$\mathbf{x} \mapsto x_1\mathbf{i} + x_2\mathbf{j} + x_3\mathbf{k}.$$

The quaternions $\mathbb{H}$ form a normed division algebra[1]. Given a quaternion $q = q_0 + q_1\mathbf{i} + q_2\mathbf{j} + q_3\mathbf{k}$, it is natural to write q as the sum of a real scalar q_0 and a pure quaternion $\mathbf{q} = q_1\mathbf{i} + q_2\mathbf{j} + q_3\mathbf{k}$ identified with a vector $\mathbf{q}$ in $\mathbb{R}^3$. In this case, $q \in \mathbb{H}$ is conveniently written as

$$q = q_0 + \mathbf{q}.$$

Given a quaternion $q = q_0 + \mathbf{q} \in \mathbb{H}$, one defines the *(quaternion) conjugate* of q as

$$\overline{q} = q_0 - \mathbf{q}.$$

Observing that $q\overline{q} = \overline{q}q \in \mathbb{R}$ for all $q \in \mathbb{H}$, a *norm* is defined on $\mathbb{H}$ by defining

$$\|q\| = (q\overline{q})^{1/2}.$$

In fact,

$$\begin{aligned} q\overline{q} &= {q_0}^2 - q_0\mathbf{q} + \mathbf{q}q_0 - \mathbf{q}^2 \\ &= {q_0}^2 - (q_1\mathbf{i} + q_2\mathbf{j} + q_3\mathbf{k})(q_1\mathbf{i} + q_2\mathbf{j} + q_3\mathbf{k}) \\ &= {q_0}^2 + {q_1}^2 + {q_2}^2 + {q_3}^2 - (q_2q_3 - q_3q_2)\mathbf{i} - (q_3q_1 - q_1q_3)\mathbf{j} - (q_1q_2 - q_2q_1)\mathbf{k} \\ &= {q_0}^2 + \|\mathbf{q}\|^2, \end{aligned}$$

[1]Every nonzero element of $\mathbb{H}$ has a multiplicative inverse.

where $\|\mathbf{q}\|$ coincides with the Euclidean norm of $\mathbf{q}$ viewed as a vector in $\mathbb{R}^3$; that is, $\|\mathbf{q}\| = ({q_1}^2 + {q_2}^2 + {q_3}^2)^{1/2}$. When $\|q\| = 1$, q is said to be a *unit quaternion.*

Given a nonzero quaternion q, the *(multiplicative) inverse* of q is given by $q^{-1} = \dfrac{\overline{q}}{q\overline{q}}$. This is quickly verified by observing that

$$q\frac{\overline{q}}{q\overline{q}} = \frac{q\overline{q}}{q\overline{q}} = 1.$$

Given two pure quaternions $\mathbf{u}, \mathbf{v} \in \mathbb{H}$, it is natural to define the quaternion dot product and cross product, which correlates nicely with the dot and cross products of $\mathbb{R}^3$. In particular, defining

$$\mathbf{u} \cdot \mathbf{v} = -\frac{1}{2}(\mathbf{uv} + \mathbf{vu}),$$

one sees that

$$\begin{aligned}\mathbf{u} \cdot \mathbf{v} &= -\frac{1}{2}(-2u_1v_1 - 2u_2v_2 - 2u_3v_3)\\ &= u_1v_1 + u_2v_2 + u_3v_3,\end{aligned}$$

which coincides with the standard dot product $\mathbf{u} \cdot \mathbf{v}$ when $\mathbf{u}$ and $\mathbf{v}$ are viewed as vectors of $\mathbb{R}^3$.

Similarly, defining

$$\mathbf{u} \times \mathbf{v} = \frac{1}{2}(\mathbf{uv} - \mathbf{vu}),$$

one obtains

$$\begin{aligned}\mathbf{u} \times \mathbf{v} &= \frac{1}{2}\left(2(q_2q_3 - q_3q_2)\mathbf{i} + 2(q_3q_1 - q_1q_3)\mathbf{j} + 2(q_1q_2 - q_2q_1)\mathbf{k}\right)\\ &= (q_2q_3 - q_3q_2)\mathbf{i} + (q_3q_1 - q_1q_3)\mathbf{j} + (q_1q_2 - q_2q_1)\mathbf{k},\end{aligned}$$

coinciding with the standard cross product when $\mathbf{u}$ and $\mathbf{v}$ are viewed as vectors of $\mathbb{R}^3$. The following useful lemma expresses the product of pure quaternions in terms of dot and cross products.

Lemma 4.0.3. *Let* $\mathbf{u}, \mathbf{v} \in \mathbb{H}$ *be pure quaternions. Then,*

$$\mathbf{uv} = -\mathbf{u} \cdot \mathbf{v} + \mathbf{u} \times \mathbf{v}.$$

Moreover,

$$\mathbf{u}\overline{\mathbf{v}} = \overline{\mathbf{u}}\mathbf{v} = \mathbf{u} \cdot \mathbf{v} - \mathbf{u} \times \mathbf{v}.$$

Proof. Proof is by direct computation:

$$\begin{aligned}-\mathbf{u} \cdot \mathbf{v} + \mathbf{u} \times \mathbf{v} &= \frac{1}{2}(\mathbf{uv} + \mathbf{vu}) + \frac{1}{2}(\mathbf{uv} - \mathbf{vu})\\ &= \frac{1}{2}(\mathbf{uv} + \mathbf{uv} + \mathbf{vu} - \mathbf{vu}\\ &= \mathbf{uv}.\end{aligned}$$

Further, when $\mathbf{u}$ is a pure quaternion, one sees that $\overline{\mathbf{u}} = -\mathbf{u}$. □

4.1 Reflections and Rotations in $\mathbb{R}^3$

Consider now the problem of rotations in Euclidean three-space. In order to specify such a rotation, one needs an angle and an axis of rotation or equivalently the plane normal to the axis of rotation.

Our first goal is to construct a 3×3 matrix A whose action on $\mathbb{R}^3$ is rotation by an angle θ about a unit vector $\mathbf{u}$. One intuitive, although complicated approach is based on Proposition 3.0.5 for rotations in $\mathbb{R}^2$.

Let $\mathbf{u} \in \mathbb{R}^3$ be a unit vector lying along a chosen (fixed) axis of rotation, such that positive angle α satisfies right-hand orientation. Using the results obtained for the plane, construct a 3×3 matrix A such that for arbitrary $\mathbf{x} \in \mathbb{R}^3$, $\mathbf{x}' = A\mathbf{x}$ represents the vector obtained by rotating $\mathbf{x}$ about the axis of rotation through angle α.

- Represent $\mathbf{u} = (u_1, u_2, u_3)$ in spherical coordinates:

$$\begin{aligned} u_1 &= \sin\varphi\cos\theta, \\ u_2 &= \sin\varphi\sin\theta, \\ u_3 &= \cos\varphi. \end{aligned}$$

- Apply a rotation in one of the coordinate planes of $\mathbb{R}^3$, fixing one component and applying a planar rotation matrix such that $\mathbf{u}$ lies in one of the coordinate planes. Label the corresponding matrix ρ_1.
- Apply another rotation in one of the coordinate planes of $\mathbb{R}^3$ such that $\mathbf{u}$ coincides with one of the basis vectors of $\mathbb{R}^3$. Label this matrix ρ_2.
- Apply the rotation by α in this coordinate plane. Label this matrix ρ_α.
- Undo the previous rotations with matrices ${\rho_2}^{-1}$ and ${\rho_1}^{-1}$.
- The desired matrix is the composition $A = {\rho_1}^{-1}{\rho_2}^{-1}\rho_\alpha\rho_2\rho_1$.

Based on Proposition 3.0.5, the following three matrices in $SO(3)$ act on $\mathbb{R}^3$ by rotating a vector $\mathbf{x}$ about the coordinate axes through an angle ω (with right-handed orientation).

$$R_{\mathfrak{i}}(\omega) := \begin{pmatrix} 1 & 0 & 0 \\ 0 & \cos\omega & -\sin\omega \\ 0 & \sin\omega & \cos\omega \end{pmatrix}, \quad R_{\mathfrak{j}}(\omega) := \begin{pmatrix} \cos\omega & 0 & \sin\omega \\ 0 & 1 & 0 \\ -\sin\omega & 0 & \cos\omega \end{pmatrix},$$

and

$$R_{\mathbf{k}}(\omega) := \begin{pmatrix} \cos\omega & -\sin\omega & 0 \\ \sin\omega & \cos\omega & 0 \\ 0 & 0 & 1 \end{pmatrix}.$$

Recalling the correspondence between rectangular and spherical coordinates in $\mathbb{R}^3$:

$$\begin{aligned} x &= \rho\cos\theta\sin\varphi, \\ y &= \rho\sin\theta\sin\varphi, \\ z &= \rho\cos\varphi. \end{aligned}$$

One method of rotating through angle ω about an arbitrary unit vector $\mathbf{u} = (1, \theta, \varphi)$ (written using spherical coordinates) is by composition of rotations:

$$\mathbf{x}' = R_{\mathbf{u}}(\omega)\mathbf{x} = R_{\mathbf{k}}(\theta)R_{\mathbf{j}}(\varphi)R_{\mathbf{k}}(\omega)R_{\mathbf{j}}(-\varphi)R_{\mathbf{k}}(-\theta)\mathbf{x}. \tag{4.1}$$

Noting that the inverse of counterclockwise rotation is clockwise rotation,

$$\mathbf{x}' = R_{\mathbf{u}}(\omega)\mathbf{x} = R_{\mathbf{k}}(\theta)R_{\mathbf{j}}(\varphi)R_{\mathbf{k}}(\omega)R_{\mathbf{j}}^{-1}(\varphi)R_{\mathbf{k}}^{-1}(\theta)\mathbf{x} = \psi_{\mathbf{u}}R_{\mathbf{k}}(\omega)\psi_{\mathbf{u}}^{-1}\mathbf{x},$$

where $\psi_{\mathbf{u}} = R_{\mathbf{k}}(\theta)R_{\mathbf{k}}(\varphi)$. In other words,

$$R_{\mathbf{u}}(\omega) = \psi_{\mathbf{u}}R_{\mathbf{k}}(\omega)\psi_{\mathbf{u}}^{-1}.$$

Reflections

Based on this intuitive approach, our second goal is to construct a 3×3 matrix B whose action on $\mathbb{R}^3$ is reflection through the plane with unit normal vector $\mathbf{u}$.

As before, the vector $\mathbf{u}$ is rotated to coincide with one of the coordinate vectors. This time, the reflection will be accomplished by one of the following coordinate axis matrices before restoring $\mathbf{u}$ to its original position:

$$X_{\mathbf{i}} := \begin{pmatrix} -1 & 0 & 0 \\ 0 & 1 & 0 \\ 0 & 0 & 1 \end{pmatrix}, \quad X_{\mathbf{j}} := \begin{pmatrix} 1 & 0 & 0 \\ 0 & -1 & 0 \\ 0 & 0 & 1 \end{pmatrix}, \quad X_{\mathbf{k}} := \begin{pmatrix} 1 & 0 & 0 \\ 0 & 1 & 0 \\ 0 & 0 & -1 \end{pmatrix}.$$

In this case, the reflection across the plane orthogonal to $\mathbf{u} = \langle 1, \theta, \varphi \rangle$ achieved via the mapping

$$\begin{aligned} \mathbf{x}' &= X_{\mathbf{u}}\mathbf{x} \\ &= R_{\mathbf{k}}(\theta)R_{\mathbf{j}}(\varphi)X_{\mathbf{k}}R_{\mathbf{j}}^{-1}(\varphi)R_{\mathbf{k}}^{-1}(\theta)\mathbf{x}. \end{aligned}$$

Rotations via Reflections

Suppose **u** and **v** are unit vectors in $\mathbb{R}^3$, separated by angle α measured from **u** to **v** as pictured in Figure 4.1. Let **x** be an arbitrary vector in the **uv**-plane and let $\mathbf{u}_\perp$, $\mathbf{v}_\perp$ denote the planes orthogonal to **u** and **v**, respectively. Let $\mathbf{x}'$ be obtained by reflecting **x** across $u_\perp$, and let $\mathbf{x}''$ be obtained from $\mathbf{x}'$ by reflection across $v_\perp$. Identifying the **uv**-plane with $\mathbb{C}$, it follows from Theorem 3.0.6 that $\mathbf{x}''$ is obtained from **x** by rotation through angle 2α in the plane generated by **u** and **v**.

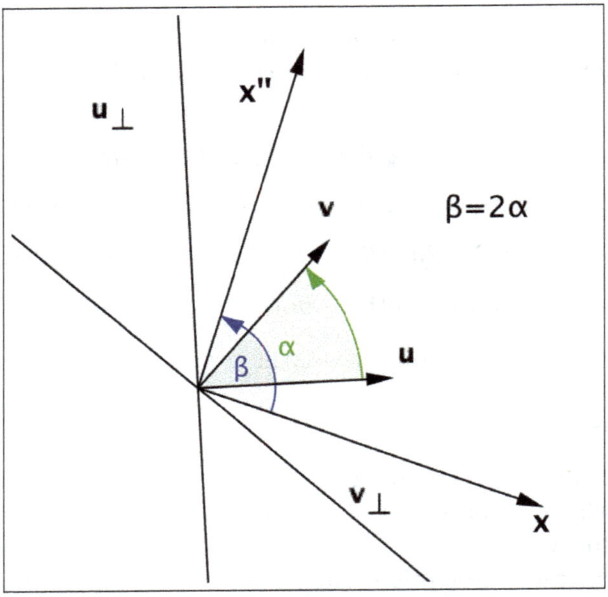

Figure 4.1 Rotation from two reflections in the **uv**-plane.

Extending to arbitrary $\mathbf{x} \in \mathbb{R}^3$, one writes $\mathbf{x} = \mathbf{x}_{uv} + \mathbf{x}'$, where $\mathbf{x}'$ is normal to the **uv**-plane. It follows that $\mathbf{x}'$ is parallel to the line of intersection of the planes $\mathbf{u}_\perp$ and $\mathbf{v}_\perp$ (i.e., the axis of rotation) and is thus invariant under reflections in these planes.

Note that when A defines a rotation, the eigenvectors of A lie along the axis of rotation (in $\mathbb{R}^3$). When A is a reflection, the reflection is through the plane orthogonal to the eigenvectors.

4.2 Quaternions and Geometry in $\mathbb{R}^3$

For convenience, the notation $\mathbf{u}$ will be taken to mean either the vector $(u_1, u_2, u_3) \in \mathbb{R}^3$ *or* the pure quaternion $u_1\mathbf{i} + u_2\mathbf{j} + u_3\mathbf{k}$, depending on context. Writing $\overline{\mathbf{q}} = x_0 - \mathbf{x}$ whenever $\mathbf{q} = x_0 + \mathbf{x}$, we see that $q\overline{q} = \|\mathbf{q}\|^2$, where $\|\mathbf{q}\|$ is the *norm* or $\mathbf{q}$. It follows that the inverse of q is given by

$$\mathbf{q}^{-1} = \frac{\overline{\mathbf{q}}}{\|\mathbf{q}\|^2}.$$

Lemma 4.2.1. *Conjugation is an* anti-involution *on* $\mathbb{H}$. *In particular, for* $\mathbf{u}, \mathbf{v} \in \mathbb{H}$, *conjugation satisfies the following:*

(1) $\overline{\mathbf{uv}} = \overline{\mathbf{v}}\,\overline{\mathbf{u}}$;
(2) $\overline{\overline{\mathbf{u}}} = \mathbf{u}$.

Proof. Write $\mathbf{v} = v_0 + \mathbf{v}'$ and $\mathbf{u} = u_0 + \mathbf{u}'$. Then,

$$\begin{aligned}
\overline{\mathbf{uv}} &= \overline{(u_0 + \mathbf{u}')(v_0 + \mathbf{v}')} \\
&= \overline{u_0 v_0 + u_0 \mathbf{v}' + \mathbf{u}' v_0 + \mathbf{u}'\mathbf{v}'} \\
&= u_0 v_0 - u_0 \mathbf{v}' - \mathbf{u}' v_0 - \mathbf{u}' \cdot \mathbf{v}' - \mathbf{u}' \times \mathbf{v}' \\
&= v_0 u_0 - \mathbf{v}' u_0 - v_0 \mathbf{u}' - \mathbf{v}' \cdot \mathbf{u}' + \mathbf{v}' \times \mathbf{u}' \\
&= v_0 u_0 - \mathbf{v}' u_0 - v_0 \mathbf{u}' + \mathbf{v}'\mathbf{u}' \\
&= (v_0 - \mathbf{v}')(u_0 - \mathbf{u}').
\end{aligned}$$

Moreover, $\overline{\overline{\mathbf{u}}} = \overline{u_0 - \mathbf{u}'} = u_0 + \mathbf{u}' = \mathbf{u}$. □

Lemma 4.2.2. *Let* $\mathbf{u}$ *be a fixed pure unit quaternion, and let* $\mathbf{x}$ *be an arbitrary pure quaternion. Then,* $\mathbf{x}' = \overline{\mathbf{u}\mathbf{x}\overline{\mathbf{u}}}$ *is a pure quaternion representing the reflection of vector* $\mathbf{x}$ *across the plane orthogonal to unit vector* $\mathbf{u}$ *in* $\mathbb{R}^3$.

Proof. Writing $\mathbf{x} = \mathbf{x}_u + \mathbf{x}_{u'}$, where $\mathbf{x}_u$ is parallel to $\mathbf{u}$ and $\mathbf{x}_{u'}$ is orthogonal to $\mathbf{u}$, it follows that

$$\begin{aligned}
\overline{\mathbf{u}\mathbf{x}\overline{\mathbf{u}}} &= \overline{\mathbf{u}(\mathbf{x}_u + \mathbf{x}_{u'})\overline{\mathbf{u}}} \\
&= \overline{\mathbf{u}\mathbf{x}_u\overline{\mathbf{u}} + \mathbf{u}\mathbf{x}_{u'}\overline{\mathbf{u}}} \\
&= \overline{\mathbf{u}\mathbf{x}_u\overline{\mathbf{u}}} + \overline{\mathbf{u}\mathbf{x}_{u'}\overline{\mathbf{u}}} \\
&= -\mathbf{x}_u + \mathbf{x}_{u'}.
\end{aligned}$$

□

The next theorems give a geometric algebra approach to rotations in $\mathbb{R}^3$.

Theorem 4.2.3 (Quaternion Rotation Formula). *Let $\mathbf{w}$ be a pure unit quaternion, let $\theta \in \mathbb{R}$, and let $\mathbf{x}$ be an arbitrary pure quaternion. Setting $\mathbf{q} = \cos(\theta/2) + \sin(\theta/2)\mathbf{w}$, the mapping $\mathbf{x} \mapsto \mathbf{q}\mathbf{x}\overline{\mathbf{q}}$ corresponds to rotation of $\mathbf{x}$ about the unit vector $\mathbf{w}$ through angle θ with right-hand orientation in $\mathbb{R}^3$.*

Proof. Let $\mathbf{w}$ be a pure unit quaternion. Let $\mathbf{u}$ and $\mathbf{v}$ be pure unit quaternions orthogonal to $\mathbf{w}$, separated by angle $\theta/2$, such that $\mathbf{w}$ is a positive scalar multiple of $\mathbf{u} \times \mathbf{v}$. Note that $\mathbf{w}$ lies in the same direction as $\mathbf{u} \times \mathbf{v}$.

Let $\mathbf{x}$ be an arbitrary pure quaternion. By Lemma 4.2.2, the mapping $\mathbf{x} \mapsto \overline{\mathbf{u}\mathbf{x}\overline{\mathbf{u}}}$ is the reflection of $\mathbf{x}$ in the plane $\mathbf{u}_\perp$. Applying the anti-involution property of quaternion conjugation, we see that the transformation

$$\begin{aligned}
\overline{\mathbf{u}\mathbf{x}\overline{\mathbf{u}}} &\mapsto \overline{\mathbf{v}\,\overline{\mathbf{u}\mathbf{x}\overline{\mathbf{u}}}\,\overline{\mathbf{v}}} \\
&= \overline{\overline{\mathbf{v}}}\,\overline{\overline{\overline{\mathbf{u}}}}\,\overline{\overline{\mathbf{x}}}\,\overline{\overline{\mathbf{u}}}\,\overline{\mathbf{v}} \\
&= \mathbf{v}\overline{\mathbf{u}}\mathbf{x}\mathbf{u}\overline{\mathbf{v}} \\
&= (-\mathbf{v}\mathbf{u})\mathbf{x}(-\mathbf{u}\mathbf{v}) \\
&= \overline{\mathbf{v}}\mathbf{u}\mathbf{x}\overline{\mathbf{u}}\mathbf{v}
\end{aligned}$$

is the composition of reflections in $\mathbf{u}_\perp$ and $\mathbf{v}_\perp$, respectively. By Theorem 3.0.6, this transformation is a rotation by $2(\theta/2) = \theta$ in the $\mathbf{uv}$-plane. Applying Lemma 4.0.3,

$$\overline{\mathbf{v}}\mathbf{u} = \mathbf{u} \cdot \mathbf{v} + \mathbf{u} \times \mathbf{v} = \cos(\theta/2) + \sin(\theta/2)\mathbf{w}.$$

Finally, we see that setting $\mathbf{q} = \cos(\theta/2) + \sin(\theta/2)\mathbf{w}$ gives

$$(\overline{\mathbf{v}}\mathbf{u})\mathbf{x}(\overline{\mathbf{u}}\mathbf{v}) = (\overline{\mathbf{v}}\mathbf{u})\mathbf{x}(\overline{\overline{\mathbf{v}}\mathbf{u}}) = \mathbf{q}\mathbf{x}\overline{\mathbf{q}}.$$

□

Matrix Approach to Reflections and Rotations

Dirac notation is often convenient for distinguishing between row vectors and column vectors. A column vector is denoted by $|v\rangle$, while a row vector is denoted by $\langle v|$. In this way, the inner product of vectors $\mathbf{u}$ and $\mathbf{v}$ is easily denoted by $\langle \mathbf{u}|\mathbf{v}\rangle$. With this notation, the outer product $\mathbf{v}\mathbf{v}^\dagger$ is represented as $|\mathbf{v}\rangle\langle\mathbf{v}|$.

Recall that an element r in a ring R is said to be *idempotent* if $r^2 = r$. This definition extends naturally to algebras. In particular, projection operators are always idempotent. Idempotents are commonly used to generate ideals in Clifford algebras.

Lemma 4.2.4. *Let* $\mathbf{u}$ *be a unit column vector in* $\mathbb{R}^n$. *The operator* $|\mathbf{u}\rangle\langle\mathbf{u}| = \mathbf{u}\mathbf{u}^\dagger$ *is a rank-one orthogonal projection onto* $\text{span}(\{\mathbf{u}\})$.

Proof. Observing that $\mathbf{u}^\dagger\mathbf{u} = 1$, there exist vectors $\mathbf{v}_1, \mathbf{v}_2 \in \mathbb{R}^3$ such that $\beta = \{\mathbf{u}, \mathbf{v}_1, \mathbf{v}_2\}$ is an orthonormal basis for $\mathbb{R}^3$. Consequently, $\mathbf{u}^\dagger\mathbf{v}_j = 0$ for $j = 1, 2$. Letting $\mathbf{x} \in \mathbb{R}^3$ be arbitrary and writing $\mathbf{x} = x_u\mathbf{u} + x_1\mathbf{v}_1 + x_2\mathbf{v}_2$, it follows that

$$\begin{aligned}(|\mathbf{u}\rangle\langle\mathbf{u}|)\mathbf{x} &= \mathbf{u}\mathbf{u}^\dagger(x_u\mathbf{u} + x_1\mathbf{v}_1 + x_2\mathbf{v}_2)\\ &= x_u\mathbf{u}\mathbf{u}^\dagger\mathbf{u} + x_1\mathbf{u}\mathbf{u}^\dagger\mathbf{v}_1 + x_2\mathbf{u}\mathbf{u}^\dagger\mathbf{v}_2\\ &= x_u\mathbf{u}(\mathbf{u}^\dagger\mathbf{u})\\ &= x_u\mathbf{u}.\end{aligned}$$

□

Corollary 4.2.5. *Let* $\mathbf{u}$ *be a unit vector in* $\mathbb{R}^n$. *The operator*

$$\mathbb{I} - |\mathbf{u}\rangle\langle\mathbf{u}|$$

is an orthogonal projection onto the hyperplane with normal $\mathbf{u}$.

Proof. Proof is by direct computation and is left as an exercise. □

Corollary 4.2.6. *Let* $\mathbf{u} \in \mathbb{R}^n$ *be a unit vector. The operator* $\mathbb{I} - 2|\mathbf{u}\rangle\langle\mathbf{u}|$ *reflects arbitrary vectors through the hyperplane orthogonal to* $\mathbf{u}$.

Proof. Let $\mathbf{x} = \mathbf{x}_u + \mathbf{x}_{u'}$ where $\mathbf{x}_u \in \text{span}(\{\mathbf{u}\})$ and $\mathbf{x}_{u'} \in \{\mathbf{u}\}^\top$. In light of Corollary 4.2.5,

$$\begin{aligned}(\mathbb{I} - 2|\mathbf{u}\rangle\langle\mathbf{u}|)\mathbf{x} &= (\mathbb{I} - 2|\mathbf{u}\rangle\langle\mathbf{u}|)(\mathbf{x}_u + \mathbf{x}_{u'})\\ &= \mathbf{x}_u + \mathbf{x}_{u'} - 2|\mathbf{u}\rangle\langle\mathbf{u}|\mathbf{x}_u\rangle - 2|\mathbf{u}\rangle\langle\mathbf{u}|\mathbf{x}_{u'}\rangle\\ &= \mathbf{x}_u + \mathbf{x}_{u'} - 2\mathbf{x}_u\\ &= \mathbf{x}_{u'} - \mathbf{x}_u.\end{aligned}$$

In other words, $\mathbb{I} - 2|\mathbf{u}\rangle\langle\mathbf{u}|$ leaves the component of $\mathbf{x}$ orthogonal to $\mathbf{u}$ fixed, but reverses the direction of the component parallel to $\mathbf{u}$. □

The action of the reflection operator is seen in Figure 4.2. With the hyperplane reflection operator $\mathbb{I} - 2|\mathbf{u}\rangle\langle\mathbf{u}|$ in hand, we are now ready to return to the problem of constructing a rotation operator for $\mathbb{R}^n$. The next theorem gives a straightforward construction of $n \times n$ rotation matrices acting on $\mathbb{R}^n$.

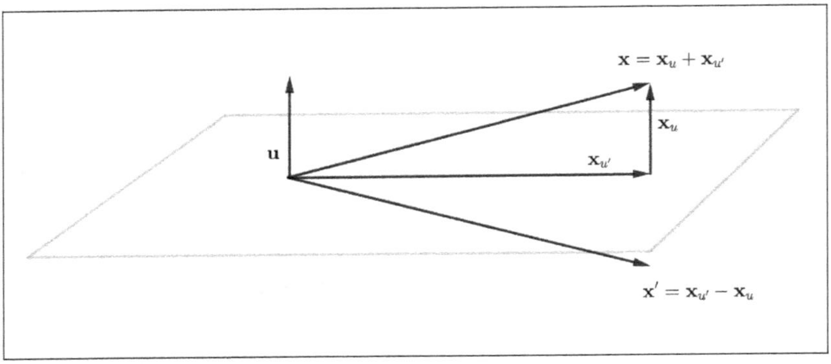

Figure 4.2 Reflection of $\mathbf{x}$ across plane normal to $\mathbf{u}$.

Theorem 4.2.7. *Let* $\mathbf{u}, \mathbf{v}$ *denote unit vectors lying in the desired plane of rotation, such that the angle measured from* $\mathbf{u}$ *to* $\mathbf{v}$ *(assuming right-hand orientation) is* $\alpha/2$*. Then, the operator on* $\mathbb{R}^n$ *that acts by rotation through angle* α *in the* $\mathbf{uv}$*-plane is*

$$\mathbb{I} - 2|\mathbf{v}\rangle\langle\mathbf{v}| - 2|\mathbf{u}\rangle\langle\mathbf{u}| + 4\langle\mathbf{v}|\mathbf{u}\rangle|\mathbf{v}\rangle\langle\mathbf{u}|.$$

Proof. For convenience, Dirac notation is replaced with row and column vectors. Considering the composition

$$\begin{aligned}(\mathbb{I} - 2\mathbf{v}\mathbf{v}^\dagger)(\mathbb{I} - 2\mathbf{u}\mathbf{u}^\dagger) &= \mathbb{I} - 2\mathbf{v}\mathbf{v}^\dagger - 2\mathbf{u}\mathbf{u}^\dagger + 4\mathbf{v}\mathbf{v}^\dagger\mathbf{u}\mathbf{u}^\dagger \\ &= \mathbb{I} - 2\mathbf{v}\mathbf{v}^\dagger - 2\mathbf{u}\mathbf{u}^\dagger + 4\langle\mathbf{v}, \mathbf{u}\rangle\mathbf{v}\mathbf{u}^\dagger,\end{aligned}$$

the result now follows from Theorem 3.0.6 after identifying the complex plane with the $\mathbf{uv}$-plane. □

Application: Rotating a Wireframe Object with Mathematica

As illustration of the quaternion rotation formula, a wireframe cube is defined by specifying a set of vertices and connecting them pairwise in an appropriate manner. The cube is then rotated about a given axis by applying the same rotation to all of the vertices.

Choosing a relatively small angle of rotation, iterations of the mapping generate "frames" suitable for animation. It is a simple matter to then animate the rotation of the cube about the axis using *Mathematica*.

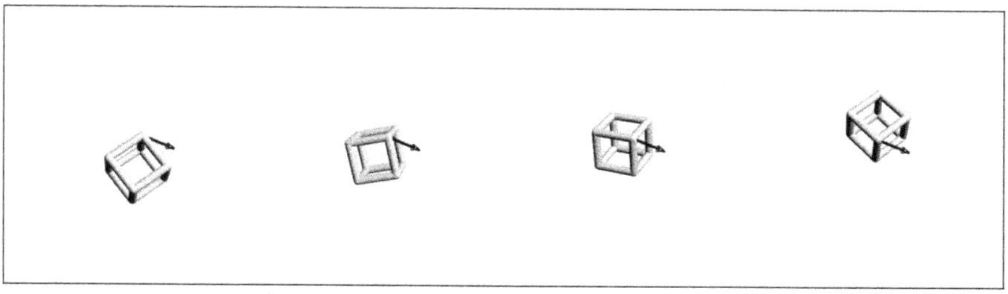

Figure 4.3 Rotating a wireframe cube.

Exercises

Exercise 4.1: Show that defining $||q|| = (\overline{q}q)^{1/2}$ establishes a norm on $\mathbb{H}$ by showing that

(1) $||q|| \geq 0$ for all $q \in \mathbb{H}$, with equality implying $q = 0$;
(2) $||aq|| = |a|||q||$ for $q \in \mathbb{H}$ and $a \in \mathbb{R}$;
(3) $||u + v|| \leq ||u|| = ||v||$ for $u, v \in \mathbb{H}$.

In other words, the mapping $||\cdot|| : \mathbb{H} \to \mathbb{R}$ is positive definite and homogeneous, and satisfies the Triangle Inequality.

Exercise 4.2: Let $\mathbf{u}$ be a pure unit quaternion, and let $\mathbf{x} \in \mathbb{H}$ be an arbitrary pure quaternion. Show that

$$\mathbf{ux\overline{u}} = \mathbf{\overline{u}xu}.$$

Exercise 4.3: Let u, v, w be pure quaternions. Derive a quaternion formula for the volume of the parallelepiped determined by the corresponding vectors $\mathbf{u}, \mathbf{v}, \mathbf{w}$ in $\mathbb{R}^3$. Hint: Recall the dot and cross product vector formulation.

Exercise 4.4: Let $\mathbf{u}$ be a fixed pure unit quaternion, and let $\mathbf{x}$ be an arbitrary pure quaternion. Show that $\mathbf{x}' = \overline{\mathbf{ux\overline{u}}}$ is a pure quaternion representing the reflection of vector $\mathbf{x}$ across the plane orthogonal to unit vector $\mathbf{u} \in \mathbb{R}^3$.

Exercise 4.5: Let $\mathbf{u}$ be a unit column vector in $\mathbb{R}^n$. Show that $|\mathbf{u}\rangle\langle\mathbf{u}|$ is idempotent.

Exercise 4.6: Let $\mathbf{u} \in \mathbb{R}^n$ be a unit vector. Show that the operator $\mathbb{I} - 2|\mathbf{u}\rangle\langle\mathbf{u}|$ reflects arbitrary vectors through the hyperplane orthogonal to $\mathbf{u}$.

Exercise 4.7: Construct a quaternion $\mathbf{q}$ so that the map $\mathbf{x} \mapsto \mathbf{q}\mathbf{x}\overline{\mathbf{q}}$ represents a rotation by angle $\pi/3$ about the line L: $x = y = z$ in $\mathbb{R}^3$. Use this quaternion to rotate $(1, 0, 0)$ about L. (For orientation, consider the unit vector parallel to L that points into the first octant.)

Exercise 4.8: Using the Pauli matrices, derive a matrix-based version of the quaternion rotation formula (Theorem 4.2.3).

Chapter 5

Euclidean Clifford Algebras

We are now ready to generalize our discussion to higher dimensions. We begin with the basis-free definition of the Clifford algebra over $\mathbb{R}^n$.

Definition 5.0.1. The Euclidean Clifford algebra of $\mathbb{R}^n$, denoted $\mathcal{C}\ell_n$, is defined as the associative algebra with geometric product defined by

$$\mathbf{x}^2 = \|\mathbf{x}\|^2.$$

With this simple definition, it is possible to derive a number of important and useful properties. First, let $\{\mathbf{e}_1, \ldots, \mathbf{e}_n\}$ be an orthonormal basis for $\mathbb{R}^n$. Expanding an arbitrary vector $\mathbf{x} \in \mathbb{R}^n$ in terms of this basis, i.e., writing $\mathbf{x} = x_1\mathbf{e}_1 + \cdots + x_n\mathbf{e}_n$, one sees that

$$\begin{aligned}(x_1\mathbf{e}_1 + \cdots + x_n\mathbf{e}_n)^2 &= (x_1\mathbf{e}_1 + \cdots + x_n\mathbf{e}_n)(x_1\mathbf{e}_1 + \cdots + x_n\mathbf{e}_n)\\ &= ({x_1}^2 + \cdots + {x_n}^2) + \sum_{1\le i<j\le n} x_i x_j(\mathbf{e}_i\mathbf{e}_j + \mathbf{e}_j\mathbf{e}_i)\\ &= \|\mathbf{x}\|^2 + \sum_{1\le i<j\le n} x_i x_j(\mathbf{e}_i\mathbf{e}_j + \mathbf{e}_j\mathbf{e}_i).\end{aligned}$$

It follows immediately that the geometric product of orthogonal vectors is anti-commutative. That is, $\mathbf{xy} = -\mathbf{yx}$ whenever $\langle \mathbf{x}, \mathbf{y}\rangle = 0$.

Continuing, let $\mathbf{x}, \mathbf{y} \in \mathbb{R}^n$ be arbitrary. Now consider the following:

$$\begin{aligned}(\mathbf{x}+\mathbf{y})^2 &= ((x_1+y_1)\mathbf{e}_1 + \cdots + (x_n+y_n)\mathbf{e}_n)^2\\ &= ((x_1+y_1)\mathbf{e}_1 + \cdots + (x_n+y_n)\mathbf{e}_n)((x_1+y_1)\mathbf{e}_1 + \cdots + (x_n+y_n)\mathbf{e}_n)\\ &= \sum_{i=1}^{n}(x_i+y_i)^2 + \sum_{1\le i<j\le n}(x_i+y_i)(x_j+y_j)(\mathbf{e}_i\mathbf{e}_j + \mathbf{e}_j\mathbf{e}_i)\\ &= \|\mathbf{x}\|^2 + \|\mathbf{y}\|^2 + 2\sum_{i=1}^{n}(x_iy_i) + \sum_{1\le i<j\le n}(x_i+y_i)(x_j+y_j)(\mathbf{e}_i\mathbf{e}_j + \mathbf{e}_j\mathbf{e}_i).\end{aligned}$$

Using anti-commutativity of $\mathbf{e}_1, \mathbf{e}_2$, this reduces to

$$(\mathbf{x}+\mathbf{y})^2 = \|\mathbf{x}\|^2 + \|\mathbf{y}\|^2 + 2\sum_{i=1}^{n} x_i y_i. \tag{5.1}$$

On the other hand, expanding the (non-commutative) binomial $(\mathbf{x}+\mathbf{y})^2$ gives

$$(\mathbf{x}+\mathbf{y})^2 = \mathbf{x}^2 + \mathbf{y}^2 + \mathbf{x}\mathbf{y} + \mathbf{y}\mathbf{x} = \|\mathbf{x}\|^2 + \|\mathbf{y}\|^2 + \mathbf{x}\mathbf{y} + \mathbf{y}\mathbf{x}. \tag{5.2}$$

Finally, setting (5.1) equal to (5.2), we have

$$\mathbf{x}\mathbf{y} + \mathbf{y}\mathbf{x} = 2\sum_{i=1}^{n}(x_i y_i) = 2\langle \mathbf{x}, \mathbf{y}\rangle.$$

Hence, the Euclidean inner product actually represents the *symmetric part* of the geometric product of two vectors. That is, one can write

$$\langle \mathbf{x}, \mathbf{y}\rangle = \frac{1}{2}(\mathbf{x}\mathbf{y} + \mathbf{y}\mathbf{x}).$$

Note that the square of a vector is a scalar. In general, the *geometric product* of two vectors is the sum of a symmetric inner product and an anti-symmetric *exterior product*:

$$\mathbf{x}\mathbf{y} = \underbrace{\langle \mathbf{x}, \mathbf{y}\rangle}_{\text{symmetric part}} + \underbrace{\mathbf{x}\wedge\mathbf{y}}_{\text{anti-symmetric part}}.$$

5.1 Grassmann (Exterior) Algebra

The geometric meaning of the inner product should already be familiar. The exterior product of two vectors is interpreted as an oriented parallelogram. For example, the exterior product $\mathbf{e}_1 \wedge \mathbf{e}_2$ is the unit square in the xy-plane with counterclockwise orientation. On the other hand, $\mathbf{e}_2 \wedge \mathbf{e}_1$ is the same plane segment with clockwise orientation. This oriented plane segment is called a *bivector*.

Given an orthonormal basis $\{\mathbf{e}_1, \dots, \mathbf{e}_n\}$ for $\mathbb{R}^n$, the exterior product $\mathbf{e}_1 \wedge \cdots \wedge \mathbf{e}_n$ will be denoted by $\mathbf{e}_{[n]}$ for convenience. This notation is explained more fully in subsequent sections.

More generally, for vectors $\mathbf{x}$ and $\mathbf{y}$ in a vector space V, the exterior product $\mathbf{x}\wedge\mathbf{y}$ is a bivector representing an oriented parallelogram determined by $\mathbf{x}$ and $\mathbf{y}$. Observe that the exterior product is anti-symmetric; i.e., $\mathbf{x}\wedge\mathbf{y} = -\mathbf{y}\wedge\mathbf{x}$. Observe also that if two vectors $\mathbf{x}$ and $\mathbf{y}$ are *parallel*, their exterior product is zero. Hence, $\mathbf{x}\wedge\mathbf{x} = 0$.

Definition 5.1.1. Letting $\mathcal{B} = \{\mathbf{e}_1, \ldots, \mathbf{e}_n\}$ be an orthonormal basis for a vector space V, the *Grassmann algebra* over V is the associative algebra generated by $\mathcal{B}$ with the exterior product.

The wedge product of k pairwise non-parallel vectors is referred to as a *k-blade*, or a *blade of grade k*. A homogeneous sum of k-blades is a *k-vector* or *multivector of grade k*. It can be shown that the collection of k-blades from the exterior algebra forms a subspace of $\bigwedge V$. This subspace is typically denoted $\bigwedge^k V$. In fact, $\bigwedge V$ is a direct sum of these subspaces, and is thus a *graded algebra*. In particular,

$$\bigwedge V = \mathbb{F} \oplus V \oplus \bigwedge^2 V \oplus \cdots \oplus \bigwedge^n V,$$

where $n = \dim V$.

Example 5.1.2. Let $\{\mathbf{e}_i : 1 \leq i \leq 3\}$ be a basis for $\mathbb{R}^3$. Two typical elements from $\bigwedge \mathbb{R}^3$ are

$$u = 4\,\mathbf{e}_1 + \mathbf{e}_3,$$

and

$$v = \mathbf{e}_2 \wedge \mathbf{e}_3 - \mathbf{e}_1 \wedge \mathbf{e}_2.$$

Then,

$$u \wedge v = 3\,\mathbf{e}_1 \wedge \mathbf{e}_2 \wedge \mathbf{e}_3.$$

Given $u \in \bigwedge V$, we define the notation $u^{\wedge k}$ to mean

$$u^{\wedge k} = \underbrace{u \wedge \cdots \wedge u}_{k \text{ times}}.$$

Example 5.1.3. Let $u = \mathbf{e}_1 \wedge \mathbf{e}_2 + 2\mathbf{e}_3 - \mathbf{e}_4 \in \bigwedge \mathbb{R}^4$. Then,

$$\begin{aligned} u^{\wedge 2} &= (\mathbf{e}_1 \wedge \mathbf{e}_2 + 2\mathbf{e}_3 - \mathbf{e}_4) \wedge (\mathbf{e}_1 \wedge \mathbf{e}_2 + 2\mathbf{e}_3 - \mathbf{e}_4) \\ &= 4\,\mathbf{e}_1 \wedge \mathbf{e}_2 \wedge \mathbf{e}_3 - 2\,\mathbf{e}_1 \wedge \mathbf{e}_2 \wedge \mathbf{e}_4. \end{aligned}$$

For convenience, we extend the inner product to basis blades $\mathbf{e}_I$ in the exterior algebra $\bigwedge V$ so that we can refer to coefficients of arbitrary elements. The coefficient of $\mathbf{e}_I$ in the expansion of $u \in \bigwedge V$ is denoted $\langle u, \mathbf{e}_I \rangle$, so that the canonical expansion of u is given by

$$u = \sum_{I \in 2^{[n]}} \langle u, \mathbf{e}_I \rangle \mathbf{e}_I.$$

Theorem 5.1.4. *Let $\mathbf{v}_1, \ldots, \mathbf{v}_k \in \mathbb{R}^n$, with canonical expansion $\mathbf{v}_i = \sum_{\ell=1}^{n} a_{i\,\ell}\, \mathbf{e}_\ell$ for $1 \le i \le k$, and let A denote the rectangular matrix*

$$A = \begin{pmatrix} a_{11} & \cdots & a_{1n} \\ & \ddots & \\ a_{k1} & \cdots & a_{kn} \end{pmatrix}.$$

For any multi-index $I \in 2^{[n]}$ of size k, let A_I denote the order k submatrix of A whose columns are indexed by I. The coefficient of $\mathbf{e}_I$ in the expansion of the exterior product $\mathbf{v}_1 \wedge \cdots \wedge \mathbf{v}_k$ is then given by the following:

$$\langle \mathbf{v}_1 \wedge \cdots \wedge \mathbf{v}_k, \mathbf{e}_I \rangle = \det(A_I).$$

Proof. Proof is by induction on $k \le n$. When $k = 1$, the result holds by definition. Suppose the result holds for some $k \ge 2$, and now consider $\mathbf{v}_1 \wedge \cdots \wedge \mathbf{v}_k \wedge \mathbf{v}_{k+1}$, where $\mathbf{v}_{k+1} = \sum_{i=1}^{n} a_i \mathbf{e}_i$. For convenience, let $B = \mathbf{v}_1 \wedge \cdots \wedge \mathbf{v}_k$. For any multi-index I of size $k+1$, the coefficient of $\mathbf{e}_I$ in the expansion of this wedge product is given by

$$\begin{aligned}
\langle B \wedge \mathbf{v}_{k+1}, \mathbf{e}_I \rangle &= \left\langle \sum_{\ell \in I} \langle B, \mathbf{e}_{I\setminus\{\ell\}} \rangle \mathbf{e}_{I\setminus\{\ell\}} (a_\ell\, \mathbf{e}_\ell), \mathbf{e}_I \right\rangle \\
&= \left\langle \sum_{j=1}^{k+1} \langle B, \mathbf{e}_{I\setminus\{I_j\}} \rangle \mathbf{e}_{I\setminus\{I_j\}} (a_{I_j}\, \mathbf{e}_{I_j}), \mathbf{e}_I \right\rangle \\
&= \left\langle \sum_{j=1}^{k+1} \langle B, \mathbf{e}_{I\setminus\{I_j\}} \rangle \mathbf{e}_I (-1)^j a_{I_j}, \mathbf{e}_I \right\rangle \\
&= \sum_{j=1}^{k+1} a_{I_j} (-1)^j |A_{I\setminus\{I_j\}}| \\
&= |A_I|.
\end{aligned}$$

When $k = n$, the summation becomes cofactor expansion of A along the nth row, which yields the following result:

$$\mathbf{v}_1 \wedge \cdots \wedge \mathbf{v}_n = \det(A)\mathbf{e}_{[n]}.$$

□

A number of results in elementary linear algebra are now obtained via exterior algebra. The first is a test for linear independence.

Corollary 5.1.5. *Let* $S = \{\mathbf{v}_1, \ldots, \mathbf{v}_k\} \subset \mathbb{R}^n$. *Then,* $\mathbf{v}_1 \wedge \cdots \wedge \mathbf{v}_n \neq 0$ *if and only if* S *is linearly independent.*

Areas of parallelograms and triangles are recovered next.

Lemma 5.1.6 (Area of parallelogram). *Let* A *denote the area of the parallelogram generated by vectors* $\mathbf{u}, \mathbf{v} \in \mathbb{R}^2$. *Then,*

$$\mathbf{u} \wedge \mathbf{v} = \pm A\, \mathbf{e}_{\{1,2\}}.$$

Proof. The proof is left as an exercise. □

Corollary 5.1.7 (Area of triangle). *Let* A *denote the area of the triangle with vertices* $\mathbf{x}_1, \mathbf{x}_2, \mathbf{x}_3 \in \mathbb{R}^2$. *Then,*

$$(\mathbf{x}_3 - \mathbf{x}_1) \wedge (\mathbf{x}_2 - \mathbf{x}_1) = \pm 2A\, \mathbf{e}_{[2]}.$$

If three points determine a triangle, they cannot be collinear. Since triangles have positive area, a test for collinearity naturally follows.

Corollary 5.1.8 (Test for collinearity). *Three points* $\mathbf{x}_1, \mathbf{x}_2, \mathbf{x}_3 \in \mathbb{R}^2$ *are collinear if and only if* $(\mathbf{x}_3 - \mathbf{x}_1) \wedge (\mathbf{x}_2 - \mathbf{x}_1) = 0$.

The exterior product of three vectors is a *trivector* representing an oriented volume element. This is the oriented parallelepiped determined by nonzero vectors $\mathbf{x}, \mathbf{y}, \mathbf{z} \in V$. Its volume is computed in Lemma 5.1.9.

Lemma 5.1.9 (Volume of parallelepiped). *Let* $\mathfrak{V}$ *denote the volume of the parallelepiped generated by vectors* $\mathbf{u}, \mathbf{v}, \mathbf{w} \in \mathbb{R}^3$. *Then,*

$$\mathbf{u} \wedge \mathbf{v} \wedge \mathbf{w} = \pm \mathfrak{V}\, \mathbf{e}_{[3]}.$$

Proof. The proof is left as an exercise. □

Note that

$$\mathbf{x} \wedge \mathbf{y} \wedge \mathbf{z} = (\mathbf{x} \cdot (\mathbf{y} \times \mathbf{z}))\, \mathbf{e}_{[3]},$$

where the left-hand side is computed in the exterior algebra while the right-hand side is computed in $\mathbb{R}^3$.

Corollary 5.1.10 (Volume of tetrahedron). *Let* $\mathfrak{V}$ *denote the volume of the tetrahedron with vertices* $\mathbf{x}_1, \mathbf{x}_2, \mathbf{x}_3, \mathbf{x}_4 \in \mathbb{R}^3$. *Then,*

$$(\mathbf{x}_4 - \mathbf{x}_1) \wedge (\mathbf{x}_3 - \mathbf{x}_1) \wedge (\mathbf{x}_2 - \mathbf{x}_1) = \pm 3\mathfrak{V}\, \mathbf{e}_{[3]}.$$

Corollary 5.1.11 (Test for coplanarity). *Four points* $\mathbf{x}_1, \mathbf{x}_2, \mathbf{x}_3, \mathbf{x}_4 \in \mathbb{R}^3$ *are coplanar if and only if* $(\mathbf{x}_4 - \mathbf{x}_1) \wedge (\mathbf{x}_3 - \mathbf{x}_1) \wedge (\mathbf{x}_2 - \mathbf{x}_1) = 0$.

5.2 Contractions

In addition to the exterior product, we define the *left contraction operator* in the exterior algebra $\bigwedge V$ as follows. Beginning with an orthonormal basis $\{\mathbf{e}_i\}$ of the n-dimensional vector space V, consider an ordered subset $\{\mathbf{e}_{\ell_j}\}_{1\leq j\leq k}$, where $k \leq n$. Then,

$$\mathbf{e}_{\ell_j} \lrcorner (\mathbf{e}_{\ell_1} \wedge \cdots \wedge \mathbf{e}_{\ell_j} \wedge \cdots \wedge \mathbf{e}_{\ell_k}) = (-1)^{j-1}(\mathbf{e}_{\ell_1} \wedge \cdots \wedge \mathbf{e}_{\ell_{j-1}} \wedge \mathbf{e}_{\ell_{j+1}} \wedge \cdots \wedge \mathbf{e}_{\ell_k}).$$

Similarly, *right contraction* is defined by

$$(\mathbf{e}_{\ell_1} \wedge \cdots \wedge \mathbf{e}_{\ell_j} \wedge \cdots \wedge \mathbf{e}_{\ell_k}) \llcorner \mathbf{e}_{\ell_j} = (-1)^{k-j}(\mathbf{e}_{\ell_1} \wedge \cdots \wedge \mathbf{e}_{\ell_{j-1}} \wedge \mathbf{e}_{\ell_{j+1}} \wedge \cdots \wedge \mathbf{e}_{\ell_k}).$$

In the event $\ell \notin I$, we define $\mathbf{e}_\ell \lrcorner \mathbf{e}_I = \mathbf{e}_I \llcorner \mathbf{e}_\ell = 0$.

Example 5.2.1. Letting u and v be defined as in Example 5.1.2,

$$\mathbf{e}_1 \lrcorner v = -\mathbf{e}_2,$$

and

$$v \llcorner \mathbf{e}_2 = -\mathbf{e}_3 - \mathbf{e}_1.$$

Contraction is clearly not associative, but we can extend it to blades as follows: let $I, J \in 2^{[n]}$, where $I = \{i_1, \ldots, i_k\}$ and $k \leq |J|$. Then, define

$$\mathbf{e}_I \lrcorner \mathbf{e}_J = \mathbf{e}_{i_1} \lrcorner (\cdots \lrcorner (\mathbf{e}_{i_{k-1}} \lrcorner (\mathbf{e}_{i_k} \lrcorner \mathbf{e}_J))).$$

In this way, we find

$$\mathbf{e}_I \lrcorner \mathbf{e}_J = \begin{cases} (-1)^{\sum_{j\in J} \sharp\{i\in I : i > j\}} \mathbf{e}_{J\setminus I} & \text{if } I \subseteq J, \\ 0 & \text{otherwise.} \end{cases}$$

5.3 Properties of the Geometric Product

Let us now return to the Euclidean geometric algebra $\mathcal{C}\ell_n$. Recall the geometric product of vectors: $\mathbf{xy} = \mathbf{x} \cdot \mathbf{y} + \mathbf{x} \wedge \mathbf{y}$.

This product can be extended associatively and additively to arbitrary multivectors of the exterior algebra $\bigwedge \mathbb{R}^n$. This is made more rigorous by replacing the inner product with contraction, so that one can define products of vectors and blades. To be specific, let $\mathbf{x} \in \mathbb{R}^n$, let B denote a blade in $\bigwedge \mathbb{R}^n$, and define

$$\mathbf{x}B := \mathbf{x} \lrcorner B + \mathbf{x} \wedge B.$$

Similarly, one defines

$$B\mathbf{x} := B \llcorner \mathbf{x} + B \wedge \mathbf{x}.$$

Definition 5.3.1. A blade B_1 will be said to *divide* blade B, denoted $B_1|B$, if there exists a blade B_2 such that $B = B_1 \wedge B_2$. Letting $\sharp B$ denote the grade of blade B, $B = B_1 \wedge B_2$ implies $\sharp B = \sharp B_1 + \sharp B_2$.

Note that a blade B can always be written as a scaled geometric product of orthonormal vectors. More specifically, if B is a blade of grade k, there exists an ordered orthonormal set $\{\mathbf{b}_1, \ldots, \mathbf{b}_k\}$ and scalar β such that $B = \beta\mathbf{b}_1 \cdots \mathbf{b}_k$. It thus makes sense to define the *norm* of a blade by

$$\|B\| = \|\beta\mathbf{b}_1 \cdots \mathbf{b}_k\| := |\beta|. \tag{5.3}$$

Given a blade $B = \mathbf{b}_1 \wedge \cdots \wedge \mathbf{b}_k = \mathbf{b}_1 \cdots \mathbf{b}_k$, the *reversion* of B, denoted $\tilde{B}$, is defined to be

$$\begin{aligned}\tilde{B} &= \mathbf{b}_k \cdots \mathbf{b}_1 \\ &= (-1)^{\frac{k(k-1)}{2}} \mathbf{b}_1 \cdots \mathbf{b}_k \\ &= (-1)^{\frac{k(k-1)}{2}} B.\end{aligned} \tag{5.4}$$

Lemma 5.3.2. *The Euclidean contraction of two blades is given by*

$$B_1 \rfloor B = \begin{cases} (-1)^{\frac{\sharp B_1(\sharp B_1 - 1)}{2}} \|B_1\|^2 B_2 & \text{if } B = B_1 \wedge B_2, \\ 0 & \text{otherwise.} \end{cases}$$

Proof. Suppose $B = B_1 \wedge B_2$, where $\sharp B_1 = k_1$ and $\sharp B_2 = k_2$. Setting $k = k_1 + k_2$, it follows that for appropriate scalar β, one can write

$$\begin{aligned}B &= \beta\mathbf{b}_1 \wedge \cdots \wedge \mathbf{b}_k \\ &= \beta\mathbf{b}_1 \cdots \mathbf{b}_k \\ &= \beta\mathbf{b}_1 \cdots \mathbf{b}_{k_1}\mathbf{b}_{k_1+1} \cdots \mathbf{b}_{k_1+k_2} \\ &= B_1 \wedge B_2,\end{aligned}$$

where $B_1 = \beta_1\mathbf{b}_1 \cdots \mathbf{b}_{k_1}$ and $B_2 = \frac{\beta}{\beta_1}\mathbf{b}_{k_1+1} \cdots \mathbf{b}_{k_1+k_2}$. Assuming orthogonality of the vectors, we can write $B_1 \wedge B_2 = B_1 B_2$. It follows that

$$\begin{aligned}B_1 \rfloor B &= \beta_1(\mathbf{b}_1 \cdots \mathbf{b}_{k_1}) \rfloor B_1 B_2 \\ &= \beta_1\mathbf{b}_1 \cdots \mathbf{b}_{k_1} \rfloor (-1)^{(k_1(k_1-1))/2} \tilde{B}_1 B_2 \\ &= (-1)^{\frac{k_1(k_1-1)}{2}} \beta_1{}^2 (\mathbf{b}_1 \cdots \mathbf{b}_{k_1}) \rfloor (\mathbf{b}_{k_1} \cdots \mathbf{b}_1) B_2 \\ &= (-1)^{\frac{k_1(k_1-1)}{2}} \beta_1{}^2 (\mathbf{b}_1 \rfloor \cdots \mathbf{b}_{k_1-1} \rfloor (\mathbf{b}_{k_1} \rfloor (\mathbf{b}_{k_1}\mathbf{b}_{k_1-1} \cdots \mathbf{b}_1)) \cdots) B_2 \\ &= (-1)^{\frac{k_1(k_1-1)}{2}} \beta_1{}^2 B_2.\end{aligned}$$

□

Corollary 5.3.3. *Let B be a blade B in the Euclidean Clifford algebra. Then,*

$$\widetilde{B}B = B\widetilde{B} = \|B\|^2.$$

Corollary 5.3.4. *Blades are invertible in the Euclidean Clifford algebra. In particular,* $B^{-1} = \dfrac{\widetilde{B}}{B\widetilde{B}}$.

Turning back to geometry, k-blades in $\mathcal{C}\ell_n$ represent k-dimensional subspaces of $\mathbb{R}^n$. When $B_1|B$, the subspace corresponding to B has a direct sum decomposition into subspaces corresponding to B_1 and its orthogonal complement within B. The contraction $B_1 \lrcorner B$ is an oriented scaled volume element in this orthogonal complement. In this way, contraction represents a "generalized" projection onto ${B_1}^{\perp}$.

Lemma 5.3.5. *Let $\mathbf{u} \in \mathbb{R}^n$ be a unit vector. Let $\mathbf{x} \in \mathbb{R}^n$ be arbitrary and show that $\mathbf{x}' = -\mathbf{uxu}$ is the reflection of $\mathbf{x}$ through the hyperplane orthogonal to $\mathbf{u}$.*

Proof. The proof is left as an exercise. □

Using geometric algebra, rotations in $\mathbb{R}^n$ can be expressed much more simply.

Theorem 5.3.6. *Let $\mathbf{u}, \mathbf{v} \in \mathbb{R}^n$ be unit vectors such that the angle between $\mathbf{u}$ and $\mathbf{v}$ is $\frac{\omega}{2}$ (measured from $\mathbf{u}$ to $\mathbf{v}$). Then, for arbitrary $\mathbf{x} \in \mathbb{R}^n$,*

$$\mathbf{x}' = \mathbf{vuxuv}$$

is the vector obtained by rotating $\mathbf{x}$ through angle ω about the origin in the uv-plane.

Proof. Taken with Lemma 5.3.5, the result is yet another corollary of Theorem 3.0.6. □

Corollary 5.3.7. *Let $\mathbf{v}_1, \ldots, \mathbf{v}_k$ be a collection of unit vectors in $\mathbb{R}^n$, and let $\mathfrak{w} = \mathbf{v}_1 \cdots \mathbf{v}_k$. It follows that for $\mathbf{x} \in \mathbb{R}^n$, the mapping*

$$\mathbf{x} \mapsto \mathfrak{w}\mathbf{x}\overline{\mathfrak{w}}$$

is a rotation when k is even, a reflection when $k = 1$, and an improper reflection when $k > 1$ is odd.

Proof. Note that when k is even, $\tilde{\mathfrak{w}} = \overline{\mathfrak{w}}$, while k odd implies $\tilde{\mathfrak{w}} = -\overline{\mathfrak{w}}$. The rest follows from Lemma 5.3.5 and Theorem 5.3.6. □

5.4 Basis Blades of $\mathcal{C}\ell_n$

Since squares of vectors are always scalars, one can always expand elements of the Clifford algebra in terms of a generating set using multi-indices. The basis-dependent definition of the Euclidean Clifford algebra of the plane appears next.

Definition 5.4.1. Let $\{\mathbf{e}_1, \ldots, \mathbf{e}_n\}$ be an orthonormal basis for $\mathbb{R}^n$. The *Euclidean Clifford algebra of* $\mathbb{R}^n$, denoted $\mathcal{C}\ell_n$, is defined as the associative algebra generated by the unit scalar 1 along with the vectors $\mathbf{e}_1, \ldots, \mathbf{e}_n \in \mathbb{R}^n$ with the geometric product such that

$$\mathbf{e}_1{}^2 = \cdots = \mathbf{e}_n{}^2 = 1,$$
$$\mathbf{e}_i\mathbf{e}_j + \mathbf{e}_j\mathbf{e}_i = 0, \quad (i \neq j).$$

For convenience, we adopt multi-index notation to represent ordered products of basis vectors. Specifically, for a subset $I \subseteq \{1, \ldots, n\}$, define the notation

$$\mathbf{e}_I := \prod_{\ell \in I} \mathbf{e}_\ell.$$

Such a subset I is referred to as a *multi-index* for the blade $\mathbf{e}_I$.

As a vector space, $\mathcal{C}\ell_n$ is spanned by

- $1 = \mathbf{e}_\varnothing$, the unit scalar,
- $\{\mathbf{e}_1, \ldots, \mathbf{e}_n\}$, the basis vectors,
- $\{\mathbf{e}_{\{1,2\}}, \ldots, \mathbf{e}_{\{n-1,n\}}\}$, the basis 2-blades (or bi-blades),

  $\vdots$
- $\{\mathbf{e}_I : |I| = k\}$, the basis k-blades $(k < n)$,

  $\vdots$
- $\mathbf{e}_{\{1,\ldots,n\}}$, the unit *pseudo scalar.*

Consequently, the dimension of $\mathcal{C}\ell_n$ is 2^n. The basis blades of $\mathcal{C}\ell_n$ are indexed by subsets of the n-set, $[n] = \{1, \ldots, n\}$. It is also convenient to let $2^{[n]}$ denote the *power set* of the n-set. In this way, basis blades are indexed by elements of $2^{[n]}$.

Fixing an ordered orthonormal basis $\{\mathbf{e}_1, \ldots, \mathbf{e}_n\}$ for $\mathbb{R}^n$, an arbitrary element $u \in \mathcal{C}\ell_n$ has the following canonical expansion:

$$u = \sum_{I \in 2^{[n]}} u_I \, \mathbf{e}_I.$$

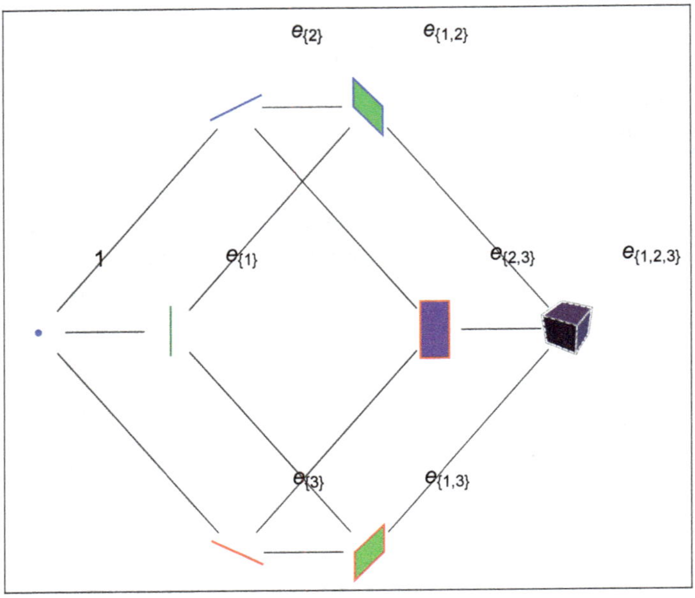

Figure 5.1 The "hypercube of blades" when $n = 3$.

For convenience, the notation $\langle u\rangle_k$ is used to indicate the *grade-k part of* u. That is,

$$\langle u\rangle_k = \left\langle \sum_{I\in 2^{[n]}} u_I\,\mathbf{e}_I \right\rangle_k = \sum_{\substack{I\in 2^{[n]}\\ |I|=k}} u_I\,\mathbf{e}_I.$$

Unless otherwise indicated, the basis blades of $\mathcal{C}\ell_n$ are ordered according to

$$\mathbf{e}_I \prec \mathbf{e}_J \Leftrightarrow \sum_{i\in I} 2^{i-1} < \sum_{j\in J} 2^{j-1},\ I, J \neq \varnothing,\ I \neq J, \tag{5.5}$$

$$\mathbf{e}_\varnothing \prec \mathbf{e}_I,\ \forall I \neq \varnothing. \tag{5.6}$$

For example, under $\prec$ the following collection is canonically ordered:

$$\{\mathbf{e}_\varnothing, \mathbf{e}_1, \mathbf{e}_2, \mathbf{e}_{12}, \mathbf{e}_3, \mathbf{e}_{13}, \mathbf{e}_{23}, \mathbf{e}_{123}, \mathbf{e}_4, \mathbf{e}_{14}, \mathbf{e}_{24}, \mathbf{e}_{124}, \mathbf{e}_{34}, \mathbf{e}_{134}, \mathbf{e}_{234}, \mathbf{e}_{1234}\}.$$

The ordering specified by $\prec$ is one of the four "admissible" monomial orders in the Grassmann algebra defined in [3]. This order is referred to as "InvLex."

Recall now the blade reversion defined in Equation (5.4). This definition extends linearly to $\mathcal{C}\ell_n$ by

$$\tilde{u} = \sum_{k=0}^{n} (-1)^{\frac{k(k-1)}{2}} \langle u \rangle_k.$$

Given an algebra $\mathcal{A}$, a homomorphism φ on $\mathcal{A}$ satisfying $\varphi \circ \varphi = \mathbb{I}$ is said to be an *involution*. A self-inverse linear map on $\mathcal{A}$ satisfying $\varphi(uv) = \varphi(v)\varphi(u)$ is said to be an *anti-involution*.

Definition 5.4.2. The *even subalgebra* of $\mathcal{C}\ell_n$ is defined by

$$\mathcal{C}\ell_n{}^+ := \{u \in \mathcal{C}\ell_n : k \cong 1 \pmod 2 \Rightarrow \langle u \rangle_k = 0\}.$$

In the next corollary, it will be convenient to use the *Clifford conjugate* $\overline{q}$ of an element $q \in \mathcal{C}\ell_3^+$. Recall that writing $u = \langle u \rangle_0 + \langle u \rangle_2 \in \mathcal{C}\ell_3^+$, the Clifford conjugate of u is given by

$$\tilde{u} = \langle u \rangle_0 - \langle u \rangle_2 .$$

Corollary 5.4.3. *Let* $\mathbf{u} = (u_x, u_y, u_z) \in \mathbb{R}^3$ *be a unit vector, let* $\omega \in \mathbb{R}$, *and define* $q = \cos\frac{\omega}{2} + \sin\frac{\omega}{2}\left(u_x\,\mathbf{e}_{\{2,3\}} + u_y\,\mathbf{e}_{\{1,3\}} + u_z\,\mathbf{e}_{\{1,2\}}\right) \in \mathcal{C}\ell_3^+$. *Let* $\mathbf{x} \in \mathbb{R}^3$ *be represented in* $\mathcal{C}\ell_3^+$ *as* $x_1\,\mathbf{e}_{\{2,3\}} + x_2\,\mathbf{e}_{\{1,3\}} + x_3\,\mathbf{e}_{\{1,2\}}$. *Then the rotation of* $\mathbf{x}$ *about* $\mathbf{u}$ *through angle* ω *with right-hand orientation is given by*

$$\mathbf{x}' = q\mathbf{x}\overline{q}.$$

Proof. The result follows from the Quaternion Rotation Formula (Theorem 4.2.3) by observing the isomorphism $\mathbb{H} \cong \mathcal{C}\ell_3^+$ defined by

$$\mathbf{i} \mapsto \mathbf{e}_{\{2,3\}}\,,$$
$$\mathbf{j} \mapsto \mathbf{e}_{\{1,3\}}\,,$$
$$\mathbf{k} \mapsto \mathbf{e}_{\{1,2\}}.$$

□

Corollary 5.4.4. *Let* $\mathbf{u} = (u_x, u_y, u_z) \in \mathbb{R}^3$ *be a unit vector, let* $\omega \in \mathbb{R}$, *and define* $q = \cos\frac{\omega}{2} - \sin\frac{\omega}{2}\left(u_x\,\mathbf{e}_{\{2,3\}} - u_y\,\mathbf{e}_{\{1,3\}} + u_z\,\mathbf{e}_{\{1,2\}}\right) \in \mathcal{C}\ell_3^+$. *Let* $\mathbf{x} \in \mathbb{R}^3$. *Then the rotation of* $\mathbf{x}$ *about* $\mathbf{u}$ *through angle* ω *with right-handed orientation is given by*

$$\mathbf{x}' = q\mathbf{x}\overline{q}.$$

Proof. Let $\mathbf{v}, \mathbf{w}$ be unit vectors in the plane orthogonal to $\mathbf{u}$ such that the angle measured from $\mathbf{v}$ to $\mathbf{w}$ is $\omega/2$. From Theorem 5.3.6, the rotated vector $\mathbf{x}'$ is given by $\mathbf{x}' = \mathbf{wvxvw}$. All that remains is to show that $q = \mathbf{wv}$ and $\overline{q} = \mathbf{vw}$.

$$\mathbf{w}\,\mathbf{v} = \langle \mathbf{w}, \mathbf{v}\rangle + \mathbf{w}\wedge\mathbf{v} = \cos\left(\frac{\omega}{2}\right) + (w_1v_2 - w_2v_1)\mathbf{e}_{\{1,2\}} + (w_1v_3 - w_3v_1)\mathbf{e}_{\{1,3\}} + (w_2v_3 - w_3v_2)\mathbf{e}_{\{2,3\}}. \quad (5.7)$$

Noting that $\mathbf{u}$ lies in the direction of $\mathbf{v}\times\mathbf{w}$ and

$$\mathbf{v}\times\mathbf{w} = \begin{vmatrix} \mathbf{i} & \mathbf{j} & \mathbf{k} \\ v_1 & v_2 & v_3 \\ w_1 & w_2 & w_3 \end{vmatrix} = \sin\left(\frac{\omega}{2}\right)\mathbf{u},$$

we have

$$\sin\left(\frac{\omega}{2}\right)(u_x, u_y, u_z) = (v_2w_3 - v_3w_2, v_3w_1 - v_1w_3, v_1w_2 - v_2w_1).$$

Hence,

$$\mathbf{w}\wedge\mathbf{v} = -\sin\left(\frac{\omega}{2}\right)\left(u_x\mathbf{e}_{\{2,3\}} - u_y\mathbf{e}_{\{1,3\}} + u_z\mathbf{e}_{\{1,2\}}\right).$$

Thus,

$$\mathbf{wv} = \cos\left(\frac{\omega}{2}\right) - \sin\left(\frac{\omega}{2}\right)\left(u_x\mathbf{e}_{\{2,3\}} - u_y\mathbf{e}_{\{1,3\}} + u_z\mathbf{e}_{\{1,2\}}\right).$$

Anti-commutativity of the exterior product then implies

$$\overline{q} = \cos\left(\frac{\omega}{2}\right) + \sin\left(\frac{\omega}{2}\right)\left(u_x\mathbf{e}_{\{2,3\}} - u_y\mathbf{e}_{\{1,3\}} + u_z\mathbf{e}_{\{1,2\}}\right) = \mathbf{vw}.$$

□

Remark 5.4.5. The elements $\mathbf{vw}$ and $\mathbf{wv}$ are elements of the *spin group* in $\mathcal{C}\ell_3$. There is a 2-to-1 mapping from this group to elements of $SO(3)$. Matrix representations of elements from the spin group are elements of the special unitary group $SU(2)$, and their irreducible representations are sometimes called *spinors*.

Decompositions in the Clifford Lipschitz Group

Let ${\mathcal{C}\ell_n}^*$ denote the multiplicative group of invertible Clifford elements. In particular,

$${\mathcal{C}\ell_n}^* = \{u \in \mathcal{C}\ell_n : u\tilde{u} \in \mathbb{R}^*\}.$$

The inverse of $u \in \mathcal{C}\ell_n$ is then seen to be $u^{-1} = \dfrac{\tilde{u}}{u\tilde{u}}$.

Definition 5.4.6. The *Clifford Lipschitz group*, Γ_n, is the subgroup of $\mathcal{C}\ell_n{}^*$ whose elements $u \in \Gamma_n$ satisfy

- $u \in \mathcal{C}\ell_n^+ \cup \mathcal{C}\ell_n^-$;
- for all $\mathbf{x} \in V$, $u\mathbf{x}\overline{u} \in V$.

Two important subgroups of the Clifford Lipschitz group are the pin and spin groups. The *pin group* $\mathrm{Pin}(n) = \{u \in \mathcal{C}\ell_n^+ \cup \mathcal{C}\ell_n^- : u\tilde{u} = \pm 1\}$ is a double covering of $O(n)$. The *spin group* $\mathrm{Spin}(n) = \{u \in \mathcal{C}\ell_n^+ \cup \mathcal{C}\ell_n^- : u\tilde{u} = 1\}$ is a double covering of $SO(n)$.

Definition 5.4.7. The *conformal orthogonal group* $\mathrm{CO}(n)$ is defined as the direct product of dilations and orthogonal transformations on $\mathbb{R}^n$.

Definition 5.4.8. An element $\mathfrak{u} \in \mathcal{C}\ell_n$ is said to be *decomposable* if $\mathfrak{u} = \mathbf{v}_1 \cdots \mathbf{v}_k$ for some linearly independent collection of vectors $\{\mathbf{v}_1, \ldots, \mathbf{v}_k\}$ in $\mathcal{C}\ell_n$. Equivalently, $\mathfrak{u}$ is decomposable if and only if it satisfies the following conditions:

(1) $\mathfrak{u} \in \mathcal{C}\ell_n^+ \cup \mathcal{C}\ell_n^-$;
(2) For all $\mathbf{x} \in V$, $\mathfrak{u}\mathbf{x}\overline{\mathfrak{u}} \in V$.

In fact, the decomposable elements of $\mathcal{C}\ell_n$ are precisely the elements of the Clifford Lipschitz group, Γ_n. Further, one quickly sees that decomposable elements $\mathfrak{u} \in \mathcal{C}\ell_n^+ \cup \mathcal{C}\ell_n^-$ satisfying $\mathfrak{u}\tilde{\mathfrak{u}} = \alpha \neq 0$ provide a double covering of the conformal orthogonal group $\mathrm{CO}(n)$.

For convenience, let $\sharp\mathfrak{u}$ denote the maximum grade among nonzero terms in the canonical basis blade expansion of $\mathfrak{u}$. The additive representation of $\mathfrak{u}$ with respect to any basis $\{\mathbf{e}_i : 1 \leq i \leq n\}$ of V is then of the form $\mathfrak{u} = \sum_{\substack{I \subseteq [n] \\ (|I| - \sharp\mathfrak{u}) \equiv 0 \pmod 2}} u_I \mathbf{e}_I$. When $k = \sharp\mathfrak{u}$, $\mathfrak{u}$ will also be referred to as a *decomposable k-element* of $\mathcal{C}\ell_n$. A problem providing motivation now is to efficiently represent such an element, which consists of as many as $\sum_{\ell=0}^{\lfloor k/2 \rfloor} \binom{n}{k-2\ell}$ nonzero terms.

As a consequence of the definition of a decomposable element, there exists a constant $\alpha \in \mathbb{R}$ and a linearly independent collection $\{\mathbf{w}_1, \ldots, \mathbf{w}_k\}$ of unit vectors in $\mathbb{R}^n$ such that

$$\alpha \mathbf{w}_1 \cdots \mathbf{w}_k = \mathfrak{u}.$$

In the context of geometric algebra, any element constructed as the product of a number of non-null vectors is commonly referred to as a *versor*. The element $\mathfrak{u}$ described above is correctly regarded as a k-versor.

Given a unit vector $\mathbf{u}$ and an arbitrary vector $\mathbf{x} \in \mathbb{R}^n$, it is well known and easily verified that computing the geometric product $\mathbf{u}\mathbf{x}\overline{\mathbf{u}}$ yields a vector $\mathbf{x}'$ obtained by reflection of $\mathbf{x}$ through the hyperplane orthogonal to $\mathbf{u}$.

By considering compositions of reflections, one similarly verifies that given a second unit vector $\mathbf{v}$, the geometric product $\mathbf{u}\mathbf{v}\mathbf{x}\mathbf{v}\mathbf{u}$ gives a vector $\mathbf{x}'$ obtained by rotating $\mathbf{x}$ in the $\mathbf{u}\mathbf{v}$-plane by twice the angle measured from $\mathbf{v}$ to $\mathbf{u}$.

Given $\mathfrak{u}=\mathbf{w}_1 \cdots \mathbf{w}_k \in \mathcal{C}\ell_n$, it will be convenient to refer to $V_\mathfrak{u}=\text{span}(\{\mathbf{w}_1, \ldots, \mathbf{w}_k\})$ as the $\mathfrak{u}$-*subspace* of $\mathbb{R}^n$. In this case, define the mapping $\varphi_\mathfrak{u} : \mathbb{R}^n \to \mathbb{R}^n$ by

$$\varphi_\mathfrak{u}(\mathbf{x}) = \mathfrak{u}\mathbf{x}\frac{\overline{\mathfrak{u}}}{\mathfrak{u}\widetilde{\mathfrak{u}}} = \mathfrak{u}\mathbf{x}\widehat{\mathfrak{u}^{-1}}.$$

Lemma 5.4.9. *The operator* $\pi_\mathfrak{u} := \mathbb{I}-\varphi_\mathfrak{u}$ *is a projection into the* $\mathfrak{u}$-*subspace of* $\mathbb{R}^n$.

Proof. Write $\mathbb{R}^n = V_\mathfrak{u} \oplus {V_\mathfrak{u}}'$, where ${V_\mathfrak{u}}'$ is the orthogonal complement of $V_\mathfrak{u}$ in $\mathbb{R}^n$. Observe that for $\mathbf{w}' \in V_\mathfrak{u}'$, anti-commutativity of $\mathbf{w}'$ with the basis vectors of $\mathfrak{u}$ guarantees that $\varphi_\mathfrak{u}(\mathbf{w}') = \mathbf{w}'$. On the other hand, for $\mathbf{w} \in V_\mathfrak{u}$, a suitable change of basis for $V_\mathfrak{u}$ allows us to write $\mathfrak{u} = \mathbf{u}_1 \cdots \mathbf{u}_{(\sharp\mathfrak{u})-1}\mathbf{w}$, so that

$$\begin{aligned}\varphi_\mathfrak{u}(\mathbf{w}) &= \frac{(-1)^{\sharp\mathfrak{u}}}{\|\mathfrak{u}\|^2}\mathbf{u}_1 \cdots \mathbf{u}_{(\sharp\mathfrak{u})-1}\mathbf{w}\mathbf{w}\mathbf{w}\mathbf{u}_{(\sharp\mathfrak{u})-1} \cdots \mathbf{u}_1 \\ &= \frac{(-1)^{\sharp\mathfrak{u}}}{\|\mathfrak{u}\|^2}\|\mathbf{w}\|^2\mathbf{u}_1 \cdots \mathbf{u}_{(\sharp\mathfrak{u})-1}\mathbf{w}\mathbf{u}_{(\sharp\mathfrak{u})-1} \cdots \mathbf{u}_1.\end{aligned}$$

Since $\mathbf{w} \in V_\mathfrak{u}$, this is clearly an element of $V_\mathfrak{u}$. Now, letting $\mathbf{x} = \mathbf{w}+\mathbf{w}' \in \mathbb{R}^n$ be arbitrary,

$$\begin{aligned}\pi_\mathfrak{u}(\mathbf{x}) &= \mathbf{x} - \varphi_\mathfrak{u}(\mathbf{x}) \\ &= (\mathbf{w} + \mathbf{w}') - \varphi_\mathfrak{u}(\mathbf{w} + \mathbf{w}') \\ &= (\mathbf{w} + \mathbf{w}') - \varphi_\mathfrak{u}(\mathbf{w}) - \varphi_\mathfrak{u}(\mathbf{w}') \\ &= \mathbf{w} + \mathbf{w}' - \mathbf{w}' - \varphi_\mathfrak{u}(\mathbf{w}) \\ &= \mathbf{w} - \mathfrak{u}\mathbf{w}\widehat{\mathfrak{u}^{-1}} \in V_\mathfrak{u}.\end{aligned}$$

□

Utilizing these basic facts allows one to develop and implement an efficient algorithm for factoring versors and blades in $\mathcal{C}\ell_n$. The same algorithm works equally well in the negative-definite[1] Clifford algebra $\mathcal{C}\ell_{0,n}$.

5.5 Versor Decomposition in Definite Signatures

When $A \in \mathrm{SO}(n)$ acts as plane rotation in $\mathbb{R}^n$, there exists a two-versor $\mathfrak{b} \in \mathcal{C}\ell_n$ such that

$$A\mathbf{x} = \mathfrak{b}\mathbf{x}\mathfrak{b}^{-1}$$

for all $\mathbf{x} \in \mathbb{R}^n$.

Beginning with such a versor, written explicitly in terms of a fixed basis in $\mathcal{C}\ell_n$, one task of interest is to obtain a factorization $\mathfrak{b} = \mathbf{b}_1\mathbf{b}_2$, where $\mathbf{b}_1, \mathbf{b}_2$ are unit vectors of $\mathbb{R}^n$. An intuitive geometric approach to accomplish this is to first apply a "probing vector." The normalized component of this vector lying in the plane of rotation represents one factor, $\mathbf{b}_1$, of the versor. This factor is rotated to its image, $\mathbf{u}$, by the action of the versor. Halfway between the probing vector's projection and the projection's image lies the second factor, $\mathbf{b}_2 = (\mathbf{b}_1 + \mathbf{u})/\|\mathbf{b}_1 + \mathbf{u}\|$, of the versor (see Figure 5.2). A nice description of the ideas behind this process can be found in the work of Aragón-Gonzales, Aragón, *et al.* [6].

By normalizing $\mathbf{b}_1$ and $\mathbf{u}$, one guarantees that the angle between $\mathbf{b}_1$ and $\mathbf{b}_2$ is $\theta/2$, where θ is the angle measured from $\mathbf{b}_1$ to $\mathbf{u}$. For arbitrary $\mathbf{x} \in \mathbb{R}^n$, it follows that $\mathbf{b}_2\mathbf{b}_1\mathbf{x}\mathbf{b}_1\mathbf{b}_2$ is rotation of $\mathbf{x}$ by angle θ in the $\mathbf{b}_1\mathbf{b}_2$-plane.

A natural extension of this geometric approach allows one to iteratively factor blades and versors in Clifford algebras of definite signature. Consider now a $2k$-versor $\mathfrak{b}$ such that $\varphi_\mathfrak{b} : \mathbf{x} \mapsto \mathfrak{b}\mathbf{x}\mathfrak{b}^{-1}$ represents the composition of k plane rotations of $\mathbf{x}$ in $\mathbb{R}^n$. An important assumption is that the linear operator $\varphi_\mathfrak{b}$ does *not* have -1 as an eigenvalue[2].

When $\mathfrak{b}$ is a versor of odd grade, one vector can be "factored out" before reverting to the iterated rotor factorization. Moreover, the group action can be generalized from $O(n)$ to $CO(n)$ by considering arbitrary scalar multiples of rotors and versors. An implementation of this approach is seen in Algorithm 1.

Example 5.5.1. Consider $\mathfrak{b} = 4 + 8\mathbf{e}_{\{1,2\}} + 6\mathbf{e}_{\{1,3\}} - 6\mathbf{e}_{\{2,3\}} \in \mathcal{C}\ell_3$. The action of $\mathbf{x} \mapsto \mathfrak{b}\mathbf{x}\bar{\mathfrak{b}}$ is the composition of a plane rotation and dilation

[1] Generators of $\mathcal{C}\ell_{0,n}$ satisfy ${\mathbf{e}_i}^2 = -1$ for $i = 1, \ldots, n$.

[2] The requirement that -1 is not an eigenvalue of $\varphi_\mathfrak{b}$ ensures that $\mathbf{b}_\ell{}' := (\mathbf{b}_\ell + \mathbf{u})/\|\mathbf{b}_\ell + \mathbf{u}\|$ is well-defined.

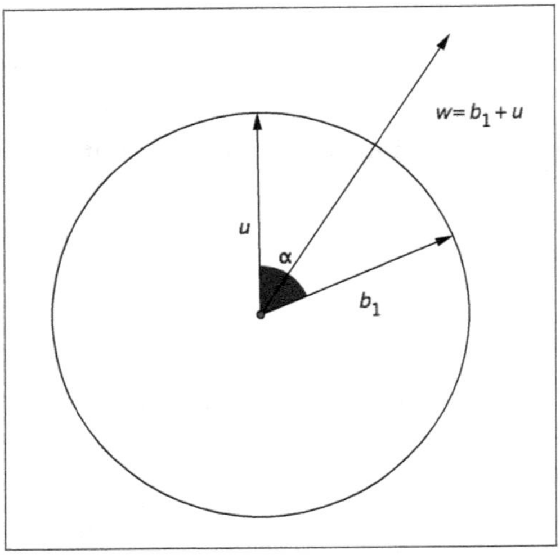

Figure 5.2 Applying a probing vector to factor a two-versor.

by factor $\mathfrak{b}\tilde{\mathfrak{b}} = 152$ in $\mathbb{R}^3$. Letting $\mathbf{p} = \mathbf{e}_1$ serve as a probing vector, we compute $\mathbf{p}' = \mathfrak{b}\mathbf{p}\widehat{\mathfrak{b}^{-1}}$ and obtain $\mathbf{p}' = -\frac{6}{19}\mathbf{e}_1 + \frac{1}{19}\mathbf{e}_2 - \frac{18}{19}\mathbf{e}_3$. Letting $\mathbf{b}_1 = (\mathbf{p} - \mathbf{p}')/\|\mathbf{p} - \mathbf{p}'\|$, we obtain the normalized projection $\mathbf{b}_1$ of $\mathbf{p}$ into the plane of rotation. In particular,

$$\mathbf{b}_1 = \frac{5}{\sqrt{38}}\mathbf{e}_1 - \frac{1}{5\sqrt{38}}\mathbf{e}_2 + \frac{9\sqrt{2}}{5\sqrt{19}}\mathbf{e}_3.$$

Computing $\mathbf{u} = \mathfrak{b}\mathbf{b}_1\widehat{\mathfrak{b}^{-1}}$, we obtain

$$\mathbf{u} = -\frac{275\mathbf{e}_1 + 293\mathbf{e}_2 + 426\mathbf{e}_3}{95\sqrt{38}}.$$

Computing the unit vector $\mathbf{b}_2$, which lies halfway between $\mathbf{b}_1$ and its image, we obtain

$$\mathbf{b}_2 = (\mathbf{b}_1 + \mathbf{u})/\|\mathbf{b}_1 + \mathbf{u}\| = \frac{50}{95}\mathbf{e}_1 - \frac{78}{95}\mathbf{e}_2 - \frac{21}{95}\mathbf{e}_3.$$

The rotation induced by $\mathfrak{b}$ now corresponds to the composition of two reflections across the orthogonal complements of $\mathbf{b}_1$ and $\mathbf{b}_2$, respectively. Note that $\mathbf{b}_2$ is the normalization of $\mathbf{w}$ in Figure 5.2. The factorization of $\mathfrak{b}$ is then given by

$$\mathfrak{b} = \sqrt{152}\,\mathbf{b}_2\mathbf{b}_1 = 4 + 8\mathbf{e}_{\{1,2\}} + 6\mathbf{e}_{\{1,3\}} - 6\mathbf{e}_{\{2,3\}}.$$

input : Additive representation of $\mathfrak{u}$, an invertible k-versor, expanded w.r.t. generators $\{\mathbf{e}_i : 1 \leq i \leq n\}$.
output: Vectors $\{\mathbf{b}_1, \ldots, \mathbf{b}_k\}$ such that $\mathfrak{u} = \alpha \mathbf{b}_k \cdots \mathbf{b}_1$.
$\ell \leftarrow 1$;
$\mathfrak{u}' \leftarrow \mathfrak{u}$;

while $\sharp\mathfrak{u}' > 1$ **do**
  Choose a random unit vector $\mathbf{x} \in \mathbb{R}^n$ *and compute its image under the action of* $\mathfrak{u}'$. ;
  Let $\mathbf{x} \in \mathbb{R}^n$ such that $\mathbf{x} \lrcorner \mathfrak{u} = 0$ and $\|\mathbf{x}\| = 1$;
  $\mathbf{x}' \leftarrow \mathfrak{u}'\mathbf{x}\widehat{\mathfrak{u}'^{-1}}$;
  $\mathbf{b}_\ell \leftarrow (\mathbf{x} - \mathfrak{u}'\mathbf{x}\widehat{\mathfrak{u}'^{-1}})/\|\mathbf{x} - \mathfrak{u}'\mathbf{x}\widehat{\mathfrak{u}'^{-1}}\|$;
  If $\mathfrak{u}'$ *is of odd grade, factor out a vector (reflection). Otherwise, factor out a 2-versor (plane rotation).*;
  if $\sharp\mathfrak{u}' \equiv 1 \pmod 2$ **then**
    $\mathfrak{u}' \leftarrow \mathfrak{u}'\mathbf{b}_\ell$;
    $\ell \leftarrow \ell + 1$;
  else
    $\mathbf{z} \leftarrow \mathfrak{u}'\mathbf{b}_\ell\mathfrak{u}'^{-1}$;
    if $\langle \mathbf{b}_\ell, \mathbf{z} \rangle \neq -1$;
    then
      $\mathbf{b}_{\ell+1} \leftarrow (\mathbf{b}_\ell + \mathbf{z})/\|\mathbf{b}_\ell + \mathbf{z}\|$;
      $\mathfrak{w} \leftarrow \mathbf{b}_{\ell+1}\mathbf{b}_\ell$;
      $\ell \leftarrow \ell + 2$;
    else
      $\mathfrak{w} \leftarrow \mathbf{b}_\ell$;
      $\ell \leftarrow \ell + 1$;
    end
    Compute lower-grade versor;
    $\mathfrak{u}' \leftarrow \mathfrak{u}'\mathfrak{w}^{-1}$;
  end
end
return $\{\mathbf{b}_1, \ldots, \mathbf{b}_{\ell-1}, \mathfrak{u}'\}$;

Algorithm 1: VersorFactor: Factor Versors in Definite Signatures

5.6 Blade Factorization

Algorithm 2 provides an efficient method for blade decomposition. Unlike the approach of Algorithm 1, it makes use of a *single term* of the canonical expansion to obtain each vector of the decomposition, as opposed to computing the full blade conjugation. That is, subspace projections are computed using a single basis blade from the expansion in place of the expansion itself. Algorithm 2 returns a collection of linearly independent vectors whose exterior product is the input blade.

The combinatorial approach developed here effectively computes geometric contractions using differences of multi-indices. Further, because multi-indices are well-ordered by $\mathbf{f}_I \preceq \mathbf{f}_J \Leftrightarrow \sum_{i\in I} 2^{i-1} \leq \sum_{j\in J} 2^{j-1}$, the following function is well defined:

$$\mathrm{FirstTerm}\left(\sum_I \alpha_I \mathbf{f}_I\right) := \min_{\{\mathbf{f}_X : \alpha_X \neq 0\}} \alpha_X \mathbf{f}_X.$$

The FirstTerm procedure thereby provides the means for choosing the term that drives the blade's decomposition.

Remark 5.6.1. Like Algorithm 2, Fontijne's algorithm [45] also utilizes a single term from the blade's canonical expansion to compute the blade's constituent vectors. The initial term selected in Fontijne's algorithm is a term whose scalar coefficient is equal in magnitude to the ∞-norm of the blade. This term is then used in conjunction with geometric contractions to extract the blade's constituent vectors.

input : Blade $\mathfrak{b}$ of grade k expressed as a sum $\sum_I \alpha_I \mathbf{e}_I$.
output: Scalar α and set of vectors $\{\mathbf{b}_1, \ldots, \mathbf{b}_k\}$ such that $\mathfrak{b} = \alpha \mathbf{b}_k \wedge \cdots \wedge \mathbf{b}_1$.
$\alpha \mathbf{e}_{M=\{m_1,\ldots,m_k\}} \leftarrow \mathrm{FirstTerm}(\mathfrak{b})$;
$\mathfrak{b}' \leftarrow \frac{1}{\alpha}\mathfrak{b}$;
for $\ell \leftarrow 1$ **to** k **do**
  $\mathfrak{u} \leftarrow \mathbf{e}_{M\setminus\{m_\ell\}}$;
  $\mathbf{b}_\ell \leftarrow \langle \mathfrak{b}'\mathfrak{u}^{-1}\rangle_1$;
end
return $\{\alpha, \mathbf{b}_1, \ldots, \mathbf{b}_k\}$;

Algorithm 2: FastBladeFactor

It is noteworthy that the output of Algorithm 2 is a set of vectors whose exterior (wedge) product is the input blade. If one wishes to recover a set of vectors whose Clifford (geometric) product is equal to the input blade, it is necessary to othogonalize the vectors output by the algorithm via Gram-Schmidt orthogonalization.

Gram-Schmidt Revisited

The following application of orthogonal projections is useful to recall. Suppose $S = \{\mathbf{v}_1, \ldots, \mathbf{v}_k\}$ is a linearly independent collection in $\mathbb{R}^n$. Using properties of the geometric product, Gram-Schmidt orthogonalization is accomplished in $\mathcal{C}\ell_n$ as follows.

(1) First, set $\mathbf{u}_1 := \mathbf{v}_1$.

(2) For each $\ell = 2, \ldots, k$, set

$$\mathbf{u}_\ell := \mathbf{v}_\ell - \sum_{j=1}^{\ell-1} \left(\frac{\mathbf{v}_\ell \mathbf{u}_j + \mathbf{u}_j \mathbf{v}_\ell}{2{\mathbf{u}_j}^2} \right) \mathbf{u}_j.$$

(3) The collection $S' = \{\mathbf{u}_1, \ldots, \mathbf{u}_k\}$ now satisfies $\operatorname{span}(S) = \operatorname{span}(S')$. If unit vectors are required, normalize the collection S' by dividing each element by its norm.

```
clGramSchmidt[L_] := Module[{B, lp, ell},
  B = {L[[1]]};
  lp = 2;
  While[lp ≤ Length[L],
   B =
    Append[B,
     clExpand[
      L[[lp]] - Sum[(clExpand[L[[lp]]⊙B[[ell]]] + clExpand[B[[ell]]⊙L[[lp]]]) /
          (2 clExpand[B[[ell]]⊙B[[ell]]]) B[[ell]], {ell, 1, lp - 1}]]];
   lp++;];
  Return[B];]
```

Figure 5.3 Clifford-algebraic Gram-Schmidt orthogonalization with *Mathematica*.

Exercises

Exercise 5.1: Let $\{\mathbf{e}_1, \mathbf{e}_2, \mathbf{e}_3\}$ be an orthonormal basis for $\mathbb{R}^3$. Show that in $\mathcal{C}\ell_3$, ${\mathbf{e}_i}^2 = 1$ and $\mathbf{e}_i\,\mathbf{e}_j = -\mathbf{e}_j\,\mathbf{e}_i$ when $i \neq j$.

Exercise 5.2: Let $\mathbf{u} \in \mathbb{R}^n$ be a unit vector. Show that $\frac{1}{2}(1+\mathbf{u})$ is idempotent in $\mathcal{C}\ell_n$.

Exercise 5.3: Show that reversion and Clifford conjugation are anti-involutions on $\mathcal{C}\ell_n$.

Exercise 5.4: Show that ${\mathcal{C}\ell_n}^+$ is a subalgebra of $\mathcal{C}\ell_n$.

Exercise 5.5: Prove that the quaternion algebra $\mathbb{H}$ is isomorphic to the even subalgebra $\mathcal{C}\ell_3^+$.

Exercise 5.6: Set $u, v, w \in \bigwedge \mathbb{R}^3$ as follows:

$$\begin{aligned} u &= 3\mathbf{e}_1 - 2\mathbf{e}_2 + 3\mathbf{e}_3 \\ v &= \mathbf{e}_1 + 5\mathbf{e}_2 - 7\mathbf{e}_3 \\ w &= 5\mathbf{e}_1 - \mathbf{e}_2 + 6\mathbf{e}_3. \end{aligned}$$

(1) Compute $u \wedge v$.
(2) Compute $u \wedge w$.
(3) Compute $u \wedge v \wedge w$.

Exercise 5.7: Rotate the vector $(3, 1, 0)$ through $\pi/3$ radians about the unit vector $\left(\frac{1}{\sqrt{2}}, 0, \frac{1}{\sqrt{2}}\right) \in \mathbb{R}^3$.

Exercise 5.8: Verify that $u = 2\mathbf{e}_{\{1,2\}} + 10\mathbf{e}_{\{1,3\}} + 3\mathbf{e}_{\{2,3\}}$ is a blade in $\bigwedge \mathbb{R}^3$, and use Algorithm 2 to obtain vectors $\mathbf{u}_1, \mathbf{u}_2$ and scalar α such that $u = \alpha \mathbf{u}_1 \wedge \mathbf{u}_2$.

Exercise 5.9: Prove Lemma 5.1.6.

Exercise 5.10: Prove Lemma 5.1.9.

Chapter 6

Clifford Algebras of Arbitrary Signature

For positive integer n, let V be an n-dimensional vector space over $\mathbb{R}$ with orthonormal basis $\beta = \{\mathbf{e}_i\}_{1\leq i\leq n}$. Suppose p and q are nonnegative integers such that $n = p+q$. Let $\mathcal{B}$ be the multiplicative group generated by $\{\pm 1\}\cup\beta$ with multiplication defined by the following:

$$\mathbf{e}_i\,\mathbf{e}_j = \begin{cases} -\mathbf{e}_j\,\mathbf{e}_i & \text{if } i \neq j, \\ 1 & \text{if } 1 \leq i = j \leq p, \\ -1 & \text{if } p+1 \leq i = j \leq p+q. \end{cases}$$

Given $I \subseteq \{1, 2, \ldots, n\}$, define the *multi-index notation* $\mathbf{e}_I$ to denote the canonically ordered product

$$\mathbf{e}_I = \prod_{i\in I} \mathbf{e}_i.$$

Lemma 6.0.1. *The set β with the operation of multiplication as defined above generates a nonabelian group $\mathcal{B}$ of order 2^{n+1}.*

Proof. Considering all products of elements of β, it is apparent that elements of $\mathcal{B}$ are of the form $\pm\mathbf{e}_I$, where $I \subseteq \{1, \ldots, n\}$ is a multi-index. There are 2^n such subsets, and two choices of sign for each element, hence the result. $\square$

Elements of $\mathcal{B}$ will be referred to as *blades.* Considering binary representations of subsets of the n-set, it becomes apparent that the Cayley graph of $\mathcal{B}/\{\pm 1\}$ is isomorphic to the n-dimensional hypercube $\mathcal{Q}_n$.

Observing the convention $\mathbf{e}_\varnothing = 1$, define the *even subgroup* of $\mathcal{B}$ by

$$\mathcal{B}^+ := \{\pm\mathbf{e}_I : |I| \text{is even}\}.$$

Lemma 6.0.2. *Let $p, q \geq 0$ be fixed such that $p+q=n$, and let $\mathcal{B}$ be the group of blades defined above. Then, the even subgroup is normal; i.e.,*

$$\mathcal{B}^+ \lhd \mathcal{B}.$$

Proof. The proof is left as an exercise. □

Definition 6.0.3. Given positive integer n and nonnegative integers p, q such that $n = p+q$, the *Clifford algebra of signature* (p, q) is the group algebra of $\mathcal{B}$ over $\mathbb{R}$, where $\mathcal{B}$ is defined as above.

Note that $\mathcal{B}$ can be written as a disjoint union of the form $\mathcal{B} = \bigcup_{k=0}^{n} \mathcal{B}_k$, where $\mathcal{B}_k = \{\pm \mathbf{e}_I : |I| = k\}$. In light of this, $\mathcal{C}\ell_{p,q}$ is a *graded algebra*; i.e., it has the canonical decomposition

$$\mathcal{C}\ell_{p,q} = \mathbb{R} \oplus \mathbb{R}\mathcal{B}_1 \oplus \cdots \oplus \mathbb{R}\mathcal{B}_{p+q}.$$

Definition 6.0.4. Given $u \in \mathcal{C}\ell_{p,q}$, define the *grade-$k$ part* of u by

$$\langle u \rangle_k := \sum_{\{I:|I|=k\}} u_I \, \mathbf{e}_I.$$

Definition 6.0.5. The *even subalgebra* of $\mathcal{C}\ell_{p,q}$ is defined by

$$\mathcal{C}\ell_{p,q}^+ := \bigoplus_{k=0}^{\lfloor n/2 \rfloor} \mathbb{R}\mathcal{B}_{2k} = \mathbb{R}\mathcal{B}^+.$$

Lemma 6.0.6. *The geometric product of arbitrary blades $\mathbf{e}_I, \mathbf{e}_J \in \mathcal{C}\ell_{p,q}$ is given by*

$$\mathbf{e}_I \, \mathbf{e}_J = (-1)^{\phi(I,J)+|I\cap J\cap Q|} \mathbf{e}_{I\triangle J},$$

where $I \triangle J = (I \cup J) \setminus (I \cap J)$, $Q = \{p+1, \ldots, p+q\}$, and $\phi(I, J) : 2^{[n]} \times 2^{[n]} : \mathbb{Z}_{\geq 0}$ is defined by

$$\phi(I, J) = \sum_{j \in J} |\{i \in I : i > j\}|.$$

Proof. The proof is left as an exercise. □

Corollary 6.0.7. *Let u, v be arbitrary elements of $\mathcal{C}\ell_{p,q}$ $(p+q=n)$. In light of Lemma 6.0.6, the product $u\,v$ can be expanded as*

$$u\,v = \sum_{I,J \in 2^{[n]}} (-1)^{|I\cap J\cap Q|+\phi(I,J)} u_I \, v_J \, \mathbf{e}_{I\triangle J}.$$

6.1 Involutions

Involutions are self-inverse homomorphisms from a group or an algebra to itself. More specifically, an involution on an algebra $\mathcal{A}$ is a bijective homomorphism $\varphi : \mathcal{A} \to \mathcal{A}$ such that $\varphi(\varphi(u)) = u$ for all $u \in \mathcal{A}$. The Clifford algebra $\mathcal{C}\ell_{p,q}$ comes with three natural involutions defined below.

Definition 6.1.1. Let $\mathbf{e}_I$ be a basis blade in the Clifford algebra $\mathcal{C}\ell_{p,q}$, where $p + q = n > 0$. Let $u \in \mathcal{C}\ell_{p,q}$ be arbitrary. The *grade involution* of u, denoted $\hat{u}$, is defined by linear extension of

$$\widehat{\mathbf{e}_I} = (-1)^{|I|}\mathbf{e}_I.$$

The *reversion* of u denoted $\tilde{u}$, is defined by linear extension of

$$\widetilde{\mathbf{e}_I} = (-1)^{|I|(|I|-1)/2}\mathbf{e}_I.$$

Finally, the *Clifford conjugate* of u, denoted $\overline{u}$, is defined by linear extension of

$$\overline{\mathbf{e}_I} = (-1)^{|I|(|I|+1)/2}\mathbf{e}_I.$$

Alternatively, for any $u \in \mathcal{C}\ell_{p,q}$, one can write

$$\begin{aligned}
\hat{u} &= \sum_{k=0}^{p+q}(-1)^k \langle u\rangle_k\,, \\
\tilde{u} &= \sum_{k=0}^{p+q}(-1)^{\frac{k(k-1)}{2}} \langle u\rangle_k\,, \\
\overline{u} &= \sum_{k=0}^{p+q}(-1)^{\frac{k(k+1)}{2}} \langle u\rangle_k\,.
\end{aligned}$$

Example 6.1.2. For any $u \in \mathcal{C}\ell_{p,q}$ where $p + q = 3$,

$$\begin{aligned}
\hat{u} &= \langle u\rangle_0 - \langle u\rangle_1 + \langle u\rangle_2 - \langle u\rangle_3\,, \\
\tilde{u} &= \langle u\rangle_0 + \langle u\rangle_1 - \langle u\rangle_2 - \langle u\rangle_3\,, \\
\overline{u} &= \langle u\rangle_0 - \langle u\rangle_1 - \langle u\rangle_2 + \langle u\rangle_3\,.
\end{aligned}$$

Note that writing $I = \{I_2, \ldots, I_{|I|}\}$, the reversion is computed by re-ordering the exterior product

$$\widetilde{\mathbf{e}_I} = \mathbf{e}_{I_{|I|}} \wedge \cdots \wedge \mathbf{e}_{I_1} = (-1)^{|I|(|I|-1)/2}\mathbf{e}_1 \wedge \cdots \wedge \mathbf{e}_{I_{|I|}} = (-1)^{|I|(|I|-1)/2}\mathbf{e}_I.$$

The Clifford conjugate is the composition of the grade and reversion involutions. That is, $\overline{u} = \widetilde{\hat{u}}$.

6.2 Pin and Spin Groups

Note that $\mathcal{C}\ell_{p,q}$ can be decomposed into *even* and *odd* parts. In other words, one can write $\mathcal{C}\ell_{p,q} = \mathcal{C}\ell_{p,q}^{+} \oplus \mathcal{C}\ell_{p,q}^{-}$, where $\mathcal{C}\ell_{p,q}^{+}$ is the even subalgebra.

Within the Clifford algebra $\mathcal{C}\ell_{p,q}$, the set of even or odd elements satisfying $u\tilde{u} = \pm 1$ forms a group called the *pin group* $\mathrm{Pin}(p,q)$. The collection of even elements satisfying $u\tilde{u} = \pm 1$ is called the *spin group* $\mathrm{Spin}(p,q)$. That is,

$$\mathrm{Pin}(p,q) = \{u \in \mathcal{C}\ell_{p,q}^{+} \cup \mathcal{C}\ell_{p,q}^{-} : u\tilde{u} = \pm 1\}$$

$$\mathrm{Spin}(p,q) = \{u \in \mathcal{C}\ell_{p,q}^{+} : u\tilde{u} = \pm 1\}, \text{and}$$

$$\mathrm{Spin}_{+}(p,q) = \{u \in \mathcal{C}\ell_{p,q}^{+} : u\tilde{u} = 1\}.$$

Notation. Let G and H be groups. We use the notation $H < G$ to indicate H is a *subgroup* of G. If H is a *normal subgroup* of G, we write $H \lhd G$.

Lemma 6.2.1.

$$\mathrm{Spin}_{+}(p,q) \lhd \mathrm{Spin}(p,q).$$

Proof. Let $u \in \mathrm{Spin}_{+}(p,q)$ and let $v \in \mathrm{Spin}(p,q)$. We need to show that $vuv^{-1} \in \mathrm{Spin}(p,q)$.

First, suppose $v \in \mathrm{Spin}(p,q)$ implies $v^{-1} = \tilde{v}$, so that

$$\begin{aligned}(vuv^{-1})(\widetilde{vuv^{-1}}) &= (vu\tilde{v})(\widetilde{vu\tilde{v}}) = (vu\tilde{v})(v\tilde{u}\tilde{v}) \\ &= vu(\tilde{v}v)\tilde{u}\tilde{v} = v(u\tilde{u})\tilde{v} = v\tilde{v} = 1. \end{aligned} \tag{6.1}$$

Hence, $vuv^{-1} \in \mathrm{Spin}(p,q)$.

On the other hand, if $v \notin \mathrm{Spin}(p,q)$, $v^{-1} = -\tilde{v}$. Thus,

$$\begin{aligned}(vuv^{-1})(\widetilde{vuv^{-1}}) &= (-vu\tilde{v})(\widetilde{-vu\tilde{v}}) = (vu\tilde{v})(v\tilde{u}\tilde{v}) \\ &= vu(\tilde{v}v)\tilde{u}\tilde{v} = v(u\tilde{u})\tilde{v} = v\tilde{v} = 1. \end{aligned} \tag{6.2}$$

Again, $vuv^{-1} \in \mathrm{Spin}_{+}(p,q)$, which therefore is a normal subgroup of $\mathrm{Spin}(p,q)$. □

Recall that the *orthogonal group* $O(n)$ is the group of $n \times n$ matrices $M \in \mathrm{GL}_n(\mathbb{R})$ satisfying $M^{\top}M = I$, and $\det M = \pm 1$. The *special orthogonal group* $SO(n)$ is the subgroup of $O(n)$ whose elements have determinant 1.

It can be shown that there is a two-to-one homomorphism from $\mathrm{Pin}(n)$ to the orthogonal group $O(n)$. Similarly, there is a two-to-one homomorphism from the spin group $\mathrm{Spin}(n)$ to the special orthogonal group $SO(n)$.

6.3 The Clifford Algebra $\mathcal{C}\ell_Q(V)$

Let V be an n-dimensional vector space over $\mathbb{R}$ equipped with a nondegenerate quadratic form Q. Associate with Q the symmetric bilinear form

$$\langle \mathbf{x}, \mathbf{y} \rangle_Q = \frac{1}{2}\left[Q(\mathbf{x}+\mathbf{y}) - Q(\mathbf{x}) - Q(\mathbf{y})\right].$$

The *exterior product* on V satisfies the canonical anti-commutation relation (CAR) $\mathbf{u} \wedge \mathbf{v} = -\mathbf{v} \wedge \mathbf{u}$ for all $\mathbf{u}, \mathbf{v} \in V$. Geometrically, the exterior product of two vectors represents an oriented parallelogram generated by the two vectors. By associative extension, the exterior product of k linearly independent vectors represents an oriented k-volume. It follows immediately from the CAR that the exterior product of linearly *dependent* vectors is zero.

The *Clifford algebra* $\mathcal{C}\ell_Q(V)$ is the real algebra obtained from associative linear extension of the Clifford vector product

$$\mathbf{x}\mathbf{y} := \langle \mathbf{x}, \mathbf{y} \rangle_Q + \mathbf{x} \wedge \mathbf{y}, \ \forall \mathbf{x}, \mathbf{y} \in V. \tag{6.3}$$

Given a nondegenerate quadratic form Q, the mapping $\|\cdot\|_Q : V \to \mathbb{R}$ defined by

$$\|\mathbf{x}\|_Q = |\langle \mathbf{x}, \mathbf{x} \rangle_Q|^{1/2}, \ (\mathbf{x} \in V)$$

is readily seen to be a seminorm, referred to henceforth as the *Q-seminorm* on V.

A vector $\mathbf{x}$ is said to be *anisotropic* if $\|\mathbf{x}\|_Q \neq 0$. A set S of Q-orthogonal vectors is said to be *Q-orthonormal* if $\|\mathbf{x}\|_Q = 1$ for all $\mathbf{x} \in S$.

Note that since Q is nondegenerate, all vectors of a Q-orthogonal basis for V must be anisotropic. Given a collection of Q-orthogonal vectors $\{\mathbf{x}_i\}$, a Q-orthonormal basis $\{\mathbf{u}_i : 1 \leq i \leq n\}$ for V is obtained by defining

$$\mathbf{u}_i := \frac{\mathbf{x}_i}{\|\mathbf{x}_i\|_Q},$$

for each $i = 1, \ldots, n$. In particular, for each $i = 1, \ldots, n$,

$$\mathbf{u}_i{}^2 = \langle \mathbf{u}_i, \mathbf{u}_i \rangle_Q = \frac{\langle \mathbf{x}_i, \mathbf{x}_i \rangle_Q}{|\langle \mathbf{x}_i, \mathbf{x}_i \rangle_Q|} = \pm 1.$$

These vectors then generate the Clifford algebra $\mathcal{C}\ell_Q(V)$.

Generally speaking, the exterior product of k linearly independent vectors is called a *k-blade* or *blade of grade k*. When the vectors are Q-orthogonal, one sees from (6.3) that the Clifford product coincides with the exterior product.

Given an arbitrary Q-orthogonal basis $\{\mathbf{e}_i : 1 \leq i \leq n\}$ for V, multi-index notation for canonical basis blades is adopted in the following manner. Denote the n-set $\{1, \dots, n\}$ by $[n]$, and denote the associated *power set* by $2^{[n]}$. The ordered product of basis vectors (i.e., algebra generators) is then conveniently denoted by

$$\prod_{i \in I} \mathbf{e}_i = \mathbf{e}_I,$$

for any subset $I \subseteq [n]$, also denoted $I \in 2^{[n]}$.

These products of generators are referred to as basis *blades* for the algebra. The *grade* of a basis blade is defined to be the cardinality of its multi-index. An arbitrary element $u \in \mathcal{C}\ell_Q(V)$ has a canonical basis blade decomposition of the form

$$u = \sum_{I \subseteq [n]} u_I \, \mathbf{e}_I,$$

where $u_I \in \mathbb{R}$ for each multi-index I. The *grade-k part* of $u \in \mathcal{C}\ell_Q(V)$ is then naturally defined by $\langle u \rangle_k := \sum_{|I|=k} u_I \mathbf{e}_I$. It is now evident that $\mathcal{C}\ell_Q(V)$ has a canonical vector space decomposition of the form

$$\mathcal{C}\ell_Q(V) = \bigoplus_{k=0}^{n} \langle \mathcal{C}\ell_Q(V) \rangle_k.$$

Example 6.3.1. Let $\{\mathbf{e}_1, \mathbf{e}_2\}$ denote an orthonormal basis for the two-dimensional Euclidean space $\mathbb{R}^2$. The associated quadratic form is $Q(x, y) = x^2 + y^2$, and a general element of the Clifford algebra $\mathcal{C}\ell_Q(\mathbb{R}^2)$ is of the form

$$a_0 + a_1 \, \mathbf{e}_1 + a_2 \, \mathbf{e}_2 + a_{\{1,2\}} \, \mathbf{e}_{\{1,2\}},$$

where $a_I \in \mathbb{R}$ for each multi index $I \in 2^{[2]}$.

An arbitrary element $u \in \mathcal{C}\ell_Q(V)$ is said to be *homogeneous of grade* k if $\langle u \rangle_k \neq 0$ and $\langle u \rangle_\ell = 0$ for all $\ell \neq k$. As the degree of a polynomial refers to the maximal exponent appearing in terms of the polynomial, an arbitrary multivector $u \in \mathcal{C}\ell_Q(V)$ is said to be *heterogeneous of grade* k if $\langle u \rangle_k \neq 0$ and $\langle u \rangle_\ell = 0$ for $\ell > k$.

It is not difficult to see that $\mathcal{C}\ell_Q(V)$ contains the following two subspaces: $\mathcal{C}\ell_Q(V)^+ = \operatorname{span}(\{\mathbf{e}_I : |I| \equiv 0 \pmod 2\})$, called the *even subalgebra* of $\mathcal{C}\ell_Q(V)$, and $\mathcal{C}\ell_Q(V)^- := \operatorname{span}(\{\mathbf{e}_I : |I| \equiv 1 \pmod 2\})$, which is a subspace, but not a subalgebra.

The *reversion* on $\mathcal{C}\ell_Q(V)$ is defined on arbitrary blade $\mathfrak{u} = \mathbf{u}_1 \wedge \cdots \wedge \mathbf{u}_{\sharp\mathfrak{u}}$ by

$$\tilde{\mathfrak{u}} := \mathbf{u}_{\sharp u} \wedge \cdots \wedge \mathbf{u}_1 = (-1)^{\sharp\mathfrak{u}(\sharp\mathfrak{u}-1)/2}\mathfrak{u}$$

and is extended linearly to all of $\mathcal{C}\ell_Q(V)$. Similarly, the *grade involution* is defined by linear extension of $\hat{\mathfrak{u}} := (-1)^{\sharp\mathfrak{u}}\mathfrak{u}$, and *Clifford conjugation* is defined as the composition of reversion and grade involution. Specifically, Clifford conjugation acts on an arbitrary blade $\mathfrak{u}$ according to $\overline{\mathfrak{u}} := (-1)^{\sharp\mathfrak{u}(\sharp\mathfrak{u}+1)/2}\mathfrak{u}$.

By utilizing reversion, the inner product $\langle \cdot, \cdot \rangle_Q$ is seen to extend to the full algebra $\mathcal{C}\ell_Q(V)$ by bilinear linear extension of

$$\langle \mathfrak{b}_1, \mathfrak{b}_2 \rangle_Q := \langle \mathfrak{b}_1 \widetilde{\mathfrak{b}_2} \rangle_0$$

for arbitrary basis blades $\mathfrak{b}_1, \mathfrak{b}_2$.

Given the Clifford product, the *left contraction* operator is now conveniently defined for vector $\mathbf{x}$ and arbitrary multivector $v \in \mathcal{C}\ell_Q(V)$ by linear extension of

$$\mathbf{x}v = \mathbf{x} \lrcorner v + \mathbf{x} \wedge v.$$

A similar definition holds for the *right contraction*; i.e., $u\mathbf{x} := u \llcorner \mathbf{x} + u \wedge \mathbf{x}$. The left and right contraction operators then extend associatively to blades and linearly to arbitrary elements $u, v \in \mathcal{C}\ell_Q(V)$. Moreover, left and right contractions are dual to the exterior product and satisfy the following:

$$\langle u \lrcorner v, w \rangle_Q = \langle v, \tilde{u} \wedge w \rangle_Q ,$$
$$\langle u \llcorner v, w \rangle_Q = \langle u, w \wedge \tilde{v} \rangle_Q .$$

Exercises

Exercise 6.1: Given arbitrary $u, v \in \mathcal{C}\ell_{p,q}$ $(p + q = n)$, let $I \in 2^{[n]}$ be a fixed multi-index and find a formula for the coefficient of $\mathbf{e}_I$ in the product uv.

Exercise 6.2: Let $u = 3 + \mathbf{e}_{\{1,2\}} + 3\mathbf{e}_{\{1,2,3\}} \in \mathcal{C}\ell_{3,0}$. Compute $\hat{u}$, $\tilde{u}$, and $\overline{u}$.

Exercise 6.3: Let $u = 1 + 2\,\mathbf{e}_1 - 4\,\mathbf{e}_{\{1,3\}} + 5\,\mathbf{e}_{\{1,2,3\}} - 2\,\mathbf{e}_{\{1,4\}} \in \mathcal{C}\ell_{3,1}$. Compute $\hat{u}, \tilde{u}, \overline{u}$, and u^2.

Exercise 6.4: Prove that grade involution is an *automorphism*.

Exercise 6.5: Show that reversion and Clifford conjugation are *anti-automorphisms*. That is, let $u, v \in \mathcal{C}\ell_{p,q}$ and show $\widetilde{uv} = \tilde{v}\tilde{u}$ and $\overline{uv} = \overline{v}\,\overline{u}$.

Exercise 6.6: Show that $\forall u \in \mathcal{C}\ell_{p,q}^{+}$, $\hat{u} = u$.

Exercise 6.7: Let $\mathbf{e}_i \in \mathcal{C}\ell_{p,q}$, and let $\alpha \in \mathbb{R}$. Show that

$$\exp(\alpha\,\mathbf{e}_i) = \begin{cases} \cosh\alpha + \sinh\alpha\,\mathbf{e}_i & 1 \le i \le p, \\ \cos\alpha + \sin\alpha\,\mathbf{e}_i & p+1 \le i \le p+q. \end{cases}$$

Exercise 6.8: Show that $SO(n) \lhd O(n)$.

Exercise 6.9: Find a formula for $\left(\sum_{i=1}^{p+q} \mathbf{e}_i\right)^k$ in $\mathcal{C}\ell_{p,q}$.

Chapter 7

Decompositions in $\mathrm{CO}_Q(V)$

Beginning with a finite-dimensional vector space V equipped with a non-degenerate quadratic form Q, we consider the decompositions of particular elements of the Clifford Lipschitz group Γ in the Clifford algebra $\mathcal{C}\ell_Q(V)$. These elements represent the conformal orthogonal group $\mathrm{CO}_Q(V)$, defined as the direct product of the orthogonal group $O_Q(V)$ with dilations.

In Euclidean Clifford algebras, it is well known that elements $\mathfrak{u} \in \Gamma$ satisfying $\mathfrak{u}\tilde{\mathfrak{u}} = \alpha \in \mathbb{R}$ represent scaled orthogonal transformations on V; i.e., $\mathbf{x} \mapsto \mathfrak{u}\mathbf{x}\overline{\mathfrak{u}}$ is a conformal orthogonal transformation on V. When $\mathfrak{u}\tilde{\mathfrak{u}} = \pm 1$, one sees that the mapping $\mathbf{x} \mapsto \mathfrak{u}\mathbf{x}\overline{\mathfrak{u}}$ is an element of the orthogonal group $\mathrm{O}(n)$. More precisely, such an element $\mathfrak{u}$ is an element of the *Pin group*. The geometric significance of these mappings is detailed in a number of works, including [22] and [68].

When an invertible element $\mathfrak{u} \in \mathcal{C}\ell_Q(V)$ can be written as an ordered Clifford product of anisotropic vectors from V, such a multiplicative representation $\mathfrak{u} = \prod_{i=1}^{k} \mathbf{v}_i$ is called a *decomposition* of $\mathfrak{u}$. The goal here is to consider decompositions of Clifford group elements, with an eye toward efficient symbolic computation. While the theoretical underpinnings have been understood and studied in various forms for decades, the advent of newer computing technologies and algorithms have shed new light on these concepts.

The basic problem considered here is not new. To wit, versor factorization algorithms can be found in the work of Christian Perwass [78], and efficient blade factorization algorithms are found in the works of Dorst and Fontijne [33], [45].

More recently, the general problem of factorization in Clifford algebras of arbitrary signature was considered by Helmstetter [57]. The Lipschitz monoid (or Lipschitz semi-group) is the multiplicative monoid generated in $\mathcal{C}\ell_Q(V)$ over a field $\Bbbk$ by all scalars in $\Bbbk$, all vectors in V, and all $1+\mathbf{xy}$ where $\mathbf{x}$ and $\mathbf{y}$ are vectors that span a totally isotropic plane. The elements of this monoid are called the Lipschitzian elements. Given a Lipschitzian element a in a Clifford algebra $\mathcal{C}\ell_Q(V)$ over a field $\Bbbk$ containing at least three scalars, Helmstetter showed that, if a is not in the subalgebra generated by a totally isotropic subspace of V, then it is a product of linearly independent vectors of V.

When the quadratic form Q is definite, the decomposable elements of $\mathcal{C}\ell_Q(V)$ are precisely the elements of the Clifford Lipschitz group. When Q is indefinite, we pass to a proper subset of Γ. In particular, an element $\mathfrak{u} \in \mathcal{C}\ell_Q(V)$ is said to be *decomposable* if there exists a collection $\{\mathbf{w}_1, \ldots, \mathbf{w}_k\}$ of linearly independent anisotropic vectors such that $\mathfrak{u} = \mathbf{w}_1 \cdots \mathbf{w}_k$ *and* if the "top form" (i.e., grade-k part) of $\mathfrak{u}$ is invertible.

7.1 Decomposable Elements of $\mathbf{CO}_Q(V)$

To maintain generality in the theoretical background, let Q denote a nondegenerate quadratic form, and let V be an n-dimensional real vector space with inner product $\langle , \rangle_Q$ induced by Q. The Clifford algebra of this space is then denoted by $\mathcal{C}\ell_Q(V)$.

The collection of all Q-orthogonal transformations on V forms a group called the *orthogonal group* of Q, denoted $\mathrm{O}_Q(V)$. Specifically, $T \in \mathrm{O}_Q(V)$ if and only if for every $\mathbf{x} \in V$, $Q(T(\mathbf{x})) = Q(\mathbf{x})$. The *conformal orthogonal group*, denoted $\mathrm{CO}_Q(V)$, is the direct product of the orthogonal group with the group of dilations. More specifically, $\tau \in \mathrm{CO}_Q(V)$ if and only if for every $\mathbf{x} \in V$, there exists a scalar λ such that $Q(\tau(\mathbf{x})) = \lambda^2 Q(\mathbf{x})$.

The concept of a blade is commonplace in Clifford algebras, where it refers to the Clifford product of a collection of pairwise-orthogonal vectors. In such cases, the exterior product coincides with the Clifford (geometric) product.

For a positive integer k, a *blade of grade* k, or k-*blade*, is a homogeneous multivector $\mathfrak{u}$ of grade k that can be written in the form $\mathfrak{u} = \mathbf{w}_1 \cdots \mathbf{w}_k$ for some Q-*orthogonal* collection $\{\mathbf{w}_1, \ldots, \mathbf{w}_k\} \subset V$.

A nonzero element $\mathfrak{u} \in \mathcal{C}\ell_Q(V)$ is said to be *invertible* if $\mathfrak{u}\tilde{\mathfrak{u}}$ is a nonzero scalar. In this case, $\mathfrak{u}^{-1} = \dfrac{\tilde{\mathfrak{u}}}{\mathfrak{u}\tilde{\mathfrak{u}}}$.

Due to complications arising from the use of indefinite quadratic forms, we tighten our definition of decomposable elements for the general case. As a result, the decomposable elements of $\mathcal{C}\ell_Q(V)$ are no longer in one-to-one correspondence with elements of the Clifford Lipschitz group.

Definition 7.1.1. An invertible element $\mathfrak{u} \in \mathcal{C}\ell_Q(V)$ of grade k is said to be *decomposable* if there exists a linearly independent collection $\{\mathbf{w}_1, \ldots, \mathbf{w}_k\}$ of anisotropic vectors in V such that $\mathfrak{u} = \mathbf{w}_1 \cdots \mathbf{w}_k$ and $\langle \mathfrak{u} \rangle_k$ is invertible[1]. In this case, $\mathfrak{u}$ is referred to as a *decomposable* k-element.

As a consequence of this definition, any decomposable element $\mathfrak{u}$ is either even or odd; i.e., $\mathfrak{u} \in \mathcal{C}\ell_Q(V)^+ \cup \mathcal{C}\ell_Q(V)^-$. Further, invertibility is guaranteed by $\mathfrak{u}\tilde{\mathfrak{u}} \in \mathbb{R}^*$. The next definition lends meaning to the notion of whether a vector can be said to "divide" a blade or decomposable element.

Definition 7.1.2. Let $\mathfrak{u}$ be a decomposable element in $\mathcal{C}\ell_Q(V)$. An anisotropic vector $\mathbf{w} \in V$ is said to *divide* $\mathfrak{u}$ if and only if there exists a decomposable element $\mathfrak{u}' \in \mathcal{C}\ell_Q(V)$ of grade $\sharp\mathfrak{u} - 1$ such that $\mathfrak{u} = \pm\mathbf{w}\mathfrak{u}'$. In this case, one writes $\mathbf{w}\big|\mathfrak{u}$.

A basic result inherent to the decomposition algorithms is the following.

Lemma 7.1.3. *If* $\mathfrak{u}$ *is a decomposable* k*-element, then the grade-*k *part of* $\mathfrak{u}$ *is a* k*-blade, and any anisotropic vector* $\mathbf{v}$ *dividing this blade also divides* $\mathfrak{u}$.

Proof. If $\mathfrak{u} = \mathbf{w}_1 \cdots \mathbf{w}_k$ is a decomposable k-element, then the grade-k part $\langle \mathfrak{u} \rangle_k$ represents an oriented k-volume in V. Any factorization of this blade thereby spans a k-dimensional subspace of V, and by decomposability there exists an anisotropic basis β for this subspace. Any vector $\mathbf{v} \in \beta$ divides the blade $\langle \mathfrak{u} \rangle_k$. Writing $\mathbf{v}$ as a linear combination of the (unknown) vectors $\{\mathbf{w}_1, \ldots, \mathbf{w}_k\}$ then gives

$$\begin{aligned}
\mathbf{v}^{-1}\mathfrak{u} &= \frac{1}{\mathbf{v}^2}(a_1\mathbf{w}_1 + \cdots + a_k\mathbf{w}_k)\mathbf{w}_1 \cdots \mathbf{w}_k \\
&= \frac{1}{\mathbf{v}^2}\sum_{j=1}^{k} a_j \mathbf{w}_1 \cdots \check{\mathbf{w}}_j \cdots \mathbf{w}_k,
\end{aligned}$$

where $\check{\mathbf{w}}_j$ indicates the omission of $\mathbf{w}_j$ from the product. Letting $\mathfrak{u}' = \mathbf{v}^{-1}\mathfrak{u}$, associativity guarantees that $\mathfrak{u} = \mathbf{v}\mathfrak{u}'$ where $\mathfrak{u}'$ is a $(k-1)$-element. Decomposability of $\mathfrak{u}'$ depends on its invertibility; i.e. $\mathfrak{u}'\tilde{\mathfrak{u}'} \in \mathbb{R}^*$ is required. This is verified by computation:

$$\mathfrak{u}'\widetilde{\mathfrak{u}'} = (\mathbf{v}^{-1}\mathfrak{u})(\widetilde{\mathbf{v}^{-1}\mathfrak{u}}) = \mathbf{v}^{-1}(\mathfrak{u}\tilde{\mathfrak{u}})\mathbf{v}^{-1} = \frac{\mathfrak{u}\tilde{\mathfrak{u}}}{\mathbf{v}^2} \in \mathbb{R}^*.$$

□

[1] Requiring invertibility of the top form makes the *decomposable* elements of $\mathcal{C}\ell_Q(V)$ a proper subset of the Lipschitz group of $\mathcal{C}\ell_Q(V)$.

The theoretical basis for an essential tool used in the decomposition algorithms is provided by the following proposition.

Theorem 7.1.4. *Given a decomposable k-element $\mathfrak{u} = \mathbf{w}_1 \cdots \mathbf{w}_k \in \mathcal{C}\ell_Q(V)$, let $n = \dim V$ and define $\varphi_\mathfrak{u} \in \mathrm{O}_Q(V)$ by*

$$\varphi_\mathfrak{u}(\mathbf{v}) = \mathfrak{u}\mathbf{v}\widehat{\mathfrak{u}^{-1}}.$$

Then $\varphi_\mathfrak{u}$ has an eigenspace $\mathcal{E}$ of dimension $n-k$ with corresponding eigenvalue 1.

Proof. If $\mathbf{v}$ is in the orthogonal complement of $\mathfrak{u}$, one sees immediately that

$$\begin{aligned}
\mathfrak{u}\mathbf{v}\widehat{\mathfrak{u}^{-1}} &= \frac{1}{\mathfrak{u}\tilde{\mathfrak{u}}}(\mathbf{w}_1 \cdots \mathbf{w}_k)\mathbf{v}(\widehat{\mathbf{w}_k \cdots \mathbf{w}_1}) \\
&= \frac{(-1)^k}{\mathfrak{u}\tilde{\mathfrak{u}}}(\mathbf{w}_1 \cdots \mathbf{w}_k)\mathbf{v}(\mathbf{w}_k \cdots \mathbf{w}_1) \\
&= \mathbf{v}.
\end{aligned}$$

Hence, $\dim \mathcal{E} \geq n-k$.

On the other hand, since $\langle \mathfrak{u} \rangle_k = \mathbf{w}_1 \wedge \cdots \wedge \mathbf{w}_k$, which is invertible by our definition of decomposability, there exists an anisotropic orthogonal collection $\{\mathbf{v}_1, \ldots, \mathbf{v}_k\}$ such that $\mathbf{v}_1 \cdots \mathbf{v}_k = \langle \mathfrak{u} \rangle_k$. Setting $\mathfrak{w} = \mathbf{v}_1 \cdots \mathbf{v}_k$, it follows that

$$\begin{aligned}
\mathfrak{w}\mathbf{v}_k\widehat{\mathfrak{w}^{-1}} &= \frac{1}{\mathfrak{w}\tilde{\mathfrak{w}}}(\mathbf{v}_1 \cdots \mathbf{v}_k)\mathbf{v}_k(\widehat{\mathbf{v}_k \cdots \mathbf{v}_1}) \\
&= \frac{(-1)^k}{\mathfrak{w}\tilde{\mathfrak{w}}}{\mathbf{v}_k}^2(\mathbf{v}_1 \cdots \mathbf{v}_k)(\mathbf{v}_{k-1} \cdots \mathbf{v}_1) \\
&= \frac{(-1)^k}{\mathfrak{w}\tilde{\mathfrak{w}}}{\mathbf{v}_k}^2(-1)^{\frac{k(k-1)}{2}+\frac{(k-1)(k-2)}{2}}(\mathbf{v}_k \cdots \mathbf{v}_1)(\mathbf{v}_1 \cdots \mathbf{v}_{k-1}) \\
&= \frac{(-1)^k}{\mathfrak{w}\tilde{\mathfrak{w}}}{\mathbf{v}_k}^2(-1)^{(k-1)}(\mathbf{v}_k \cdots \mathbf{v}_1)(\mathbf{v}_1 \cdots \mathbf{v}_{k-1}) \\
&= -\frac{{\mathbf{v}_k}^2}{\mathfrak{w}\tilde{\mathfrak{w}}}\mathbf{v}_k(\mathbf{v}_{k-1} \cdots \mathbf{v}_1)(\mathbf{v}_1 \cdots \mathbf{v}_{k-1}) \\
&= -\mathbf{v}_k.
\end{aligned}$$

The corresponding result is similarly obtained for $\mathbf{v}_1$. For $1 < j < k$, one can consider

$$
\begin{aligned}
\mathfrak{w}\mathbf{v}_j\widehat{\mathfrak{w}^{-1}} &= \frac{1}{\mathfrak{w}\tilde{\mathfrak{w}}}(\mathbf{v}_1\cdots\mathbf{v}_k)\mathbf{v}_j(\widehat{\mathbf{v}_k\cdots\mathbf{v}_1})\\
&= \frac{(-1)^k}{\mathfrak{w}\tilde{\mathfrak{w}}}(\mathbf{v}_1\cdots\mathbf{v}_{j-1}\mathbf{v}_j\cdots\mathbf{v}_k)\mathbf{v}_j(\mathbf{v}_k\cdots\mathbf{v}_j\mathbf{v}_{j-1}\cdots\mathbf{v}_1)\\
&= \frac{(-1)^k}{\mathfrak{w}\tilde{\mathfrak{w}}}(\mathbf{v}_1\cdots\mathbf{v}_{j-1})(\mathbf{v}_k\cdots\mathbf{v}_j)\mathbf{v}_j(\mathbf{v}_j\cdots\mathbf{v}_k)(\mathbf{v}_{j-1}\cdots\mathbf{v}_1)\\
&= \frac{(-1)^k{\mathbf{v}_j}^2}{\mathfrak{w}\tilde{\mathfrak{w}}}(\mathbf{v}_1\cdots\mathbf{v}_{j-1})(\mathbf{v}_k\cdots\mathbf{v}_{j+1}\mathbf{v}_j)(\mathbf{v}_{j+1}\cdots\mathbf{v}_k)(\mathbf{v}_{j-1}\cdots\mathbf{v}_1)\\
&= \frac{(-1)^{k+(k-j)^2}{\mathbf{v}_j}^2}{\mathfrak{w}\tilde{\mathfrak{w}}}(\mathbf{v}_1\cdots\mathbf{v}_{j-1})(\mathbf{v}_j\cdots\mathbf{v}_k)(\mathbf{v}_k\cdots\mathbf{v}_{j+1})(\mathbf{v}_{j-1}\cdots\mathbf{v}_1)\\
&= \left(\frac{(-1)^{k+(k-j)^2}}{\mathfrak{w}\tilde{\mathfrak{w}}}\prod_{\ell=j}^{k}{\mathbf{v}_\ell}^2\right)(\mathbf{v}_1\cdots\mathbf{v}_{j-1})\mathbf{v}_j(\mathbf{v}_{j-1}\cdots\mathbf{v}_1)\\
&= \left(\frac{(-1)^{k+(k-j)^2+(j-1)^2}}{\mathfrak{w}\tilde{\mathfrak{w}}}\prod_{\ell=j}^{k}{\mathbf{v}_\ell}^2\right)\mathbf{v}_j(\mathbf{v}_{j-1}\cdots\mathbf{v}_1)(\mathbf{v}_1\cdots\mathbf{v}_{j-1})\\
&= (-1)^{k(k+1)+2(j^2-kj-j)+1}\mathbf{v}_j\\
&= -\mathbf{v}_j.
\end{aligned}
$$

It follows that $\mathrm{span}(\{\mathbf{v}_1,\ldots,\mathbf{v}_k\})$ is an eigenspace of the transformation $\mathbf{x}\mapsto\mathfrak{w}\mathbf{x}\widehat{\mathfrak{w}^{-1}}$ corresponding to eigenvalue -1. Letting $\mathbf{v}\in\mathrm{span}(\{\mathbf{v}_1,\ldots,\mathbf{v}_k\})$, it is not difficult to see that writing $\mathfrak{u}=\mathfrak{w}+\mathfrak{u}'$ implies

$$
\mathfrak{u}\mathbf{v}\widehat{\mathfrak{u}^{-1}} = \frac{1}{\mathfrak{u}\tilde{\mathfrak{u}}}\left(\mathfrak{w}\mathbf{v}\overline{\mathfrak{w}}+\mathfrak{u}'\mathbf{v}\overline{\mathfrak{w}}+\mathfrak{w}\mathbf{v}\overline{\mathfrak{u}'}+\mathfrak{u}'\mathbf{v}\overline{\mathfrak{u}'}\right).
$$

Observe that $\mathbf{v}\overline{\mathfrak{w}}$ and $\mathfrak{w}\mathbf{v}$ are blades of grade $k-1$ orthogonal to $\mathbf{v}$, while the highest grade terms of $\mathfrak{u}'$ are of grade $k-2$. Consequently, the "cross terms" contribute no components parallel to $\mathbf{v}$. In other words, $\mathfrak{u}\mathbf{v}\widehat{\mathfrak{u}^{-1}}=\mathbf{v}$ implies

$$
\mathfrak{w}\mathbf{v}\overline{\mathfrak{w}}+\mathfrak{u}'\mathbf{v}\overline{\mathfrak{u}'}=(\mathfrak{u}\tilde{\mathfrak{u}})\mathbf{v}. \tag{7.1}
$$

Note that writing $\mathfrak{u}=\mathfrak{w}+\mathfrak{u}'$ gives

$$
\mathfrak{u}\tilde{\mathfrak{u}}=\mathfrak{w}\tilde{\mathfrak{w}}+\mathfrak{u}'\widetilde{\mathfrak{u}'}.
$$

Further, if $\mathbf{v}$ divides $\mathfrak{w}$, one sees that $\mathfrak{u}'\mathbf{v}\overline{\mathfrak{u}'}=\lambda\mathbf{v}$ implies $\lambda=\mathfrak{u}'\widetilde{\mathfrak{u}'}$. Finally, a little algebra applied to (7.1) yields

$$
\begin{aligned}
\mathfrak{u}'\mathbf{v}\overline{\mathfrak{u}'} &= (\mathfrak{w}\tilde{\mathfrak{w}}+\mathfrak{u}\tilde{\mathfrak{u}})\mathbf{v}\\
&= (2\mathfrak{w}\tilde{\mathfrak{w}}+\mathfrak{u}'\widetilde{\mathfrak{u}'})\mathbf{v}.
\end{aligned}
$$

This implies $\mathfrak{u}'\widetilde{\mathfrak{u}'}=2\mathfrak{w}\tilde{\mathfrak{w}}+\mathfrak{u}'\widetilde{\mathfrak{u}'}$. Since $\mathfrak{w}$ is anisotropic, this is a contradiction. It follows that $\mathbf{v}\in V_{\mathfrak{u}}$ implies $\varphi_{\mathfrak{u}}(\mathbf{v})\neq\mathbf{v}$, so that $\dim\mathcal{E}\leq n-k$. $\square$

Given $\mathfrak{u} = \mathbf{w}_1 \cdots \mathbf{w}_k$, it will be convenient to refer to $V_{\mathfrak{u}} =$ span $(\{\mathbf{w}_1, \ldots, \mathbf{w}_k\})$ as the $\mathfrak{u}$-*subspace* of V. As seen in Theorem 7.1.4, when the orthogonal complement of the $\mathfrak{u}$-subspace is nontrivial, any unit vector of $V_{\mathfrak{u}^\star}$ is an eigenvector of $\varphi_{\mathfrak{u}}$ having eigenvalue 1. This observation allows one to define a $\mathfrak{u}$-subspace projection by

$$\pi_{\mathfrak{u}}(\mathbf{x}) := \left(\mathbf{x} - \mathfrak{u}\mathbf{x}\widehat{\mathfrak{u}^{-1}}\right).$$

Corollary 7.1.5. *Let* $\mathbf{x} \in V$ *be arbitrary. Then* $\mathbf{x} - \varphi_{\mathfrak{u}}(\mathbf{x}) \in V_{\mathfrak{u}}$. *In other words, the operator* $\pi_{\mathfrak{u}} := \mathbb{I} - \varphi_{\mathfrak{u}}$ *is a projection into the subspace determined by* $\mathfrak{u}$.

Proof. Write $V = V_{\mathfrak{u}} \oplus V'_{\mathfrak{u}}$, where $V'_{\mathfrak{u}}$ is the orthogonal complement of $V_{\mathfrak{u}}$ in V. Then, letting $\mathbf{x} = \mathbf{w} + \mathbf{w}' \in V$ be arbitrary,

$$\begin{aligned}
\pi_{\mathfrak{u}}(\mathbf{x}) &= (\mathbf{x} - \varphi_{\mathfrak{u}}(\mathbf{x})) \\
&= \left((\mathbf{w} + \mathbf{w}') - \mathfrak{u}(\mathbf{w} + \mathbf{w}')\widehat{\mathfrak{u}^{-1}}\right) \\
&= \left(\mathbf{w}\mathbf{w}' - \mathbf{w}' - \mathfrak{u}\mathbf{w}\widehat{\mathfrak{u}^{-1}}\right) \\
&= \left(\mathbf{w} - \mathfrak{u}\mathbf{w}\widehat{\mathfrak{u}^{-1}}\right) \in V_{\mathfrak{u}}.
\end{aligned}$$

It is clear that the null space of $\pi_{\mathfrak{u}}$ is $V'_{\mathfrak{u}}$, so that the range is $\mathfrak{u}$. □

Now that all tools are in hand, it is possible to formalize a decomposition algorithm for decomposable elements of $\mathcal{C}\ell_Q(V)$. Algorithm 3 makes use of the projection operator defined in Corollary 7.1.5 to obtain component vectors of decomposable elements. When the algorithm is applied to a blade, the result is an orthogonal collection of vectors whose product is the blade.

7.2 Decomposition Examples with *Mathematica*

All results appearing here were obtained using *Mathematica* with the CLIFFMATH package running on a MacBook Pro.

Example 7.2.1. To compare the algorithms involving general element decomposition, blade decomposition, and fast blade factoring, consider the randomly generated grade-5 element of $\mathcal{C}\ell_8$ seen in Figure 7.1. First, the element is decomposed using Algorithm 3.

The grade-5 part of the element is a 5-blade. Applying Algorithm 3 to this blade results in the factorization seen in Figure 7.2. Applying Algorithm 2 to the blade results in the non-orthogonal factors seen in Figure

input : $\mathfrak{b}$, a decomposable k-element.
output: $\{\mathbf{b}_k, \ldots, \mathbf{b}_1\}$ such that $\mathfrak{b} = \mathbf{b}_k \cdots \mathbf{b}_1$.
;
$\ell \leftarrow 1$;
$\mathfrak{u} \leftarrow \mathfrak{b}/\|\mathfrak{b}\|$;
while $\sharp\mathfrak{u} > 1$ **do**
  Choose random anisotropic vector $\mathbf{x} \in V$ *such that* $\mathbf{x} \lrcorner \mathfrak{u} \neq 0$ *and compute its image under the action of* $\varphi_{\mathfrak{u}}$. ;
  Let $\mathbf{x} \in V$ such that $\mathbf{x} \lrcorner \mathfrak{u} \neq 0$ and $\mathbf{x}^2 \neq 0$;
  $\mathbf{x}' \leftarrow \mathfrak{u}\mathbf{x}\widehat{\mathfrak{u}^{-1}}$;
  if $(\mathbf{x} - \mathbf{x}')^2 \neq 0$ **then**
    $\mathbf{b}_{\sharp\mathfrak{u}} \leftarrow (\mathbf{x} - \mathbf{x}')/\|\mathbf{x} - \mathbf{x}'\|$;
    if $\mathfrak{u}\,{\mathbf{b}_{\sharp\mathfrak{u}}}^{-1}$ *is decomposable* **then**
      $\mathfrak{u} \leftarrow \mathfrak{u}\,{\mathbf{b}_{\sharp\mathfrak{u}}}^{-1}$;
    end
  end
end
return $\{\mathbf{b}_k, \ldots, \mathbf{b}_2, \|\mathfrak{b}\|\mathfrak{u}\}$;

Algorithm 3: CliffordDecomp

7.2 in 1/18 of the time. All decompositions were subsequently verified to reproduce the original elements.

Example 7.2.2. In Figure 7.3, runtimes are compared for decomposition of grade-4 blades and general elements in $\mathcal{C}\ell_6$ and $\mathcal{C}\ell_7$. In each case, five hundred elements of grade 4 were randomly generated.

Exercise

Exercise 7.1: Verify that $u = 2\mathbf{e}_{\{1,2\}} + 10\mathbf{e}_{\{1,3\}} + 3\mathbf{e}_{\{2,3\}}$ is a blade in $\mathcal{C}\ell_{2,1}$, and use Algorithm 3 to obtain vectors $\mathbf{u}_1, \mathbf{u}_2, \mathbf{u}_3$ such that $u = \mathbf{u}_1\mathbf{u}_2\mathbf{u}_3$.

```
6784 e{1} + 6678 e{2} + 10 984 e{3} + 7576 e{4} - 6205 e{5} - 102 e{6} - 5149 e{7} + 1202 e{8} - 676 e{1,2,3} - 1752 e{1,2,4} +
 3229 e{1,2,5} - 2116 e{1,2,6} - 1015 e{1,2,7} + 2368 e{1,2,8} - 7568 e{1,3,4} + 6862 e{1,3,5} + 1036 e{1,3,6} +
 9822 e{1,3,7} - 4980 e{1,3,8} - 2208 e{1,4,5} - 4776 e{1,4,6} - 5872 e{1,4,7} + 1608 e{1,4,8} + 3779 e{1,5,6} +
 5302 e{1,5,7} - 4159 e{1,5,8} - 11 535 e{1,6,7} + 11 564 e{1,6,8} + 5533 e{1,7,8} - 5368 e{2,3,4} + 2145 e{2,3,5} +
 4456 e{2,3,6} + 11 825 e{2,3,7} - 8856 e{2,3,8} - 4177 e{2,4,5} - 2312 e{2,4,6} - 3317 e{2,4,7} - 1372 e{2,4,8} +
 1736 e{2,5,6} + 1840 e{2,5,7} - 1356 e{2,5,8} - 9764 e{2,6,7} + 11 044 e{2,6,8} + 7064 e{2,7,8} - 4316 e{3,4,5} -
 8776 e{3,4,6} - 14 732 e{3,4,7} + 6824 e{3,4,8} + 6963 e{3,5,6} + 12 360 e{3,5,7} - 10 073 e{3,5,8} - 19 315 e{3,6,7} +
 18 832 e{3,6,8} + 6919 e{3,7,8} - 115 e{4,5,6} + 2226 e{4,5,7} - 3565 e{4,5,8} - 9345 e{4,6,7} + 12 092 e{4,6,8} +
 6359 e{4,7,8} + 7762 e{5,6,7} - 9970 e{5,6,8} - 7278 e{5,7,8} + 6650 e{6,7,8} - 1610 e{1,2,3,4,5} + 3104 e{1,2,3,4,6} +
 4254 e{1,2,3,4,7} - 4088 e{1,2,3,4,8} - 3010 e{1,2,3,5,6} - 6230 e{1,2,3,5,7} + 5180 e{1,2,3,5,8} + 4058 e{1,2,3,6,7} -
 2344 e{1,2,3,6,8} + 2132 e{1,2,3,7,8} + 1986 e{1,2,4,5,6} + 1756 e{1,2,4,5,7} - 462 e{1,2,4,5,8} + 1862 e{1,2,4,6,7} -
 4152 e{1,2,4,6,8} - 3238 e{1,2,4,7,8} - 4402 e{1,2,5,6,7} + 5526 e{1,2,5,6,8} + 3862 e{1,2,5,7,8} - 4022 e{1,2,6,7,8} +
 278 e{1,3,4,5,6} - 3172 e{1,3,4,5,7} + 4634 e{1,3,4,5,8} + 6850 e{1,3,4,6,7} - 9640 e{1,3,4,6,8} - 4190 e{1,3,4,7,8} -
 7006 e{1,3,5,6,7} + 9558 e{1,3,5,6,8} + 7726 e{1,3,5,7,8} - 7430 e{1,3,6,7,8} + 4216 e{1,4,5,6,7} - 5796 e{1,4,5,6,8} -
 4144 e{1,4,5,7,8} + 3380 e{1,4,6,7,8} + 1116 e{1,5,6,7,8} - 3440 e{2,3,4,5,6} - 7810 e{2,3,4,5,7} + 7070 e{2,3,4,5,8} +
 5968 e{2,3,4,6,7} - 4896 e{2,3,4,6,8} + 1150 e{2,3,4,7,8} - 1290 e{2,3,5,6,7} + 2150 e{2,3,5,6,8} + 2230 e{2,3,5,7,8} -
 1894 e{2,3,6,7,8} + 5882 e{2,4,5,6,7} - 7734 e{2,4,5,6,8} - 5470 e{2,4,5,7,8} + 5046 e{2,4,6,7,8} + 776 e{2,5,6,7,8} +
 8126 e{3,4,5,6,7} - 11 122 e{3,4,5,6,8} - 8550 e{3,4,5,7,8} + 7730 e{3,4,6,7,8} + 908 e{3,5,6,7,8} + 748 e{4,5,6,7,8}
```

```
In[20]:= Timing[Bf = CliffordDecomp[B]]
Out[20]= {8.438983,
  {0.0630904 e{1} + 0.264971 e{2} - 0.390084 e{3} + 0.147402 e{4} - 0.188183 e{5} + 0.748969 e{6} + 0.136306 e{7} +
   0.370096 e{8}, 0.171923 e{1} - 0.380025 e{2} - 0.579994 e{3} + 0.143699 e{4} + 0.554634 e{5} - 0.320797 e{6} +
   0.193767 e{7} - 0.144584 e{8}, 0.423971 e{1} + 0.232783 e{2} + 0.30033 e{3} + 0.431616 e{4} -
   0.282986 e{5} + 0.31425 e{6} + 0.348015 e{7} - 0.435456 e{8}, 0.653833 e{1} + 0.148705 e{2} +
   0.184215 e{3} + 0.519516 e{4} + 0.304638 e{5} - 0.363147 e{6} - 0.137426 e{7} + 0.0546965 e{8},
  31 906.1 e{1} + 10 450.9 e{2} - 7684.13 e{3} - 3218.17 e{4} + 42 554.1 e{5} + 3429.73 e{6} + 42 580. e{7} - 14 537.7 e{8}}}
```

Figure 7.1 Decomposable grade-5 element in $\mathcal{C}\ell_8$ and its decomposition.

```
In[24]:= Timing[Blf = CliffordDecomp[Bl]]
Out[24]= {2.055691, {0.436433 e{1} + 0.0787571 e{2} + 0.161208 e{3} -
    0.830435 e{4} + 0.225138 e{5} + 0.000738118 e{6} + 0.170457 e{7} + 0.0892718 e{8},
   0.180493 e{1} - 0.321771 e{2} - 0.856856 e{3} + 0.219418 e{5} + 0.192729 e{6} + 0.113192 e{7} + 0.177712 e{8},
   0.486322 e{1} - 0.110234 e{2} - 0.752037 e{5} + 0.391758 e{6} - 0.154873 e{7} - 0.0912149 e{8},
   0.246575 e{1} - 0.546527 e{2} - 0.621213 e{6} - 0.119359 e{7} - 0.490262 e{8},
   +5505.05 e{1} + 11 762.3 e{6} + 26 704.6 e{7} - 24 174.3 e{8}}}
```

```
In[26]:= Timing[Blff = FastBladeFactor[Bl]]
Out[26]= {0.107119, {e{1} + (344 e{6})/161 + (781 e{7})/161 - (101 e{8})/23, -e{2} - (139 e{6})/805 + (1586 e{7})/805 - (331 e{8})/115,
   e{3} - (993 e{6})/805 - (878 e{7})/805 + (33 e{8})/115, -e{4} + (43 e{6})/23 + (89 e{7})/23 - (74 e{8})/23, e{5} - (1552 e{6})/805 - (2127 e{7})/805 + (292 e{8})/115, -1610}}
```

Figure 7.2 Decomposition and fast decomposition of the 5-blade (grade-5 part) in Figure 7.1.

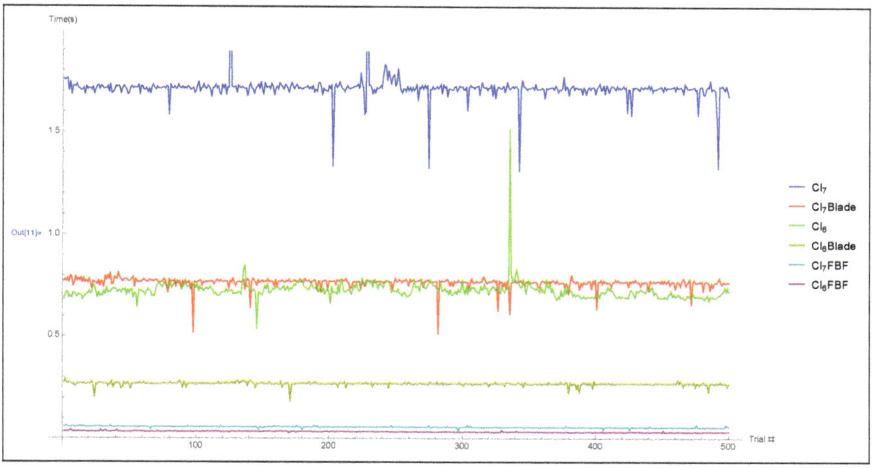

Figure 7.3 Computation time required for decomposition of grade-4 elements and 4-blades in $\mathcal{C}\ell_6$ and $\mathcal{C}\ell_7$.

Chapter 8

From Geometry to Combinatorics

As first seen in Figure 5.1, hypercubes play an important role in Clifford algebras and their "combinatorially interesting" subalgebras. In particular, each basis blade $\mathbf{e}_I$ of a Clifford algebra $\mathcal{C}\ell_{p,q}$ is in one-to-one correspondence with the canonical subspace $\text{span}\{\mathbf{e}_{\{i\}} : i \in I\}$ of its generating vector space V. In this way, computations in the algebra represent computations involving subspaces; i.e., intersections, unions, complements. By constructing subalgebras with different multiplicative properties, different types of combinatorial computations can be performed. The CLIFFMATH package for *Mathematica* facilitates symbolic computations in the following Clifford subalgebras.

- The Clifford algebra $\mathcal{C}\ell_{p,q,r}$. Generated by anticommutative generators $\{\mathbf{e}_{\{i\}} : 1 \leq i \leq n = p+q+r\}$ and unit scalar 1 satisfying
$$\mathbf{e}_{\{i\}}{}^2 = \begin{cases} 1 & 1 \leq i \leq p, \\ -1 & p+1 \leq i \leq p+q, \\ 0 & p+q+1 \leq i \leq p+q+r. \end{cases}$$
- The "sym-Clifford" algebra $\mathcal{C}\ell_{p,q,r}{}^{\text{sym}}$. Generated by pairwise commutative $\{\varsigma_{\{i\}} : 1 \leq i \leq p+q+r\}$ along with unit scalar 1 satisfying the squaring rules above.
- The n-particle zeon algebra $\mathcal{C}\ell_n{}^{\text{nil}}$. Generated by pairwise commutative $\{\zeta_{\{i\}} : 1 \leq i \leq n\}$ along with unit scalar 1 subject to $\zeta_{\{i\}}{}^2 = 0$ for $i = 1, \ldots, n$.
- The "idem-Clifford" algebra $\mathcal{C}\ell_n{}^{\text{idem}}$. This algebra is generated by pairwise-commutative idempotent generators $\{\varepsilon_{\{i\}} : 1 \leq i \leq n\}$ along with unit scalar 1. In particular, $\varepsilon_{\{i\}}{}^2 = \varepsilon_{\{i\}}$ for $i = 1, \ldots, n$.

Each of the algebras above can be viewed as the quotient of a group algebra or a semigroup algebra. The group or semigroup underlying each algebra is finitely generated by the algebra's generators and is comprised of the algebra's basis blades. Consequently, the Cayley graphs of the groups and semigroups underlying the algebras are isomorphic to hypercubes or generalizations thereof. The multiplicative structure of an algebra generated by $\{\gamma_j : 1 \leq j \leq 4\}$ and spanned by basis blades $\{\gamma_I : I \in 2^{[4]}\}$ is generally modeled by a four-dimensional hypercube as seen in Figure 8.1.

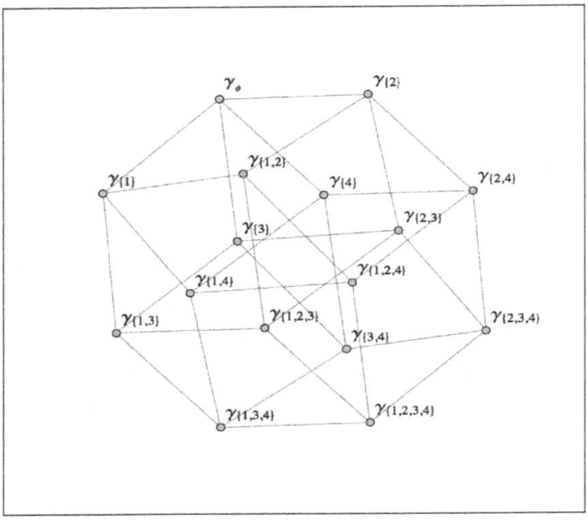

Figure 8.1 Four-dimensional hypercube.

In this chapter, we will focus on the particular groups underlying the Clifford algebras $\mathcal{C}\ell_{p,q}$ and the semigroups underlying $\mathcal{C}\ell_n{}^{\mathrm{nil}}$. The Cayley graphs of these groups and semigroups are generalizations of hypercubes, and irreducible representations of the algebras can then be characterized by considering irreducible representations of the (semi)groups.

All group and semigroup representations considered in this chapter are complex. A *representation* of a given group, G, is a homomorphism $\rho : G \to \mathrm{GL}_n(\mathbb{C})$. The *degree* of this representation is n, and the *representation space* is the space $\mathbb{C}^n$ on which the elements of $\mathrm{GL}_n(\mathbb{C})$ act.

Given a representation ρ and a subspace W of $\mathbb{C}^n$, we say W is G-*invariant* if $\rho(g)W \subseteq W$ for every $g \in G$. If the only invariant spaces are $\{0\}$ and V, the representation is said to be *irreducible*. The *character* of a representation, $\chi : G \to \mathbb{C}$, is defined by $\chi(g) = \mathrm{tr}(\rho(\mathrm{g}))$. The classification of irreducible complex representations of Clifford subalgebras appearing here was first considered in joint work with Cody Cassiday [19].

The reader is directed to Section 1.3 for a review of essential concepts and terminology from representation theory.

8.1 Representations of $\mathcal{C}\ell_{p,q}$

Let $B = \{\mathbf{e}_1, \ldots, \mathbf{e}_n\}$, and let p and q be nonnegative integers such that $p + q = n$. Let $\mathcal{B}_{p,q}$ be the *multiplicative group generated by* B along with the elements $\{\mathbf{e}_\varnothing, \mathbf{e}_\alpha\}$, subject to the following generating relations: for all $x \in B \cup \{\mathbf{e}_\varnothing, \mathbf{e}_\alpha\}$,

$$\mathbf{e}_\varnothing\, x = x\, \mathbf{e}_\varnothing = x,$$
$$\mathbf{e}_\alpha\, x = x\, \mathbf{e}_\alpha,$$
$${\mathbf{e}_\varnothing}^2 = {\mathbf{e}_\alpha}^2 = \mathbf{e}_\varnothing,$$

and

$$\mathbf{e}_i\mathbf{e}_j = \begin{cases} \mathbf{e}_\alpha\, \mathbf{e}_j\mathbf{e}_i & \text{if } i \neq j, \\ \mathbf{e}_\varnothing & \text{if } i = j \leq p, \\ \mathbf{e}_\alpha & \text{if } p+1 \leq i = j \leq n. \end{cases}$$

The group $\mathcal{B}_{p,q}$ is referred to herein as the *blade group*[1] of signature (p, q).

As always, let $2^{[n]}$ denote the power set of the n-set, $[n] = \{1, 2, \ldots, n\}$, used as indices of the generators in B. Elements of $2^{[n]}$ are assumed to be canonically ordered by

$$I \prec J \Leftrightarrow \sum_{i\in I} 2^{i-1} < \sum_{j\in J} 2^{j-1}. \tag{8.1}$$

Note that the ordering is inherited from the binary subset representation of integers.

Remark 8.1.1. The order of the blade group $\mathcal{B}_{p,q}$ is 2^{p+q+1} as seen by noting the form of its elements, i.e., $\mathcal{B}_{p,q} = \{\mathbf{e}_I, \mathbf{e}_\alpha\, \mathbf{e}_I : I \in 2^{[n]}\}$.

To simplify multiplication within $\mathcal{B}_{p,q}$, some additional mappings will be useful. For fixed positive integer j, define the map $\mu_j : 2^{[n]} \to \mathbb{N}_0$ by

$$\mu_j(I) = |\{i \in I : i > j\}|. \tag{8.2}$$

In other words, $\mu_j(I)$ is the *counting measure* of the set $\{i \in I : i > j\}$.

[1]The elements of this group are analogous to the "basis blades" of a Clifford (Grassmann) algebra.

Definition 8.1.2. The *product signature map* $\vartheta : 2^{[n]} \times 2^{[n]} \to \{\mathbf{e}_\varnothing, \mathbf{e}_\alpha\}$ is defined by

$$\vartheta(I, J) = \mathbf{e}_\alpha{}^{\left(\mu_p(I\cap J)+\sum_{j\in J}\mu_j(I)\right)}. \tag{8.3}$$

Applying multi-index notation to the generators B according to the ordered product

$$\mathbf{e}_I = \prod_{i\in I} \mathbf{e}_i \tag{8.4}$$

for arbitrary $I \in 2^{[n]}$, the multiplicative group $\mathcal{B}_{p,q}$ is now seen to be determined by the multi-indexed set $\{\mathbf{e}_I, \mathbf{e}_\alpha \mathbf{e}_I : I \in 2^{[n]}\}$ along with the associative multiplication defined by

$$\mathbf{e}_I\, \mathbf{e}_J = \vartheta(I, J)\mathbf{e}_{I\triangle J}, \tag{8.5}$$

where $I\triangle J = (I \cup J) \setminus (I \cap J)$ denotes set-symmetric difference. Inverses in $\mathcal{B}_{p,q}$ are given by

$$\mathbf{e}_I{}^{-1} = \vartheta(I, I)\mathbf{e}_I$$

since

$$\mathbf{e}_I\vartheta(I, I)\mathbf{e}_I = \vartheta(I, I)^2\mathbf{e}_{I\triangle I} = \mathbf{e}_\varnothing.$$

Elements of the form $\mathbf{e}_I$ are called *positive*, while elements of the form $\mathbf{e}_\alpha\mathbf{e}_I$ are called *negative*. Positive elements of $\mathcal{B}_{p,q}$ are now canonically ordered by

$$\mathbf{e}_I \prec \mathbf{e}_J \Leftrightarrow I \prec J$$

using the ordering on $2^{[n]}$ given by (8.1).

An element $\mathbf{e}_I \in \mathcal{B}_{p,q}$ is said to be *even* if $|I| = 2k$ for some nonnegative integer k. Otherwise, $\mathbf{e}_I$ is said to be *odd*.

Lemma 8.1.3. *The collection of even elements $\mathcal{B}_{p,q}{}^+$ of $\mathcal{B}_{p,q}$ forms a normal subgroup of $\mathcal{B}_{p,q}$; i.e.,*

$$\mathcal{B}_{p,q}{}^+ \lhd \mathcal{B}_{p,q}.$$

Proof. The proof is left as an exercise. □

Binary Clifford Multiplication

For any positive integer n, the collection $\{0,1\}^n$ of bit strings of length n is in one-to-one correspondence with integers $\{0,\ldots,2^n-1\}$. A familiar bijection is given by

$$(b_1 \cdots b_n) \mapsto \sum_{k=0}^{n-1} b_{n-k} 2^k.$$

Similarly, it is well known that the subsets of $[n] = \{1,\ldots,n\}$ are in one-to-one correspondence with integers $\{0,\ldots,2^n-1\}$ via the binary representation of subsets:

$$I = \{i_1,\ldots,i_k\} \mapsto \sum_{\ell=1}^{k} 2^{i_\ell}. \tag{8.6}$$

We now adopt the notational convention that for integer $0 \le j \le 2^n - 1$, $\underline{j} \subseteq [n]$ denotes the subset of $[n]$ associated with j by (8.6).

Definition 8.1.4. Given binary digits b_1, b_2 the *exclusive or* operator, denoted $\veebar$, is defined by

$$\begin{aligned} 0 \veebar 0 &= 0 \\ 1 \veebar 0 &= 1 \\ 0 \veebar 1 &= 1 \\ 1 \veebar 1 &= 0. \end{aligned}$$

The $\veebar$ operator is bitwise-extended to bit strings of length n. That is, given bit strings $b_1 \cdots b_n$ and $c_1 \cdots c_n$, one defines

$$(b_1\, b_2 \cdots b_n) \veebar (c_1\, c_2 \cdots c_n) = ((b_1 \veebar c_1)(b_2 \veebar c_2) \cdots (b_n \veebar c_n)).$$

To simplify notation, the binary representation of multi-index I will be denoted as b_I, that is, $b_I = (b_1 \cdots b_n)$, where

$$b_j = \begin{cases} 1 & j \in I \\ 0 & j \notin I. \end{cases}$$

Example 8.1.5. Given $I = \{1,3,4,7\}$ and $J = \{1,2,3,4,5,6,7\}$, we see that

$$\begin{aligned} I \triangle J &= \{2,5,6\} \\ b_I &= (1011001) \\ b_J &= (1111111) \\ b_I \veebar b_J &= (0100110) = b_{I\triangle J}. \end{aligned}$$

In terms of binary strings and the exclusive OR operator, Clifford multiplication satisfies

$$\mathbf{e}_{b_I}\mathbf{e}_{b_J} = \vartheta(I,J)\mathbf{e}_{b_I \veebar b_J}.$$

Representations of $\mathcal{B}_{p,q}$

A fundamental result in group representation theory [93] is that a representation ρ with character χ is irreducible if and only if χ satisfies

$$(\chi|\chi) = \frac{1}{|G|}\sum_{g\in G}\chi(g)\overline{\chi(g)} = 1.$$

Two representations ρ and r of a group G are said to be *isomorphic* if there exists an invertible mapping $f : \mathbb{C}^n \to \mathbb{C}^n$ such that

$$f \circ \rho = r \circ f.$$

Lemma 8.1.6. *The group $\mathcal{B}_{p,q}$ has at least 2^{p+q} distinct degree-1 irreducible representations.*

Proof. First, any representation ρ of degree 1 must satisfy $\rho(\mathbf{e}_\varnothing) = 1$. We claim that in the case $p+q > 1$, this implies $\rho(\mathbf{e}_\alpha) = \rho(\mathbf{e}_\varnothing) = 1$. To see this, note that ${\mathbf{e}_\alpha}^2 = \mathbf{e}_\varnothing$ clearly implies $\rho(\mathbf{e}_\alpha) = \pm 1$. Suppose $\rho(\mathbf{e}_\alpha) = -1$, and consider the following cases:

(1) The case $p = 0$ or $q = 0$. In this case, either ${\mathbf{e}_i}^2 = \mathbf{e}_\varnothing$ for all $1 \le i \le p+q$ or ${\mathbf{e}_i}^2 = \mathbf{e}_\alpha$ for all $1 \le i \le p+q$. In either case, since $p+q > 1$, one considers the product $\mathbf{e}_i\,\mathbf{e}_j$ for $1 \le i \ne j \le p+q$. By anticommutativity, $(\mathbf{e}_i\,\mathbf{e}_j)^2 = \mathbf{e}_\alpha$, but $\rho(\mathbf{e}_\alpha) = -1$ guarantees a contradiction.

(2) The case $p, q \ge 1$. In this case, one considers a pair, i, j, satisfying $1 \le i \le p$ and $p+1 \le j \le p+q$. Then ${\mathbf{e}_i}^2 = \mathbf{e}_\varnothing$, ${\mathbf{e}_j}^2 = \mathbf{e}_\alpha$, and $(\mathbf{e}_i\,\mathbf{e}_j)^2 = \mathbf{e}_\varnothing$, again leading to a contradiction.

Let $J \in 2^{[p+q]}$ denote a multi-index. A degree-1 representation ρ_J is defined by setting $\rho_J(\mathbf{e}_\varnothing) = \rho_J(\mathbf{e}_\alpha) = 1$, and for $1 \le i \le p+q$, setting

$$\rho_J(\mathbf{e}_i) = \begin{cases} 1 & i \in J, \\ -1 & i \notin J. \end{cases}$$

By considering the number of distinct subsets J, it follows immediately that the total number of representations created this way is 2^{p+q}. These representations are clearly irreducible and distinct, i.e., pairwise non-isomorphic. □

Recall that given a group G, the *conjugacy class* of an element $g \in G$ is the set

$$\mathfrak{Cl}(g) = \{hgh^{-1} : h \in G\}.$$

A well-known result in representation theory says that the number of irreducible representations of a group G is equal to the number of its conjugacy classes [93]. This provides a useful tool for establishing the following result.

Theorem 8.1.7. *Given the group $\mathcal{B}_{p,q}$ the number of conjugacy classes and subsequently the number of irreducible representations is given by the formula*

$$\kappa = 2^{p+q} + 1 + c,$$

where $c = p + q \pmod 2$.

Proof. If $p + q = 1$, the formula trivially works.

Suppose $p + q \neq 1$ and denote the center of $\mathcal{B}_{p,q}$ by $Z(\mathcal{B}_{p,q})$. If $g \in \mathcal{B}_{p,q} \backslash Z(\mathcal{B}_{p,q})$, then it is easily seen that

$$\begin{aligned} \mathfrak{Cl}(g) &= \{hgh^{-1} : h \in \mathcal{B}_{p,q}\} \\ &= \{\mathbf{e}_I g \mathbf{e}_I^{-1} : I \in 2^{[p+q]}\}. \end{aligned}$$

Observing that ${\mathbf{e}_I}^{-1} \in \{\mathbf{e}_I, \mathbf{e}_\alpha \mathbf{e}_I\}$, $\mathbf{e}_I g \in \{g\mathbf{e}_I, \mathbf{e}_\alpha g \mathbf{e}_I\}$, and ${\mathbf{e}_I}^2 \in \{\mathbf{e}_\alpha, \mathbf{e}_\varnothing\}$, it follows that $\mathfrak{Cl}(g) = \{g, \mathbf{e}_\alpha g\}$.

Further, if $g \in Z(\mathcal{B}_{p,q})$, then $\mathbf{e}_\alpha g \in Z(\mathcal{B}_{p,q})$ and $\mathfrak{Cl}(g) = \{g\}$. Hence,

$$\kappa = \frac{|\mathcal{B}_{p,q} \backslash Z(\mathcal{B}_{p,q})|}{2} + |Z(\mathcal{B}_{p,q})|.$$

In order to finish the proof, it is necessary to determine the order of the center. Let $\mathbf{e}_I \in \mathcal{B}_{p,q}$ be arbitrary. If $I = \varnothing$, one can see that $\mathbf{e}_\varnothing \in Z(\mathcal{B}_{p,q})$.

Now suppose $I \neq \varnothing$, and let $I = \{i_1, \ldots, i_h\}$ for h even. Then,

$$\begin{aligned} \mathbf{e}_{i_h} \mathbf{e}_I &= (\mathbf{e}_\alpha)^{h-1} \mathbf{e}_I \mathbf{e}_{i_h} \\ &= \mathbf{e}_\alpha \mathbf{e}_I \mathbf{e}_{i_h}. \end{aligned}$$

Whence, $\mathbf{e}_I \notin Z(\mathcal{B}_{p,q})$.

Assume now that $I = \{i_1, \ldots, i_h\} \neq \{1, 2, \ldots, p+q\}$ for h odd. Then there is a natural number ℓ such that $\ell \notin I$ and

$$\mathbf{e}_\ell \mathbf{e}_I = \mathbf{e}_\alpha \mathbf{e}_I \mathbf{e}_\ell.$$

Thus, $\mathbf{e}_I \notin Z(\mathcal{B}_{p,q})$.

Finally, suppose $I = \{1, 2, \dots, p+q\} = [p+q]$ for $p+q$ odd. It is claimed that

$$\mathbf{e}_J\mathbf{e}_{[p+q]} = \mathbf{e}_{[p+q]}\mathbf{e}_J$$

for every indexing set $J \subseteq I$. To see this, note that if $J = \varnothing$ the result trivially holds. By way of induction on the cardinality of J, assume $J = \{j\}$, set
$\mu^+ = |\{i \in [p+q] : i < j\}|$, and set $\mu^- = |\{i \in [p+q] : i > j\}|$. Then,

$$\begin{aligned}\mathbf{e}_{[p+q]}\mathbf{e}_j &= (\mathbf{e}_\alpha)^{\mu^- + \mu^+}\mathbf{e}_j\mathbf{e}_{[p+q]}\\ &= (\mathbf{e}_\alpha)^{p+q-1}\mathbf{e}_j\mathbf{e}_{[p+q]}\\ &= \mathbf{e}_j\mathbf{e}_{[p+q]}.\end{aligned}$$

Suppose $\mathbf{e}_{[p+q]}\mathbf{e}_J = \mathbf{e}_J\mathbf{e}_{[p+q]}$ for some multi-index cardinality $|J|$. The task now is to show $\mathbf{e}_{[p+q]}\mathbf{e}_J\mathbf{e}_h = \mathbf{e}_J\mathbf{e}_h\mathbf{e}_{[p+q]}$ for some natural number $h \notin J$. From the inductive hypothesis we know

$$(\mathbf{e}_{[p+q]}\mathbf{e}_J)\mathbf{e}_h = (\mathbf{e}_J\mathbf{e}_{[p+q]})\mathbf{e}_h$$

and from the basis step, we know $\mathbf{e}_{[p+q]}\mathbf{e}_h = \mathbf{e}_h\mathbf{e}_{[p+q]}$.

Combining these two facts and using associativity of the group operation,

$$\mathbf{e}_{[p+q]}\mathbf{e}_J\mathbf{e}_h = \mathbf{e}_J\mathbf{e}_{[p+q]}\mathbf{e}_h = \mathbf{e}_J\mathbf{e}_h\mathbf{e}_{[p+q]}.$$

Hence, by induction, $\mathbf{e}_J\mathbf{e}_{[p+q]} = \mathbf{e}_{[p+q]}\mathbf{e}_J$ for all $J \in 2^{[p+q]}$.

If $p+q$ is even, then $Z(\mathcal{B}_{p,q}) = \{\mathbf{e}_\alpha, \mathbf{e}_\varnothing\}$. In this case,

$$\begin{aligned}\kappa &= \frac{|\mathcal{B}_{p,q} \backslash Z(\mathcal{B}_{p,q})|}{2} + |Z(\mathcal{B}_{p,q})|\\ &= (2^{p+q} - 1) + 2\\ &= 2^{p+q} + 1.\end{aligned}$$

If $p+q$ is odd, then $Z(\mathcal{B}_{p,q}) = \{\mathbf{e}_\alpha, \mathbf{e}_\varnothing, \mathbf{e}_{[p+q]}, \mathbf{e}_\alpha\mathbf{e}_{[p+q]}\}$, which gives

$$\begin{aligned}\kappa &= \frac{|\mathcal{B}_{p,q} \backslash Z(\mathcal{B}_{p,q})|}{2} + |Z(\mathcal{B}_{p,q})|\\ &= (2^{p+q} - 2) + 4\\ &= 2^{p+q} + 2.\end{aligned}$$

□

Example 8.1.8. Consider the group $\mathcal{B}_{0,1}$. In this case, two non-faithful irreducible representations are constructed as in the proof of Lemma 8.1.6. Two more faithful irreducible representations are found in agreement with Theorem 8.1.7. The four representations are listed in the left table of Figure 8.2. Four irreducible representations of $\mathcal{B}_{1,0}$ are similarly constructed in the right table of Figure 8.2. No faithful irreducible representations exist in this case. It is not difficult to verify that the representations are distinct for each group.

$\mathcal{B}_{0,1}$	$\mathbf{e}_\varnothing$	$\mathbf{e}_\alpha$	$\mathbf{e}_1$	$\mathbf{e}_\alpha\,\mathbf{e}_1$
$\rho_\varnothing$	1	1	1	1
$\rho_{\{1\}}$	1	1	-1	-1
δ_1	1	-1	$\imath$	$-\imath$
δ_2	1	-1	$-\imath$	$\imath$

$\mathcal{B}_{1,0}$	$\mathbf{e}_\varnothing$	$\mathbf{e}_\alpha$	$\mathbf{e}_1$	$\mathbf{e}_\alpha\,\mathbf{e}_1$
$\rho_\varnothing$	1	1	1	1
$\rho_{\{1\}}$	1	1	-1	-1
δ_1	1	-1	1	-1
δ_2	1	-1	-1	1

Figure 8.2 Irreducible degree-1 representations of $\mathcal{B}_{0,1}$ (left) and $\mathcal{B}_{1,0}$ (right).

The above example could have been completed using real representations, although none would be faithful. When $p+q > 1$, $\mathcal{B}_{p,q}$ is non-Abelian, and hence has no faithful degree-1 representation regardless of representation space.

Another well-known result in group representation theory is the following, found in [93].

Lemma 8.1.9. *Let G be a finite group having κ irreducible representations. For each $i = 1, \ldots, \kappa$, let n_i denote the degree of the ith irreducible representation of G. Then,*

$$|G| = \sum_{i=1}^{\kappa} {n_i}^2.$$

Given the group $\mathcal{B}_{p,q}$ there are always 2^{p+q} distinct irreducible representations of degree 1. The remaining irreducible (complex) representations are now enumerated in the next theorem.

Theorem 8.1.10. *If $p + q = 2k > 1$, then $\mathcal{B}_{p,q}$ has one irreducible representation of degree 2^k. If $p + q = 2k + 1$, then $\mathcal{B}_{p,q}$ has two irreducible representations of degree 2^k. Moreover, all of these irreducible representations are faithful except when p is odd and q is even.*

Proof. This will be treated in two cases. First is the case of $p+q$ even. Since $p+q$ is even, one can write $p+q=2k$ for some $k\in\mathbb{N}$. Define $\tau:\mathcal{B}_{p,q}\to \mathrm{GL}_{2^k}(\mathbb{C})$ by

$$\tau(\mathbf{e}_j)=\begin{cases}\sigma_{\mathrm{x}}{}^{\otimes(j-1)}\otimes\sigma_{\mathrm{z}}\otimes\sigma_0{}^{\otimes(k-j)} & 1\le j\le k,\ j\le p\\ \sigma_{\mathrm{x}}{}^{\otimes(j-1)}\otimes\imath\sigma_{\mathrm{y}}\otimes\sigma_0{}^{\otimes(k-j)} & 1\le j\le k,\ j> p\\ \sigma_{\mathrm{x}}{}^{\otimes(j-k-1)}\otimes\sigma_{\mathrm{z}}\otimes\sigma_0{}^{\otimes(2k-j)} & k+1\le j\le 2k,\ j\le p\\ \sigma_{\mathrm{x}}{}^{\otimes(j-k-1)}\otimes\imath\sigma_{\mathrm{y}}\otimes\sigma_0{}^{\otimes(2k-j)} & k+1\le j\le 2k,\ j> p.\end{cases}\tag{8.7}$$

Setting $\tau(\mathbf{e}_\varnothing)=\sigma_0{}^{\otimes k}$ and $\tau(\mathbf{e}_\alpha)=-\sigma_0{}^{\otimes k}$, this extends by multiplication to all of $\mathcal{B}_{p,q}$. More specifically, for any multi-index I, $\tau(\mathbf{e}_I)=\prod_{\ell\in I}\tau(\mathbf{e}_\ell)$, and $\tau(\mathbf{e}_\alpha\mathbf{e}_I)=-\tau(\mathbf{e}_I)$.

This representation is clearly well defined. To verify that this representation is irreducible, let ξ be the character of τ. Let $\mathbf{e}_I\in\mathcal{B}_{p,q}$ be arbitrary. Letting $\sigma_{(\ell)}\in\{\sigma_0,\sigma_{\mathrm{x}},\sigma_{\mathrm{y}},\sigma_{\mathrm{z}}\}$ for each $\ell=1,\ldots,k$, one writes

$$\begin{aligned}\xi(\mathbf{e}_I)&=\mathrm{tr}(\tau(\mathbf{e}_I))\\&=\mathrm{tr}\left(u\bigotimes_{\ell=1}^{k}\sigma_{(\ell)}\right)\\&=u\prod_{\ell\in I}\mathrm{tr}(\sigma_{(\ell)})\end{aligned}$$

for some unit $u\in\{\pm1,\pm\imath\}$. Since $\mathrm{tr}(\sigma_{\mathrm{x}})=\mathrm{tr}(\sigma_{\mathrm{y}})=\mathrm{tr}(\sigma_{\mathrm{z}})=0$, and $\mathrm{tr}(\sigma_0)=2$, it follows that $\xi(\mathbf{e}_I)=\xi(\mathbf{e}_\alpha\mathbf{e}_I)=0$ unless $I=\varnothing$, in which case $\xi(\mathbf{e}_\varnothing)=2^k$ and $\xi(\mathbf{e}_a\mathbf{e}_\varnothing)=-2^k$. Now,

$$\begin{aligned}(\xi|\xi)&=\frac{1}{|\mathcal{B}_{p,q}|}\sum_{g\in\mathcal{B}_{p,q}}\xi(g)\overline{\xi(g)}\\&=\frac{1}{2^{2k+1}}((\xi(\mathbf{e}_\varnothing))^2+(\xi(\mathbf{e}_\alpha\mathbf{e}_\varnothing))^2)\\&=\frac{1}{2^{2k+1}}(2)(2^{2k})=1.\end{aligned}$$

Thus, τ is irreducible.

To see that the representation (8.7) is faithful, consider the kernel:

$$\ker(\tau)=\{\mathbf{e}_I:\tau(\mathbf{e}_I)=\sigma_0{}^{\otimes k}\}.$$

Noting that $\tau(\mathcal{B}_{p,q})$ is a subgroup of $\mathrm{GL}_{2^k}(\mathbb{C})$, we begin by showing that the center of this subgroup consists only of elements having the form $\pm u\sigma_0{}^{\otimes k}$, where $u\in\{\pm1,\pm\imath\}$. To begin, the center of $\tau(\mathcal{B}_{p,q})$ is

$$Z(\tau(\mathcal{B}_{p,q}))=\{\tau(\mathbf{e}_E):\tau(\mathbf{e}_E)\tau(\mathbf{e}_J)=\tau(\mathbf{e}_J)\tau(\mathbf{e}_E),\ \forall J\in 2^{[p+q]}\}.$$

Suppose an element of $Z(\tau(\mathcal{B}_{p,q}))$ is of the form $M = u\,\sigma_{(1)} \otimes \ldots \otimes \sigma_{(k)}$, where for some index h, $\sigma_{(h)} \neq \sigma_0$ but $\sigma_{(h+1)} = \cdots = \sigma_{(k)} = \sigma_0$. If $\sigma_{(h)} = \sigma_z$ or σ_y, it follows that

$$\tau(\mathbf{e}_{\{h,h+k\}}) = u({\sigma_0}^{\otimes(h-1)} \otimes \sigma_x \otimes {\sigma_0}^{(k-h)}),$$

which will anti-commute with M.

If $\sigma_{(h)} = \sigma_x$, an element anti-commuting with M is given by

$$\begin{aligned}
&\tau(\mathbf{e}_{\{1,k+1,2,k+2,\ldots,h-1,k+h-1,h\}}) \\
&= \left(\prod_{j=1}^{h-1}(-\imath)({\sigma_0}^{\otimes(j-1)} \otimes \sigma_x \otimes {\sigma_0}^{\otimes(k-j)})\right)\left({\sigma_x}^{\otimes(h-1)} \otimes \sigma_z \otimes {\sigma_0}^{\otimes(k-h)}\right) \\
&= u\left({\sigma_x}^{\otimes(h-1)} \otimes {\sigma_0}^{\otimes(k-h+1)}\right)\left({\sigma_x}^{\otimes(h-1)} \otimes \sigma_z \otimes {\sigma_0}^{\otimes(k-h)}\right) \\
&= u\,{\sigma_0}^{\otimes(h-1)} \otimes \sigma_z \otimes {\sigma_0}^{\otimes(k-h)}.
\end{aligned}$$

This proves that every element of $Z(\tau(\mathcal{B}_{p,q}))$ is of the form $u\,{\sigma_0}^{\otimes k}$ for $u \in \{\pm 1, \pm i\}$.

By construction, $\tau(\mathbf{e}_\varnothing)$ and $\tau(\mathbf{e}_\alpha)$ are elements of $Z(\tau(\mathcal{B}_{p,q}))$. Suppose E is a non-empty indexing set, and to the contrary suppose $\tau(\mathbf{e}_E) \in Z(\tau(\mathcal{B}_{p,q}))$, so that $\tau(\mathbf{e}_E) = u\,{\sigma_0}^{\otimes k}$. It is not difficult to see that $\mathbf{e}_E \notin Z(\mathcal{B}_{p,q})$, so there exists an integer m such that

$$\mathbf{e}_E\mathbf{e}_m = \mathbf{e}_\alpha\mathbf{e}_m\mathbf{e}_E.$$

Applying τ reveals

$$\tau(\mathbf{e}_E\mathbf{e}_m) = u\,{\sigma_0}^{\otimes k}\tau(\mathbf{e}_m) \neq -u\,\tau(\mathbf{e}_m){\sigma_0}^{\otimes k} = \tau(\mathbf{e}_\alpha\mathbf{e}_m\mathbf{e}_E).$$

This contradicts the homomorphism property of τ. We conclude then that $\mathbf{e}_E \notin Z(\tau(\mathcal{B}_{p,q}))$, which means

$$Z(\tau(\mathcal{B}_{p,q})) = \{\tau(\mathbf{e}_\varnothing), \tau(\mathbf{e}_\alpha)\}.$$

However, only one of these, $\tau(\mathbf{e}_\varnothing)$, is ${\sigma_0}^{\otimes k}$. Thus, $\ker(\tau) = \{\mathbf{e}_\varnothing\}$. Since the kernel is trivial, τ is faithful.

Now suppose $p + q = 2k + 1$ is odd; more specifically, suppose p is even and q is odd. Let $\tau : \mathcal{B}_p^q \to \mathrm{GL}_{2^k}(\mathbb{C})$ be defined by

$$\tau(\mathbf{e}_j) = \begin{cases}
{\sigma_x}^{\otimes(j-1)} \otimes \sigma_z \otimes {\sigma_0}^{\otimes(k-j)} & 1 \leq j \leq k,\ j \leq p \\
{\sigma_x}^{\otimes(j-1)} \otimes \bar{\imath}\sigma_y \otimes {\sigma_0}^{\otimes(k-j)} & 1 \leq j \leq k,\ j > p \\
{\sigma_x}^{\otimes(j-k-1)} \otimes \sigma_z \otimes {\sigma_0}^{\otimes(2k-j)} & k+1 \leq j \leq 2k,\ j \leq p \\
{\sigma_x}^{\otimes(j-k-1)} \otimes \bar{\imath}\sigma_y \otimes {\sigma_0}^{\otimes(2k-j)} & k+1 \leq j \leq 2k,\ j > p \\
{\sigma_x}^{\otimes(k-1)} \otimes \sigma_z & j = 2k+1,\ j \leq p \\
{\sigma_x}^{\otimes(k-1)} \otimes \bar{\imath}\sigma_y & j = 2k+1,\ j > p.
\end{cases} \tag{8.8}$$

Setting $\tau(\mathbf{e}_\varnothing) = {\sigma_0}^{\otimes k}$ and $\tau(\mathbf{e}_\alpha \mathbf{e}_i) = -\tau(\mathbf{e}_i)$, this extends by multiplication to all of $\mathcal{B}_{p,q}$.

To check irreducibility of τ, let ξ denote the character of τ. It is already known that $\xi(\mathbf{e}_I) = 0$ in all but the extreme case $I = \varnothing$. Further, since $\mathbf{e}_{[p+q]}$ is in the center of the group, its image under τ must be of the form $u\,{\sigma_0}^{\otimes k}$ for some scalar $u \in \{\pm 1, \pm \imath\}$. It thereby follows that

$$\begin{aligned}
(\xi|\xi) &= \frac{1}{|\mathcal{B}_{p,q}|} \sum_{g \in \mathcal{B}_p^q} \xi(g)\overline{\xi(g)} \\
&= \frac{1}{2^{2k+2}} \left((\xi(\mathbf{e}_\varnothing))^2 + (\xi(\mathbf{e}_\alpha))^2 \right) \\
&\quad + \frac{1}{2^{2k+2}} \left(\xi(\mathbf{e}_{[2k+1]})\overline{\xi(\mathbf{e}_{[2k+1]})} + \xi(\mathbf{e}_\alpha \mathbf{e}_{[2k+1]})\overline{\xi(\mathbf{e}_\alpha \mathbf{e}_{[2k+1]})} \right) \\
&= \frac{1}{(2^{2k+2})} \left(2^{2k} + 2^{2k} + 2^{2k} + 2^{2k} \right) \\
&= 1\,.
\end{aligned}$$

Hence, τ is irreducible.

Recall that when $p+q$ is odd, $Z(\mathcal{B}_{p,q}) = \{\mathbf{e}_\alpha, \mathbf{e}_\varnothing, \mathbf{e}_{[p+q]}, \mathbf{e}_\alpha \mathbf{e}_{[p+q]}\}$. In light of the proof that τ was faithful for $p+q$ even, showing that $\mathbf{e}_{[p+q]} \notin \ker(\tau)$ is sufficient to show that τ is faithful. Computing $\mathbf{e}_{[p+q]}$, one finds

$$\mathbf{e}_{[p+q]} \mapsto (\sigma_\mathrm{z} \otimes \imath\sigma_\mathrm{y})^{\otimes(p/2)} \otimes (\imath\sigma_\mathrm{y} \otimes \sigma_\mathrm{z})^{\otimes((q-1)/2)} \otimes \bar{\imath}\sigma_\mathrm{y}. \tag{8.9}$$

It follows that $\ker(\tau)$ is trivial.

Finally, in the case p is odd and q is even, the construction of (8.8) is again used. This representation is again irreducible, and

$$\mathbf{e}_{[p+q]} \mapsto (\sigma_\mathrm{z} \otimes \imath\sigma_\mathrm{y})^{\otimes((p-1)/2)} \otimes \sigma_\mathrm{z} \otimes (\sigma_\mathrm{z} \otimes \imath\sigma_\mathrm{y})^{\otimes(q/2)}, \tag{8.10}$$

so that the representation is faithful.

Recalling that the order of $\mathcal{B}_{p,q}$ is equal to the sum of the squares of degrees of irreducible representations, there remains one irreducible representation of $\mathcal{B}_{p,q}$ in the case $p+q$ is odd: the complex conjugate of τ. This representation is given explicitly by

$$\overline{\tau}(\mathbf{e}_j) = \begin{cases}
{\sigma_\mathrm{x}}^{\otimes(j-1)} \otimes \sigma_\mathrm{z} \otimes {\sigma_0}^{\otimes(k-j)} & 1 \le j \le k,\ j \le p \\
{\sigma_\mathrm{x}}^{\otimes(j-1)} \otimes \bar{\imath}\sigma_\mathrm{y} \otimes {\sigma_0}^{\otimes(k-j)} & 1 \le j \le k,\ j > p \\
{\sigma_\mathrm{x}}^{\otimes(j-k-1)} \otimes \sigma_\mathrm{z} \otimes {\sigma_0}^{\otimes(2k-j)} & k+1 \le j \le 2k,\ j \le p \\
{\sigma_\mathrm{x}}^{\otimes(j-k-1)} \otimes \bar{\imath}\sigma_\mathrm{y} \otimes {\sigma_0}^{\otimes(2k-j)} & k+1 \le j \le 2k,\ j > p \\
{\sigma_\mathrm{x}}^{\otimes(k-1)} \otimes \sigma_\mathrm{z} & j = 2k+1,\ j \le p \\
{\sigma_\mathrm{x}}^{\otimes(k-1)} \otimes \bar{\imath}\sigma_\mathrm{y} & j = 2k+1,\ j > p,
\end{cases}$$

where $\overline{\tau}(\mathbf{e}_\varnothing) = {\sigma_0}^{\otimes k}$ and $\overline{\tau}(\mathbf{e}_\alpha \mathbf{e}_i) = -\tau(\mathbf{e}_i)$. This extends by multiplication to all of $\mathcal{B}_{p,q}$.

To see that $\overline{\tau}$ is not isomorphic to τ, one considers the action of $\overline{\tau}$ on $\mathbf{e}_{[p+q]}$. In particular, (8.9) and (8.10) imply

$$\overline{\tau}(\mathbf{e}_{[p+q]}) = \begin{cases} {\sigma_0}^{\otimes k} = \tau(\mathbf{e}_{[p+q]}) & \text{when } q \equiv 0 \pmod 4, \\ -\imath({\sigma_0}^{\otimes k}) = -\tau(\mathbf{e}_{[p+q]}) & \text{when } q \equiv 1 \pmod 4, \\ -{\sigma_0}^{\otimes k} = \tau(\mathbf{e}_{[p+q]}) & \text{when } q \equiv 2 \pmod 4, \\ \imath({\sigma_0}^{\otimes k}) = -\tau(\mathbf{e}_{[p+q]}) & \text{when } q \equiv 3 \pmod 4. \end{cases}$$

Suppose there exists an invertible linear transformation $f \in \mathrm{GL}(\mathbb{C}^{2^k})$ satisfying $f \circ \tau = \overline{\tau} \circ f$. Then, the cases $q \equiv 1 \pmod 4$ and $q \equiv 3 \pmod 4$ imply

$$f \circ \left(\imath\left({\sigma_0}^{\otimes k}\right)\right) = -\imath\left({\sigma_0}^{\otimes k}\right) \circ f \Rightarrow f(\mathbf{v}) = -\mathbf{v},\ \forall \mathbf{v} \in \mathbb{C}^{2^k},$$

which contradicts $f \circ \overline{\tau}(\mathbf{e}_\varnothing) = {\sigma_0}^{\otimes k}$. Similarly, in the cases $q \equiv 0 \pmod 4$ and $q \equiv 2 \pmod 4$,

$$f \circ \left({\sigma_0}^{\otimes k}\right) = \left({\sigma_0}^{\otimes k}\right) \circ f \Rightarrow f(\mathbf{v}) = \mathbf{v},\ \forall \mathbf{v} \in \mathbb{C}^{2^k},$$

contradicting $f \circ \tau = \overline{\tau} \circ f$, since $\tau \neq \overline{\tau}$. □

It becomes evident in the case $p \equiv 1 \pmod 2$ and $q \equiv 0 \pmod 2$ that in order to obtain a faithful representation of $\mathcal{B}_{p,q}$, one must pass to a larger representation space. It is not difficult to show that a faithful representation is given by defining $\tau : \mathcal{B}_{p,q} \to \mathrm{GL}_{2^{k+1}}(\mathbb{C})$ by

$$\tau(\mathbf{e}_j) = \begin{cases} {\sigma_\mathrm{x}}^{\otimes(j-1)} \otimes \sigma_\mathrm{z} \otimes {\sigma_0}^{\otimes(k-j+1)} & 1 \le j \le k,\ j \le p \\ \imath\left({\sigma_\mathrm{x}}^{\otimes(j-1)} \otimes \sigma_\mathrm{z} \otimes {\sigma_0}^{\otimes(k-j+1)}\right) & 1 \le j \le k,\ j > p \\ {\sigma_\mathrm{x}}^{\otimes(j-k-1)} \otimes \sigma_\mathrm{y} \otimes {\sigma_0}^{\otimes(2k-j)} & k+1 \le j \le 2k,\ j \le p \\ \imath\left({\sigma_\mathrm{x}}^{\otimes(j-k-1)} \otimes \sigma_\mathrm{y} \otimes {\sigma_0}^{\otimes(2k-j)}\right) & k+1 \le j \le 2k,\ j > p \\ {\sigma_\mathrm{x}}^{\otimes(k+1)} & j = 2k+1,\ j \le p \\ \imath\left({\sigma_\mathrm{x}}^{\otimes(k+1)}\right) & j = 2k+1,\ j > p. \end{cases}$$

Setting $\tau(\mathbf{e}_\varnothing) = {\sigma_0}^{\otimes(k+1)}$ and $\tau(\mathbf{e}_\alpha \mathbf{e}_i) = -\tau(\mathbf{e}_i)$, this is again extended by multiplication to all of $\mathcal{B}_{p,q}$.

8.2 The "Nil-Clifford" (Zeon) Algebra $\mathcal{C}\ell_n{}^{\text{nil}}$

As subalgebras of Clifford algebras of appropriate signature, zeon algebras were first defined in [98], where they were applied to graph theory. The name "zeon algebra" came later, being first used by Feinsilver [43]. In addition to their theoretical properties and applications, they have been applied to real-world problems, including satellite communications and wireless networks [27, 73].

The *n-particle zeon algebra*, denoted $\mathcal{C}\ell_n{}^{\text{nil}}$, is defined as the real abelian algebra generated by the collection $\{\zeta_i\}$ $(1 \le i \le n)$ along with the scalar $1 = \zeta_\varnothing$ subject to the following multiplication rules:

$$\zeta_i\,\zeta_j = \zeta_j\,\zeta_i \ \text{ for } i \neq j, \text{ and}$$
$$\zeta_i{}^2 = 0 \ \text{ for } 1 \le i \le n.$$

As a vector space, this 2^n-dimensional algebra has a canonical basis of *basis blades* of the form $\{\zeta_I : I \subseteq [n]\}$, where $\zeta_I = \prod_{\iota \in I} \zeta_\iota$. The null-square property of the generators $\{\zeta_i : 1 \le i \le n\}$ guarantees that the product of two basis blades satisfies the following:

$$\zeta_I \zeta_J = \begin{cases} \zeta_{I\cup J} & I \cap J = \varnothing, \\ 0 & \text{otherwise.} \end{cases}$$

Letting $2^{[n]}$ denote the power set of the *n-set*, $[n] = \{1, 2, \ldots, n\}$, it thereby follows that an element $u \in \mathcal{C}\ell_n{}^{\text{nil}}$ has the following canonical expansion:

$$u = \sum_{I \in 2^{[n]}} u_I\, \zeta_I\,,$$

where $u_I \in \mathbb{R}$ for each multi-index I. The *grade-k part* of u is defined by $\langle u \rangle_k = \sum_{|I|=k} u_I \zeta_I$. To simplify things, the term "zeon" will be used in reference to an arbitrary element of $\mathcal{C}\ell_n{}^{\text{nil}}$.

It is often convenient to separate the scalar part (i.e., grade-0 part) of a zeon from the rest of it. To this end, for $z \in \mathcal{C}\ell_n{}^{\text{nil}}$ we write $\Re z = \langle z \rangle_0$, the *real part* of z, and $\mathfrak{D} z = z - \Re z$, the *dual part*[2] of z. A nonzero zeon u is said to be *trivial* if $u = \Re u$. Otherwise, it is said to be *nontrivial*.

Example 8.2.1. In $\mathcal{C}\ell_3{}^{\text{nil}}$ set

$$u = 2 + 5\zeta_{\{1\}} - \zeta_{\{2\}} + 3\zeta_{\{1,3\}} - 6\zeta_{\{1,2,3\}}.$$

[2]These are referred to as the body and soul of z in Neto's works [74–76].

Then,

$$\begin{aligned}
\langle u\rangle_0 &= 2\\
\langle u\rangle_1 &= 5\zeta_{\{1\}} - \zeta_{\{2\}}\\
\langle u\rangle_2 &= 3\zeta_{\{1,3\}}\\
\langle u\rangle_3 &= -6\zeta_{\{1,2,3\}}.
\end{aligned}$$

Further, $\Re u = 2$, and $\mathfrak{D}u = 5\zeta_{\{1\}} - \zeta_{\{2\}} + 3\zeta_{\{1,3\}} - 6\zeta_{\{1,2,3\}}$.

Definition 8.2.2. For nilpotent $u \in {\mathcal{C}\ell_n}^{\text{nil}}$, let $\kappa(u)$ denote the *index of nilpotency* of u; i.e., $u^{\kappa(u)} = 0$ and $u^{\kappa(u)-1} \neq 0$.

Example 8.2.3. Let $u = \zeta_{1,2} \in {\mathcal{C}\ell_n}^{\text{nil}}$ for any $n \geq 2$. Then, $\kappa(u) = 2$, since $u \neq 0$, but $u^2 = 0$. On the other hand, $\kappa(\zeta_{\{}1\} + \zeta_{\{}2\}) = 3$ because

$$(\zeta_{\{1\}} + \zeta_{\{2\}})^2 = 2\zeta_{\{1,2\}},$$

while $(\zeta_{\{1\}} + \zeta_{\{2\}})^3 = 0$.

Null Blade Semigroups $\mathfrak{G}_n$ *and* $\mathfrak{Z}_n$

Two semigroups underlie the zeon algebra. The semigroup $\mathfrak{Z}_n$ directly underlies the zeon algebra in the sense that ${\mathcal{C}\ell_n}^{\text{nil}}$ is isomorphic to a quotient of the real semigroup algebra $\mathbb{R}\mathfrak{Z}_n$.

The more fundamental semigroup, denoted here by $\mathfrak{G}_n$ underlies the Grassmann exterior algebra $\bigwedge \mathbb{R}^n$ in the sense that $\bigwedge \mathbb{R}^n$ is isomorphic to a quotient of the real semigroup algebra $\mathbb{R}\mathfrak{G}_n$.

By modifying the multiplication in $\mathcal{B}_{p,q}$ such that generators square to zero, one obtains a non-Abelian *semigroup* generated by *null squares*. The principal difference from this point forward is a lack of multiplicative inverses for elements in the algebraic structures.

Definition 8.2.4. Let $\mathfrak{G}_n$ denote the *null blade semigroup* defined as the semigroup generated by the collection $G = \{\gamma_i : 1 \leq i \leq n\}$ along with $\{\gamma_\varnothing, \gamma_\alpha, 0_\gamma\}$ satisfying the following generating relations: for all $x \in G \cup \{\gamma_\varnothing, \gamma_\alpha, 0_\gamma\}$,

$$\begin{aligned}
\gamma_\varnothing\, x &= x\, \gamma_\varnothing = x,\\
\gamma_\alpha\, x &= x\, \gamma_\alpha,\\
0_\gamma\, x &= x\, 0_\gamma = 0_\gamma,\\
{\gamma_\varnothing}^2 &= {\gamma_\alpha}^2 = \gamma_\varnothing,
\end{aligned}$$

and

$$\gamma_i\gamma_j = \begin{cases} 0_\gamma & \text{if and only if } i = j, \\ \gamma_\alpha\,\gamma_j\gamma_i & i \neq j. \end{cases}$$

Define the *anti-symmetric product signature map* $\phi : 2^{[n]} \times 2^{[n]} \to \{\gamma_\varnothing, \gamma_\alpha\}$ by

$$\phi(I, J) = \gamma_\alpha{}^{\sum_{j\in J} \mu_j(I)}.$$

Remark 8.2.5. Note that the product signature map defined by (8.3) can be extended to $\mathcal{G} \times \mathcal{G}$ and written in terms of ϕ as

$$\vartheta(I, J) = \gamma_\alpha{}^{\mu_p(I\cap J)+\phi(I,J)}.$$

Hence, ϑ has a decomposition into *signature-dependent* and *signature-independent* parts.

Applying multi-index notation to the generators $G = \{\gamma_i : 1 \leq i \leq n\}$ according to the ordered product

$$\gamma_I = \prod_{i\in I} \gamma_i$$

for arbitrary $I \in 2^{[n]}$, the multiplicative semigroup $\mathfrak{G}_n$ is now seen to be determined by the multi-indexed set $\{0_\gamma\} \cup \{\gamma_\alpha\gamma_I, \gamma_I : I \in 2^{[n]}\}$ along with the associative multiplication defined by

$$\gamma_I\,\gamma_J = \begin{cases} \gamma_\alpha{}^{\sum_{i\in I} \mu_i(J)}\gamma_{I\cup J} & I \cap J = \varnothing, \\ 0 & \text{otherwise.} \end{cases}$$

Note that the order of the null blade semigroup is $|\mathfrak{G}_n| = 2^{n+1} + 1$. The next combinatorial algebra can now be defined.

Definition 8.2.6. For fixed positive integer n, the *null blade algebra* is defined as the real semigroup algebra $\mathbb{R}\mathfrak{G}_n / \langle 0_\gamma, \gamma_\alpha + \gamma_\varnothing\rangle$, denoted $\mathcal{B}\ell_{\wedge n}$ for convenience.

It now becomes clear that the null blade algebra $\mathcal{B}\ell_{\wedge n}$ is canonically isomorphic to the Grassmann (exterior) algebra $\bigwedge \mathbb{R}^n$.

Theorem 8.2.7. *For any natural number n, there are three irreducible representations of $\mathfrak{G}_n$.*

Proof. First, each $\mathfrak{J}$-class of $\mathfrak{G}_n$ will be classified, and the regular classes will be identified. Then the maximal subgroups of a choice of distinct idempotents are computed. Finally, the number of irreducible representations can be obtained.

Let $\gamma_I \in \mathfrak{G}_n \setminus \{0_\gamma\}$ be arbitrary. Then,

$$\begin{aligned}\mathfrak{G}_n\gamma_I\mathfrak{G}_n &= \{s_1\,\gamma_I\,s_2 : s_1, s_2 \in \mathfrak{G}_n\}\\ &= \{\gamma_E, \gamma_\alpha\gamma_E, 0_\gamma : I \subseteq E\}.\end{aligned}$$

Similarly,

$$\mathfrak{G}_n(0_\gamma)\mathfrak{G}_n = \{0_\gamma\}.$$

Hence, every $w \in \mathfrak{G}_n$, the set of all things $\mathfrak{J}$ -equivalent to w is simply $\{w, \gamma_\alpha w\}$. The number of $\mathfrak{J}$-classes is then 2^n+1. However, only the regular $\mathfrak{J}$-classes are of interest. The only idempotent elements of $\mathfrak{G}_n$ are 0_γ and $\gamma_\varnothing$. It follows that the regular $\mathfrak{J}$-classes are $\{\gamma_\varnothing, \gamma_\alpha\}$ and $\{0_\gamma\}$. The two maximal subgroups are

$$G_{\gamma_\varnothing} = \{\text{invertible elements of } \gamma_\varnothing\mathfrak{G}_n\gamma_\varnothing\} = \{\gamma_\varnothing, \gamma_\alpha\},$$

and

$$G_{0_\gamma} = \{\text{invertible elemenets of } 0_\gamma\mathfrak{G}_n 0_\gamma\} = \{0_\gamma\}.$$

The trivial group, G_{0_γ}, has one conjugacy class, while $G_{\gamma_\varnothing}$ is an Abelian group of order 2, which consequently has two conjugacy classes. Thus, the number of irreducible representations of $\mathfrak{G}_n$ is three. □

Definition 8.2.8. Let $\mathfrak{Z}_n$ denote the *zeon semigroup* defined as the semigroup generated by the collection $C = \{\zeta_i : 1 \leq i \leq n\}$ along with $\{\zeta_\varnothing, 0_\zeta\}$ satisfying the following generating relations: for all $x \in C \cup \{\zeta_\varnothing, 0_\zeta\}$,

$$\begin{aligned}\zeta_\varnothing\, x &= x\,\zeta_\varnothing = x,\\ 0_\zeta\, x &= x\, 0_\zeta = 0_\zeta,\\ {\zeta_\varnothing}^2 &= 0_\zeta,\end{aligned}$$

and

$$\zeta_i\zeta_j = \begin{cases} 0_\zeta & \text{if and only if } i = j,\\ \zeta_j\zeta_i & i \neq j.\end{cases}$$

The zeon semigroup is of particular interest, as its associated semigroup algebra is canonically isomorphic to the *zeon algebra*. Using nearly the same proof as above, it becomes apparent that $\mathfrak{Z}_n$ has two copies of the trivial group as maximal subgroups, and thus has two irreducible representations, regardless of n.

The irreducible representations of both $\mathfrak{Z}_n$ and $\mathfrak{G}_n$ are almost immediately obvious. For arbitrary n, define degree-1 representations θ, ρ_0, and ρ_1 of $\mathfrak{G}_n$ by

$$\theta(s) = 1, \forall s \in \mathfrak{G}_n.$$

$$\rho_0(s) = \begin{cases} 1 & s = \gamma_\varnothing, \\ -1 & s = \gamma_\alpha, \\ 0 & \text{otherwise.} \end{cases} \qquad \rho_1(s) = \begin{cases} 1 & s = \gamma_\varnothing, \\ 1 & s = \gamma_\alpha, \\ 0 & \text{otherwise.} \end{cases}$$

Note that these representations are clearly not faithful.

In $\mathfrak{Z}_n$, the irreducible representations are simply θ and the degree-1 representation ρ given by

$$\rho(s) = \begin{cases} 1 & s = \zeta_\varnothing, \\ 0 & \text{otherwise.} \end{cases}$$

Given an arbitrary natural number n, there is a faithful representation τ of $\mathfrak{G}_n$ of order 2^n given by

$$\tau(\gamma_i) = {\sigma_{\mathrm{x}}}^{\otimes(i-1)} \otimes \eta \otimes {\sigma_0}^{\otimes(n-i)},$$

where $\eta = \sigma_{\mathrm{z}} + \imath\sigma_{\mathrm{y}} = \begin{pmatrix} 1 & 1 \\ -1 & -1 \end{pmatrix}$.

Similarly, a faithful representation ψ of $\mathfrak{Z}_n$ is given by

$$\psi(\zeta_i) = {\sigma_0}^{\otimes(i-1)} \otimes \eta \otimes {\sigma_0}^{\otimes(n-i)}.$$

8.3 Group and Semigroup Algebras

Beginning with a finite multiplicative semigroup, S, the *semigroup algebra* of S over $\mathbb{R}$, is the real algebra $\mathbb{R}S$ whose additive group is the Abelian group of formal $\mathbb{R}$-linear combinations of elements of S, i.e.,

$$\mathbb{R}S = \left\{ \sum_{s\in S} \alpha_s s : \alpha_s \in \mathbb{R} \right\}$$

and whose multiplication operation is defined by linear extension of the group multiplication operation of S. This definition restricts in a natural way to group algebras.

Given a (complex) representation ρ of a finite semigroup S, let $\widetilde{\rho}$ denote the representation of the $|S|$-dimensional semigroup algebra $\mathbb{C}S$ given by

$$x = \sum_{s\in G} \alpha_s s \Rightarrow \widetilde{\rho}(x) = \sum_{s\in S} \alpha_s \rho(s),$$

where $\alpha_s \in \mathbb{C}$ for each $s \in S$.

It is known that ρ is irreducible if and only if $\widetilde{\rho}$ is irreducible [82], so that the irreducible representations of S are in one-to-one correspondence with the irreducible representations of $\mathbb{C}S$. In particular, if $\widetilde{\rho}$ is an irreducible representation of $\mathbb{C}S$, an irreducible representation of S is obtained by restricting $\widetilde{\rho}$ to the elements of S.

Classifying the irreducible representations of $\mathcal{B}_{p,q}$ thereby classifies the irreducible representations of the group algebra $\mathbb{R}\mathcal{B}_{p,q}$. Similarly, classifying the irreducible representations for $\mathfrak{G}_n$ and $\mathfrak{Z}_n$ classifies the irreducible representations of the semigroup algebras $\mathbb{R}\mathfrak{G}_n$ and $\mathbb{R}\mathfrak{Z}_n$.

Quotient Algebras

A subring I of a commutative ring R is said to be an *ideal* (or a two-sided ideal) if, in addition to being a subring, it has the property that $ra \in I$ whenever $a \in I$ and $r \in R$. An ideal is said to be *principal* if it is generated by a single element. Letting R be a commutative ring, it is not difficult to show that the set

$$\langle a \rangle = \{ra : r \in R\}$$

is an ideal. This ideal is called the *principal ideal generated by a*.

In non-commutative rings, a distinction can be made between left and right ideals, but this is not necessary for our discussion. Given a ring R with ideal I, a *coset* of I is determined for each $r \in R$ by

$$r + I := \{r + a : a \in I\}.$$

Two cosets $r + I$, $s + I$ are equal if and only if $(r - s) \in I$. The collection of all such cosets is denoted by R/I. Observe that the zero coset of I is I itself; i.e., $0 + I = I$. Addition is defined naturally on R/I by

$$(r + I) + (s + I) = (r + s) + I.$$

One can also show that multiplication is well-defined on R/I by

$$(r + I)(s + I) = rs + I.$$

The relation on R defined by $r \sim s \Leftrightarrow (r - s) \in I$ is an equivalence relation called *congruence modulo I*.

With the multiplication and addition seen above, the set R/I is seen to be a ring called a *quotient ring* (or *factor ring*). Extending the ring R to an algebra $\mathcal{A}$ by allowing scalar multiplication on R, one obtains the *quotient algebra* $\mathcal{A}/I$.

Recall that given an algebra homomorphism, $\varphi : \mathcal{A} \to \mathcal{B}$, the *kernel* of φ is defined by

$$\ker \varphi = \{a \in \mathcal{A} : \varphi(a) = 0_{\mathcal{B}}\}.$$

Here, $0_{\mathcal{B}}$ denotes the additive identity element of $\mathcal{B}$. The *image* of φ is defined by

$$\operatorname{Im} \varphi := \varphi(\mathcal{A}) = \{\varphi(a) : a \in \mathcal{A}\}.$$

Lemma 8.3.1. *The kernel of an algebra homomorphism $\varphi : \mathcal{A} \to \mathcal{B}$ is an ideal of $\mathcal{A}$.*

Proof. The proof is left as an exercise. □

This leads us to recall one of the most important theorems in algebra, stated here without proof.

Theorem 8.3.2 (First isomorphism theorem). *Let $\mathcal{A}, \mathcal{B}$ be algebras and let $\varphi : \mathcal{A} \to \mathcal{B}$ be an algebra homomorphism. Then,*

$$\mathcal{A}/\ker \varphi \cong \operatorname{Im} \varphi.$$

The homomorphisms from group algebras and semigroup algebras to the Clifford algebra, Grassmann algebra, and other combinatorially motivated subalgebras of Clifford algebras are constructed as in the following example. In the blade group, elements of the form $\mathbf{e}_I$ are regarded as "positive elements" and are mapped to their corresponding elements in the Clifford (sub)algebra. Similarly, $\mathbf{e}_\varnothing \mapsto \mathbf{e}_\varnothing$, while $\mathbf{e}_\alpha x \mapsto -x$. In this way, elements of the form $\mathbf{e}_\alpha \mathbf{e}_I$ are regarded as "negative elements" of the group or semigroup. The kernel of such a homomorphism is seen to be the principal ideal generated by $\mathbf{e}_\varnothing + \mathbf{e}_\alpha$. Because $\mathbf{e}_\varnothing$ and $\mathbf{e}_\alpha$ are in the center of the group, the ideal is clearly two-sided.

In the null blade semigroup algebra, the homomorphism maps the absorbing unit 0_γ to $0 \in \bigwedge \mathbb{R}^n$. In this way, any element of the form $0_\gamma x$ maps to 0. With this additional information, the kernel of the homomorphism is seen to be the ideal $\langle \gamma_\varnothing + \gamma_\alpha, 0_\gamma \rangle$, generated by both $\gamma_\varnothing + \gamma_\alpha$ and 0_γ. As in the previous case, commutativity of 0_γ with all elements of the semigroup algebra guarantees two-sidedness of the ideal.

Construction of algebra homomorphisms for the remaining cases proceeds along similar lines. Returning to the fourth column of Table 8.2, the result is the following:

- The Clifford algebra $\mathcal{C}\ell_{p,q}$ $(p+q>1)$ is canonically isomorphic to the blade group quotient algebra $\mathbb{R}\mathcal{B}_{p,q}/\langle \mathbf{e}_\alpha + \mathbf{e}_\varnothing \rangle$. Considering the degree-1 representations, $\rho_J(\mathbf{e}_\varnothing) = \rho_J(\mathbf{e}_\alpha) = 1$ for all $J \in 2^{[p+q]}$. It then becomes clear that passing to the quotient has no effect on the number of irreducible representations. On the other hand, the higher-dimensional irreducible representations satisfy $\tilde{\tau}(\mathbf{e}_\varnothing + \mathbf{e}_\alpha) = 0$ *a priori*, so that representations of the group algebra are precisely the representations of the quotient algebra.
- The Grassmann exterior algebra, $\bigwedge \mathbb{R}^n$, is canonically isomorphic to the null blade semigroup algebra $\mathcal{B}\ell_{\wedge n} = \mathbb{R}\mathfrak{G}_n / \langle 0_\gamma, \gamma_\alpha + \gamma_\varnothing \rangle$. This algebra is isomorphic to the algebra of fermion creation (or annihilation) operators.
- The n-particle *zeon algebra* $\mathcal{C}\ell_n{}^{\text{nil}}$ is canonically isomorphic to the Abelian null blade semigroup algebra $\mathbb{R}\mathfrak{Z}_n / \langle 0_\zeta \rangle$. This algebra is isomorphic to an algebra of *commuting* lowering or raising (annihilation or creation) operators.

Finally, we recover all algebras appearing in the fourth column of Table 8.2 as subalgebras of Clifford algebras of appropriate signature. The necessary signatures and isomorphisms are summarized in Table 8.3.

Table 8.1 Groups and Semigroups

(Semi) Group	Generator Commutation	Generator Squares
$\mathcal{B}_{p,q}$	$\mathbf{e}_i\mathbf{e}_j = \mathbf{e}_\alpha \mathbf{e}_j \mathbf{e}_i$	$\{\underbrace{\mathbf{e}_\varnothing, \ldots, \mathbf{e}_\varnothing}_{p}, \underbrace{\mathbf{e}_\alpha, \ldots, \mathbf{e}_\alpha}_{q}\}$
$\mathfrak{G}_n$	$\gamma_i\gamma_j = \gamma_\alpha\gamma_j\gamma_i$	${\gamma_i}^2 = 0_\gamma$, $i = 1, \ldots, n$
$\mathfrak{Z}_n$	Abelian	${\zeta_i}^2 = 0_\zeta$, $i = 1, \ldots, n$

Table 8.2 Group and Semigroup Algebras

Group or Semigroup	Algebra	Quotient Algebra	Isomorphic Algebra
$\mathcal{B}_{p,q}$	$\mathbb{R}\mathcal{B}_{p,q}$	$\mathbb{R}\mathcal{B}_{p,q}/\langle \mathbf{e}_\alpha + \mathbf{e}_\varnothing \rangle$	$\mathcal{C}\ell_{p,q}$
$\mathfrak{G}_n$	$\mathbb{R}\mathfrak{G}_n$	$\mathbb{R}\mathfrak{G}_n / \langle 0_\gamma, \gamma_\alpha + \gamma_\varnothing \rangle$	$\bigwedge \mathbb{R}^n$
$\mathfrak{Z}_n$	$\mathbb{R}\mathfrak{Z}_n$	$\mathbb{R}\mathfrak{Z}_n / \langle 0_\zeta \rangle$	$\mathcal{C}\ell_n{}^{\text{nil}}$

Table 8.3 gives details for constructing subalgebras of Clifford algebras isomorphic to Grassmann and zeon algebras. In particular, generators of subalgebras are defined in terms of the algebra's generators.

Table 8.3 Clifford subalgebras

Subalgebra	Clifford Algebra	Isomorphism
$\bigwedge \mathbb{R}^n$	$\mathcal{C}\ell_{n,n}$	$\gamma_i \mapsto \mathbf{e}_{\{i\}} + \mathbf{e}_{\{n+i\}}$
${\mathcal{C}\ell_n}^{\text{nil}}$	$\bigwedge \mathbb{R}^n$	$\zeta_{\{i\}} \mapsto \gamma_{\{2i-1,2i\}}$
${\mathcal{C}\ell_n}^{\text{nil}}$	$\mathcal{C}\ell_{2n,2n}$	$\zeta_{\{i\}} \mapsto \mathbf{e}_{\{2i-1,2i\}} + \mathbf{e}_{\{2i-1,2(n+i)\}} - \mathbf{e}_{\{2i,2(n+i)-1\}}$ $+\mathbf{e}_{\{2(n+i)-1,2(n+i)\}}$
${\mathcal{C}\ell_n}^{\text{nil}}$	$\mathcal{C}\ell_{0,0,2n}$	$\zeta_{\{i\}} \mapsto \mathbf{e}_{\{2i-1,2i\}}$

8.4 Classification of Clifford Algebras

While the results stated above refer only to complex representation spaces, representations of the quotient group algebra $\mathbb{R}\mathcal{B}_{p,q}/\langle \mathbf{e}_\alpha + \mathbf{e}_\varnothing \rangle$ are covered by known results on matrix representations of Clifford algebras [18, 22, 67, 79, 80].

A *central simple algebra* over a field $\Bbbk$ is a finite-dimensional associative $\Bbbk$-algebra A such that A is a simple ring[3] and the center of A is equal to $\Bbbk$.

The following three properties determine the class of the algebra $\mathcal{C}\ell_{p,q}$:

i. $(p-q) \pmod 2$: n is even/odd: central simple or not
ii. $(p-q) \pmod 4$: $\omega^2 = \pm 1$: if not central simple, center is $\mathbb{R} \oplus \mathbb{R}$ or $\mathbb{C}$
iii. $(p-q) \pmod 8$: the Brauer class of the algebra (n even) or even subalgebra (n odd) is $\mathbb{R}$ or $\mathbb{H}$.

Each of these properties depends only on $(p-q) \pmod 8$. The complete classification table is given in Table 8.4. The double ring ${}^2\mathbb{F}$ of a field has componentwise multiplication defined by $(a_1, b_1)(a_2, b_2) = (a_1a_2, b_1b_2)$, making it a 2-dimensional algebra. This notation is extended naturally to the division algebra of quaternions by ${}^2\mathbb{H}$.

Table 8.5 gives a visual representation of this classification for $p+q \leq 5$. Rows represent $n = p+q$ and columns represent $p-q$.

[3] A non-zero ring is said to be *simple* if it has no two-sided ideal other than the zero ideal and itself.

Table 8.4 Classification of Clifford algebras

$(p-q) \pmod 8$	ω^2	$\mathcal{C}\ell_{p,q}$ $(n = p+q)$
0	$+$	$\mathbb{R}(2n/2)$
1	$+$	$\mathbb{R}(2(n-1)/2) \oplus \mathbb{R}(2(n-1)/2)$
2	$-$	$\mathbb{R}(2n/2)$
3	$-$	$\mathbb{C}(2(n-1)/2)$
4	$+$	$\mathbb{H}(2(n-2)/2)$
5	$+$	$\mathbb{H}(2(n-3)/2) \oplus \mathbb{H}(2(n-3)/2)$
6	$-$	$\mathbb{H}(2(n-2)/2)$
7	$-$	$\mathbb{C}(2(n-1)/2)$

Table 8.5 Classification of Clifford algebras II

	5	4	3	2	1	0	-1	-2	-3	-4	-5
0						$\mathbb{R}$					
1					${}^2\mathbb{R}$		$\mathbb{C}$				
2				$\mathbb{R}(2)$		$\mathbb{R}(2)$		$\mathbb{H}$			
3			$\mathbb{C}(2)$		${}^2\mathbb{R}(2)$		$\mathbb{C}(2)$		${}^2\mathbb{H}$		
4		$\mathbb{H}(2)$		$\mathbb{R}(4)$		$\mathbb{R}(4)$		$\mathbb{H}(2)$		$\mathbb{H}(2)$	
5	${}^2\mathbb{H}(2)$		$\mathbb{C}(4)$		${}^2\mathbb{R}(4)$		$\mathbb{C}(4)$		${}^2\mathbb{H}(2)$		$\mathbb{C}(4)$
ω^2	+	+	-	-	+	+	-	-	+	+	-

Exercises

Exercise 8.1: Verify the isomorphisms appearing in Table 8.3.

Exercise 8.2: Construct the zeon generator representations z_1, z_2 of $\zeta_{\{1\}}$ and $\zeta_{\{2\}}$, respectively, in $\mathcal{C}\ell_3{}^{\text{nil}}$ and use them to construct the matrix representation of $\zeta_{\{1,2\}}$.

Exercise 8.3: Construct matrix representations of generators $\{\mathbf{e}_{\{1\}}, \mathbf{e}_{\{2\}}\}$ of $\mathcal{C}\ell_{0,2}$.

Exercise 8.4: Prove Lemma 8.1.3.

PART III
Algebraic Combinatorics & Zeons

Chapter 9

Algebraic and Combinatorial Properties of Zeons

The *n-particle zeon algebra*, denoted ${\mathcal{C}\ell_n}^{\text{nil}}$, is defined as the real abelian algebra generated by the collection $\{\zeta_i\}$ $(1 \le i \le n)$ along with the scalar $1 = \zeta_\varnothing$ subject to the following multiplication rules:

$$\zeta_i\,\zeta_j = \zeta_j\,\zeta_i \;\text{ for } i \neq j, \text{ and}$$
$${\zeta_i}^2 = 0 \;\text{ for } 1 \le i \le n.$$

As a vector space, this 2^n-dimensional algebra has a canonical basis of *basis blades* of the form $\{\zeta_I : I \subseteq [n]\}$, where $\zeta_I = \prod_{\iota \in I} \zeta_\iota$. The null-square property of the generators $\{\zeta_i : 1 \le i \le n\}$ guarantees that the product of two basis blades satisfies the following:

$$\zeta_I \zeta_J = \begin{cases} \zeta_{I\cup J} & I \cap J = \varnothing, \\ 0 & \text{otherwise.} \end{cases} \tag{9.1}$$

Letting $2^{[n]}$ denote the power set of the *n-set*, $[n] = \{1, 2, \ldots, n\}$, it thereby follows that an element $u \in {\mathcal{C}\ell_n}^{\text{nil}}$ has the following canonical expansion:

$$u = \sum_{I \in 2^{[n]}} u_I\, \zeta_I\,,$$

where $u_I \in \mathbb{R}$ for each multi-index I. The *grade-k part* of u is defined by $\langle u \rangle_k = \sum_{|I|=k} u_I \zeta_I$. To simplify things, the term "zeon" will be used in reference to an arbitrary element of ${\mathcal{C}\ell_n}^{\text{nil}}$.

In recent years, combinatorial properties and applications of zeons have been studied in numerous works, many of which are summarized in the monograph [91].

9.1 Group and Semigroup Properties

The first work exploring the basic algebraic properties of the abelian multiplicative group formed by the zeon algebra's non-nilpotent elements was [32]. Observing that ${\mathcal{C}\ell_n}^{\rm nil}$ is an abelian algebra, its elements form a commutative multiplicative semigroup. The first useful result is a formula for expanding arbitrary products of zeons.

Lemma 9.1.1. *If $z_1, \ldots, z_k \in {\mathcal{C}\ell_n}^{\rm nil}$ with $z_j = \sum_I a_{j,I}\zeta_I$ for $1 \leq j \leq k$, and $\prod_{j=1}^k z_j = \sum_I b_I \zeta_I$, then*

$$b_I = \sum_{(I_1,\ldots,I_k)} \prod_{j=1}^{k} a_{j,I_j}$$

where the sum is taken over all k-tuples $(I_1, \ldots, I_k)$ of pairwise disjoint (possibly empty) subsets of I such that $I = \bigcup_j I_j$.

Proof. The proof is by induction on $k \geq 2$. When $k = 2$, the result follows immediately from $\gamma = \left(\sum_K a_{1,K}\zeta_K\right)\left(\sum_J a_{2,J}\zeta_J\right)$ in light of (9.1).

That is,

$$b_I = \sum_{J\subset I} a_{1,J}a_{2,I\setminus J}.$$

Now fix k with $2 < k \leq n$, and suppose the result holds for $k-1$. If $z_1 \cdots z_{k-1} = \sum_I c_I\zeta_I$, then

$$b_I = \sum_{J\subset I} c_J a_{k,I\setminus J} = \sum_{J\subset I}\left(\sum_{(J_1,\ldots,J_{k-1})}\prod_{j=1}^{k-1} a_{j,J_j}\right) a_{k,I\setminus J} = \sum_{(I_1,\ldots,I_k)}\prod_{j=1}^{k} a_{j,I_j}$$

for all $I \subset \{1, \ldots, n\}$, and by induction the result is proven. □

As a consequence we obtain a result that implies the nilpotency of all noninvertible elements of ${\mathcal{C}\ell_n}^{\rm nil}$.

Lemma 9.1.2. *If $z_1, \ldots, z_{n+1} \in {\mathcal{C}\ell_n}^{\rm nil}$ and $\Re z_j = 0$ for all j, then $\prod_{j=1}^{n+1} z_j = 0$. In particular, if $z \in {\mathcal{C}\ell_n}^{\rm nil}$ and $\Re z = 0$, then $z^{n+1} = 0$.*

Proof. For each j put $z_j = \sum_I a_{j,I}\zeta_I$, and put $\prod_{j=1}^{n+1} z_k = \sum_I a_I\zeta_I$, and fix $I \subset \{1,\ldots,n\}$. By Lemma 9.1.1,

$$a_I = \sum_{(I_1,\ldots,I_{n+1})} \prod_{j=1}^{n+1} a_{j,I_j}.$$

Since $\Re z_j = 0$ for all j, the sum is taken over partitions $(I_1,\ldots,I_{n+1})$ of I. But since I contains at most n elements, there is no such partition. Thus the sum is empty, and $a_I = 0$. Since I was arbitrary, it follows that $\prod_{j=1}^{n+1} z_j = 0$. □

As shown in [32], $u \in \mathcal{C}\ell_n{}^{\text{nil}}$ is invertible if and only if $\Re u \neq 0$. Moreover, the multiplicative inverse of u is unique. The result is paraphrased here for review.

Proposition 9.1.3. *Let $u \in \mathcal{C}\ell_n{}^{\text{nil}}$, and let $\kappa(\mathfrak{D}u)$ denote the index of nilpotency*[1] *of $\mathfrak{D}u$. It follows that u is invertible if and only if $\Re u \neq 0$, and setting*

$$u' = \sum_{j=1}^{\kappa(\mathfrak{D}u)-1} (-1)^{j-1}(\Re u)^{-j}(\mathfrak{D}u)^{j-1},$$

one sees that $uu' = u'u = 1$. Moreover, the inverse of u is unique.

Proof. Note that by writing $u = \Re u + \mathfrak{D}u$, where $\mathfrak{D}u = \sum_{\varnothing \neq I \in 2^{[n]}} u_I\zeta_I$ is nilpotent of index $\kappa(\mathfrak{D}u)$, one sees immediately u is not invertible if $\Re u = 0$. *Claim*: For positive integer k,

$$u\left(\sum_{j=1}^{k}(-1)^{j-1}(\Re u)^{-j}(\mathfrak{D}u)^{j-1}\right) = 1 + (-1)^{k-1}(\Re u)^{-k-1}(\mathfrak{D}u)^k.$$

Proof of the claim proceeds by induction on k. When $k = 1$, one finds

$$u(\Re u)^{-1} = (\Re u + \mathfrak{D}u)(\Re u)^{-1} = 1 - (\Re u)^{-1}\mathfrak{D}u.$$

[1] In particular, κ is the least positive integer such that $(\mathfrak{D}u)^\kappa = 0$.

Assuming the result holds for some $k \geq 1$, one finds

$$\begin{aligned}
&u\left(\sum_{j=1}^{k+1}(-1)^{j-1}(\Re u)^{-j}(\mathfrak{D}u)^{j-1}\right)\\
&\quad= u\left(\sum_{j=1}^{k}(-1)^{j-1}(\Re u)^{-j}(\mathfrak{D}u)^{j-1}\right) + (-1)^k u\left((\Re u)^{-(k+1)}(\mathfrak{D}u)^k\right)\\
&\quad= 1 + (-1)^{k-1}(\Re u)^{-k}(\mathfrak{D}u)^k + (-1)^k(u^{-k}(\mathfrak{D}u)^k + u^{-(k+1)}(\mathfrak{D}u)^{k+1})\\
&\quad= 1 + (-1)^k(\Re u)^{-(k+1)}(\mathfrak{D}u)^{k+1}.
\end{aligned}$$

This establishes the claim. It follows immediately that when $(\mathfrak{D}u)$ is nilpotent of index k,

$$\begin{aligned}
u\left(\sum_{j=1}^{k}(-1)^{j-1}(\Re u)^{-j}(\mathfrak{D}u)^{j-1}\right) &= 1 + (-1)^{k-1}(\Re u)^{-k}(\mathfrak{D}u)^k\\
&= 1.
\end{aligned}$$

To see that the inverse is unique, suppose $uu' = u\gamma = 1$. Then, $u' = (u\gamma)u' = (\gamma u)u' = \gamma(uu') = \gamma$. □

Definition 9.1.4. For $n \in \mathbb{N}$, ${\mathcal{C}\ell_n}^{\text{nil}\star}$ is defined to be the collection of invertible elements in ${\mathcal{C}\ell_n}^{\text{nil}}$. That is,

$${\mathcal{C}\ell_n}^{\text{nil}\star} = \{u \in {\mathcal{C}\ell_n}^{\text{nil}} : \Re u \neq 0\}.$$

Clearly ${\mathcal{C}\ell_n}^{\text{nil}\star}$ is closed under (commutative) zeon multiplication, and every element is invertible by Proposition 9.1.3. Hence, the invertible zeons ${\mathcal{C}\ell_n}^{\text{nil}\star}$ form an abelian group under multiplication.

Considering this characterization of the invertible elements, some interesting algebraic properties of ${\mathcal{C}\ell_n}^{\text{nil}}$ become apparent.

Lemma 9.1.5. *Every element of* ${\mathcal{C}\ell_n}^{\text{nil}}$ *is either nilpotent or invertible.*

Proof. Let $u \in {\mathcal{C}\ell_n}^{\text{nil}}$. If $\Re u \neq 0$, then u^{-1} exists. If $\Re u = 0$, then a simple application of the multinomial theorem shows that $u^{n+1} = 0$. □

Lemma 9.1.6. *There is a unique maximal ideal in* ${\mathcal{C}\ell_n}^{\text{nil}}$*, namely the collection of singular (noninvertible) elements.*

Proof. Let us denote by ${\mathcal{C}\ell_n}^{\text{nil}\circ}$ the set of all singular elements of ${\mathcal{C}\ell_n}^{\text{nil}}$. As a consequence of Lemma 9.1.1, the mapping $u \mapsto \Re u$ is an algebra homomorphism with kernel ${\mathcal{C}\ell_n}^{\mathrm{nil}_0}$. Thus, ${\mathcal{C}\ell_n}^{\mathrm{nil}_0}$ is an ideal in ${\mathcal{C}\ell_n}^{\mathrm{nil}}$. Furthermore, if another ideal W of ${\mathcal{C}\ell_n}^{\mathrm{nil}}$ contains ${\mathcal{C}\ell_n}^{\mathrm{nil}_0}$ as a proper subset, then W contains an invertible element, and therefore $W = {\mathcal{C}\ell_n}^{\mathrm{nil}}$. Hence, ${\mathcal{C}\ell_n}^{\mathrm{nil}_0}$ is maximal. □

This splits the group of invertible elements into two connected components (with the standard topology on $\mathbb{R}^{2^n}$): the subgroup of elements with positive real part, and those with negative real part. Moreover, the spectrum of any invertible element is nonempty; it contains exactly one element: the real part. The map taking an element of ${\mathcal{C}\ell_n}^{\mathrm{nil}}$ to its spectrum (considered as an element of $\mathbb{R}$) is an algebra homomorphism.

For convenience, a collection of pairwise-disjoint subsets of $2^{[n]}$ is denoted by $\{I_1|\ldots|I_k\}$. Such a collection is said to be *independent* if its elements are pairwise disjoint.

Lemma 9.1.7. *Let $u \in {\mathcal{C}\ell_n}^{\mathrm{nil}}$ and write $u = \sum_{I \in 2^{[n]}} u_I \zeta_I$. For positive integer k, let $\gamma = u^k$ be written $\gamma = \sum_{I \in 2^{[n]}} \gamma_I \zeta_I$. For fixed multi-index I, the corresponding coefficient of ζ_I in γ is given by*[2]

$$\gamma_I = \sum_{j=0}^{k} \frac{k!}{j!} {u_\varnothing}^j \sum_{\substack{\pi \in \mathcal{P}(I) \\ |\pi| = k-j}} u_\pi,$$

where $u_\pi = \prod_{b \in \pi} u_b$ is a product over blocks of the partition π.

Proof. Applying the multinomial theorem, the null-square property of zeon generators yields

$$\begin{aligned}\left(\sum_{I \in 2^{[n]}} u_I \zeta_I\right)^k &= \sum_{\substack{0 \le \ell_\varnothing, \ldots, \ell_{[n]} \\ \ell_0 + \cdots + \ell_{2^{[n]}} = k}} \binom{k}{\ell_\varnothing, \ldots, \ell_{[n]}} \prod_{I \in 2^{[n]}} {u_I}^{\ell_I} {\zeta_I}^{\ell_I} \\ &= \sum_{j=0}^{k} \binom{k}{j} {u_\varnothing}^j \sum_{\substack{\{I_1|\ldots|I_{k-j}\} \\ \text{independent}}} (k-j)! \zeta_{I_1 \cup \cdots \cup I_{k-j}} \prod_{\ell=1}^{k-j} u_{I_\ell} \\ &= \sum_{j=0}^{k} \frac{k!}{j!} {u_\varnothing}^j \sum_{\substack{\{I_1|\ldots|I_{k-j}\} \\ \text{independent}}} \zeta_{I_1 \cup \cdots \cup I_{k-j}} \prod_{\ell=1}^{k-j} u_{I_\ell}.\end{aligned}$$

[2]By convention, define ${u_\varnothing}^0 = 1$ when $u_\varnothing = 0$.

Evaluating the coefficient of a particular basis blade ζ_J is thereby accomplished by considering a sum over partitions of the multi-index J. More specifically, letting $\mathcal{P}(J)$ denote the collection of partitions of J,

$$\left\langle \left(\sum_{I\in 2^{[n]}} u_I\zeta_I \right)^k, \zeta_J \right\rangle = \sum_{j=0}^{k} \frac{k!}{j!} {u_\varnothing}^j \sum_{\substack{\pi\in\mathcal{P}(J)\\ |\pi|=k-j}} \prod_{\ell=1}^{k-j} u_{\pi_\ell}$$

$$= \sum_{j=0}^{k} \frac{k!}{j!} {u_\varnothing}^j \sum_{\substack{\pi\in\mathcal{P}(J)\\ |\pi|=k-j}} u_\pi.$$

□

9.2 Partitions and Counting Numbers

Observing that the null-square property of zeon generators guarantees that the product $\zeta_I\zeta_J$ is nonzero only if $I\cap J=\varnothing$ (that is, $\{I|J\}$ is a partition of $I\cup J$), it is not surprising that combinatorial properties of zeons can be used to generate Stirling numbers of the second kind, Bell numbers, and Bessel numbers.

In this section, we consider only partitions of the n-set, and blocks of partitions are identified with multi-indices of basis blades in ${\mathcal{C}\ell_n}^{\text{nil}}$. The following definition is based on Berezin's definition dealing with second quantization [11].

Definition 9.2.1. Let $u \in {\mathcal{C}\ell_n}^{\text{nil}}$. Then the *zeon Berezin integral* of u is defined by

$$\int u\, d\zeta_n \cdots d\zeta_1 = \int \left(\sum_{I\in 2^{[n]}} u_I\, \zeta_I \right) d\zeta_n \cdots d\zeta_1 = u_{\{1,2,\ldots n\}}.$$

In other words, $\displaystyle\int u\, d\zeta_n \cdots d\zeta_1$ is the "top-form" coefficient in the canonical expansion of u.

Proposition 9.2.2. *Let*

$$A = \sum_{\varnothing\neq J\in 2^{[n]}} \zeta_J \in {\mathcal{C}\ell_n}^{\text{nil}}.$$

Then, for $1\le k\le n$,

$$\frac{1}{k!}\int A^k\, d\zeta_n\cdots d\zeta_1 = \left\{ {n \atop k} \right\},$$

where $\left\{ {n \atop k} \right\}$ *denotes the* Stirling number of the second kind, *defined as the number of partitions of* $[n]$ *into* k *nonempty subsets.*

Proof. The proof is left as an exercise. $\square$

Proposition 9.2.3. *Let*

$$A = \sum_{\varnothing \neq J \in 2^{[n]}} \zeta_J \in \mathcal{C}\ell_n{}^{\mathrm{nil}}.$$

Then,

$$\int e^A \, d\zeta_n \cdots d\zeta_1 = B_n,$$

where B_n *denotes the* n*th Bell number.*

Proof. The proof is left as an exercise. $\square$

Proposition 9.2.4. *Let* $\chi_{\mathrm{ov}}, \chi_{\mathrm{cr}} : 2^{[n]} \times 2^{[n]} \to \{0, 1\}$ *be defined by*

$$\chi_{\mathrm{ov}}(I, J) = \begin{cases} 0 & \text{if } I \between J \\ 1 & \text{otherwise} \end{cases}$$

and

$$\chi_{\mathrm{cr}}(I, J) = \begin{cases} 0 & \text{if } I \pitchfork J \\ 1 & \text{otherwise.} \end{cases}$$

Let $A_{\between}, A_{\pitchfork} \in \mathcal{C}\ell_n{}^{\mathrm{nil}} \otimes \mathcal{C}\ell_{\binom{2^n}{2}}{}^{\mathrm{nil}}$ *be defined by*

$$A_{\between} = \sum_{\varnothing \prec I \preceq [n]} \zeta_I \otimes \prod_{\varnothing \prec J \preceq [n]} \left(\chi_{\mathrm{ov}}(I, J) \upsilon_{f(I,J)} + (1 - \chi_{\mathrm{ov}}(I, J)) \upsilon_{\varnothing}\right) \quad (9.2)$$

and

$$A_{\pitchfork} = \sum_{\varnothing \prec I \preceq [n]} \zeta_I \otimes \prod_{\varnothing \prec J \preceq [n]} \left(\chi_{\mathrm{cr}}(I, J) \upsilon_{f(I,J)} + (1 - \chi_{\mathrm{cr}}(I, J)) \upsilon_{\varnothing}\right), \quad (9.3)$$

where $f : 2^{[n]} \times 2^{[n]} \to [\binom{2^n}{2}]$ *is a symmetric integer-labeling of pairs of multi-indices, and* $\upsilon_{f(I,J)}$ *is a nilpotent generator of* $\mathcal{C}\ell_{\binom{2^n}{2}}{}^{\mathrm{nil}}$*. Then,*

$$\int \vartheta(e^{A_{\between}}) \, d\zeta_n \cdots d\zeta_1 = B_n^*$$

$$\int \vartheta(e^{A_{\pitchfork}}) \, d\zeta_n \cdots d\zeta_1 = C_n,$$

where $\vartheta : \mathcal{C}\ell_n{}^{\mathrm{nil}} \otimes \mathcal{C}\ell_{\binom{2^n}{2}}{}^{\mathrm{nil}} \to \mathcal{C}\ell_n{}^{\mathrm{nil}}$ *is the canonical projection,* B_n^* *denotes the* n*th Bessel number, and* C_n *is the* n*th Catalan number.*

Proof. The proof is based on the observation that non-overlapping (non-crossing) partitions are constructed from non-overlapping (non-crossing) blocks. By construction of the sum in (9.2), if multi-indices I and J are overlapping, their associated multi-vectors ζ_I and ζ_J are multiplied by the same abelian nilpotent generator of $\mathcal{C}\ell_{\binom{2^n}{2}}{}^{\text{nil}}$. In this way, the products of multi-vectors associated with overlapping blocks are always zero. The sum in (9.3) is constructed similarly. The remainder of the proof follows those of Propositions 9.2.2 and 9.2.3. □

Functions Defined on Partitions

Let $f : 2^{[n]} \to \mathbb{R}$ be a function on the power set of $[n]$ with $f(\varnothing) = 1$. Define the function $h : \mathcal{P}([n]) \to \mathbb{R}$ by

$$h(\pi) = \prod_{b\in\pi} f(b).$$

Here each partition element $\pi \in \mathcal{P}([n])$ is assumed to be canonically ordered.

Since π is used to denote a partition of $[n]$, σ will be used to denote a permutation of the blocks in π. In other words, $\pi \in \mathcal{P}([n])$ and $\sigma \in \mathrm{S}_k$.

Theorem 9.2.5. *Let* $0 < k \leq n$. *Then*

$$\int_{\mathrm{B}} \left(\sum_{I\in 2^{[n]}} f(I)\,\gamma_I \right)^k d\gamma_1 \cdots d\gamma_n = \sum_{\substack{\pi\in\mathcal{P}([n])\\ |\pi|=k}} \sum_{\sigma\in\mathrm{S}_k} h(\sigma(\pi)),$$

where S_k *is the symmetric group on* k *elements; i.e., we sum over all permutations of blocks of each* $\pi \in \mathcal{P}([n])$ *such that* $|\pi| = k$.

Proof. Begin with the following expansion:

$$\begin{aligned}\left(\sum_{I\in 2^{[n]}} f(I)\,\gamma_I \right)^k &= \left(\sum_{I_1\in 2^{[n]}} f(I_1)\,\gamma_{I_1} \right) \cdots \left(\sum_{I_k\in 2^{[n]}} f(I_k)\,\gamma_{I_k} \right)\\ &= \sum_{I_1,\ldots,I_k\in 2^{[n]}} f(I_1)\cdots f(I_k)\,\gamma_{I_1}\cdots\gamma_{I_k}.\end{aligned}$$

Because ${\gamma_i}^2 = 0$ for $1 \leq i \leq n$, this sum is equal to

$$\sum_{\substack{I_1,\ldots,I_k\in 2^{[n]}\\ \text{pairwise disjoint}}} f(I_1)\cdots f(I_k)\,\gamma_{I_1}\cdots\gamma_{I_k}.$$

From this follows

$$\int_{\mathrm{B}} \left(\sum_{I\in 2^{[n]}} f(I)\,\gamma_I \right)^k d\gamma_1 \cdots d\gamma_n$$

$$= \sum_{\substack{I_1,\ldots,I_k\in 2^{[n]}\\ \text{pairwise disjoint}}} f(I_1)\cdots f(I_k)\,\gamma_{i_1}\cdots\gamma_{i_k}\Big|_{I_1\cup\cdots\cup I_k=[n]}.$$

The sum is restricted to those multi-indices whose disjoint union is all of $[n]$. Since the sum is over collections of k-blocks, the Berezin integral is the sum over all k-block partitions of $[n]$. Further, it is clear that the blocks recur in all possible permutations in the expansion. □

9.3 Roots of Invertible Zeons

In this section, kth roots of invertible zeon elements are considered. More specifically, conditions for existence of roots are established, numbers of existing roots are determined, and computational methods (recursive and closed formulas) for constructing roots are developed.

Existence and Recursive Formulations

As will be shown, invertible zeons have roots of all odd orders and roots of all even orders when their scalar parts are positive. A recursive algorithm establishes their existence and provides a convenient method for their computation.

Theorem 9.3.1. *Let $w \in \mathcal{C}\ell_n{}^{\mathrm{nil}\star}$, and let $k \in \mathbb{N}$. Then, $\exists u \in \mathcal{C}\ell_n{}^{\mathrm{nil}\star}$ such that $u^k = w$, provided $w_\varnothing > 0$ when k is even. Further, writing $w = \varphi + \zeta_{\{n\}}\psi$, where $\varphi, \psi \in \mathcal{C}\ell_{n-1}{}^{\mathrm{nil}}$, u is computed recursively by*

$$u = w^{1/k} = \varphi^{1/k} + \zeta_{\{n\}}\frac{1}{k}\varphi^{-(k-1)/k}\psi.$$

Proof. Assuming $w \in \mathcal{C}\ell_n{}^{\mathrm{nil}\star}$ guarantees $w_\varnothing \neq 0$, so the scalar part of w has odd roots of all orders. Even-order roots $w_\varnothing{}^{1/k}$ exist for positive values of $w_\varnothing$.

Proof is by induction on n. When $n = 1$, let $w = a + b\zeta_{\{1\}}$, where $w_\varnothing = a$. Applying the binomial theorem and null-square properties of zeon generators, one finds

$$\left(a^{1/k} + \frac{b}{ka^{(k-1)/k}}\zeta_{\{1\}}\right)^k = a + ka^{(k-1)/k}\frac{b}{ka^{(k-1)/k}}\zeta_{\{1\}} = a + b\zeta_{\{1\}}.$$

Next, suppose the result holds for some $n-1 \geq 1$ and let $w \in \mathcal{C}\ell_n{}^{\text{nil}}$ be written $w = \varphi + \zeta_{\{n\}}\psi$, where $\varphi, \psi \in \mathcal{C}\ell_{n-1}{}^{\text{nil}}$. In particular, this implies $\varphi \in \mathcal{C}\ell_n{}^{\text{nil}\star}$. Let $\alpha = \varphi^{1/k}$, and let $u = \alpha + \frac{1}{k}\zeta_{\{n\}}\alpha^{-(k-1)}\psi$. Then

$$\begin{aligned} u^k = \left(\alpha + \zeta_{\{n\}}\frac{1}{k}\alpha^{-(k-1)}\psi\right)^k &= \varphi + k\alpha^{(k-1)}\frac{1}{k}\zeta_{\{n\}}\alpha^{-(k-1)}\psi \\ &= \varphi + \zeta_{\{n\}}\psi \\ &= w. \end{aligned}$$

□

Cardinality of Zeon Roots

Whenever an element $u \in \mathcal{C}\ell_n{}^{\text{nil}}$ has a kth root, it is only natural to wonder how many kth roots u has. For example, an element of the form $u = a\zeta_I \in \mathcal{C}\ell_n{}^{\text{nil}_\circ}$, where $a \neq 0$ and $|I| \geq 2$, has infinitely many square roots of the form $x = b\zeta_J + \frac{a}{b}\zeta_{I\setminus J}$, where $\varnothing \neq J \subsetneq I$ and $b \neq 0$, since

$$x^2 = \left(b\zeta_J + \frac{a}{2b}\zeta_{I\setminus J}\right)^2 = 2b\frac{a}{2b}\zeta_{J\cup(I\setminus J)} = a\zeta_I = u.$$

In fact, for nilpotent elements of $\mathcal{C}\ell_n{}^{\text{nil}}$, existence of a kth root guarantees existence of infinitely many kth roots.

Lemma 9.3.2. *Suppose $\alpha \in \mathcal{C}\ell_n{}^{\text{nil}}$ is nilpotent and let $k \geq 2$ be an integer. If there exists $u \in \mathcal{C}\ell_n{}^{\text{nil}}$ such that $u^k = \alpha$, then there exist infinitely many such elements u.*

Proof. For any scalar t, $u^k = \alpha$ implies

$$\begin{aligned} (u + t\zeta_{[n]})^k &= u^k + \sum_{\ell=0}^{k-1}\binom{k}{\ell}u^\ell(t\zeta_{[n]})^{k-\ell} \\ &= \alpha. \end{aligned}$$

□

When u is a nilpotent zeon monomial, the structure of kth roots can be determined.

Lemma 9.3.3 (Roots of nilpotent monomials). *Let $u = a\zeta_I \in \mathcal{C}\ell_n{}^{\text{nil}_\circ}$, where $a \neq 0$ and $I \neq \varnothing$. For positive integer k ($2 \leq k \leq |I|$), the element u has infinitely many kth roots of the form*

$$x = \sum_{\ell=1}^{k} b_\ell \zeta_{J_\ell},$$

where $I = J_1 \sqcup \cdots \sqcup J_k$ *is any* k*-block partition of* I *and* $\prod_{\ell=1}^{k} b_\ell = \frac{a}{k!}$. *Moreover, no* k*th roots exist when* $k > |I|$.

Proof. Applying Lemma 9.1.7,

$$x^k = \left(\sum_{\ell=1}^{k} b_\ell \zeta_{J_\ell}\right)^k = k! \prod_{\ell=1}^{k} b_\ell \zeta_{J_\ell} = k! \frac{a}{k!} \zeta_{J_1 \cup \cdots \cup J_\ell} = a\zeta_I = u.$$

It is not difficult to see that the monomial $a\zeta_I$ has no kth roots when $k > |I|$ because the necessary partitioning of I is impossible. □

Of particular interest is determining numbers of kth roots of invertible elements of $\mathcal{C}\ell_n{}^{\text{nil}}$. When u is invertible, the number of kth roots depends on the parity of k.

Theorem 9.3.4. *Let* $\alpha \in \mathcal{C}\ell_n{}^{\text{nil}\star}$, *and let* $k \in \mathbb{N}$. *Then, assuming* $\alpha_\varnothing > 0$ *when* k *is even,*

$$\sharp\{u : u^k = \alpha\} = \begin{cases} 1 & \textit{when } k \equiv 1 \pmod 2, \\ 2 & \textit{when } k \equiv 0 \pmod 2. \end{cases}$$

Proof. Note that the choice of scalar term $u_\varnothing$ is unique when k is odd. Now suppose $u^k = \alpha = v^k$, and observe that $u - v$ is nilpotent. Writing $u = a + \beta$ for some nilpotent β, it follows that $v = a + \gamma$ for nilpotent γ. Observe that the product $\alpha\delta$ of an invertible element α and a nilpotent δ is zero if and only if $\delta = 0$, since $0 = \alpha^{-1}0 = \delta$. Hence, assuming $u^k = v^k$, one finds

$$\begin{aligned} u^k - v^k &= (u - v)(u^{k-1} + u^{k-2}v + \cdots + v^{k-1}) \\ &= (u - v)\left[(a^{k-1} + \delta_1) + (a^{k-1} + \delta_2) + \cdots + (a^{k-1} + \delta_k)\right] \\ &= (u - v)\left[ka^{k-1} + \delta\right], \end{aligned}$$

where $\delta = \delta_1 + \cdots + \delta_k$ is nilpotent by the ideal property of $\mathcal{C}\ell_n{}^{\text{nil}_\text{o}}$ established in Lemma 9.1.6. It is clear that $ka^{k-1} + \delta$ is invertible, so $(u - v)(ka^{k-1} + \delta) = 0$ implies $(u - v) = 0$.

In the case of even k, there are two possible choices for the scalar term, $\pm u_\varnothing$. In one of these cases, $u^k = v^k$ implies $u - v$ is nilpotent and the proof proceeds as above. In the other case, $u^k = v^k$ implies $u + v$ is nilpotent. Considering this case in detail, one writes $u = a + \beta$ and $v = -a + \gamma$ for nilpotent elements β and γ. For even values of k, a little algebra thereby yields

$$\begin{aligned}u^k - v^k &= (u+v)\left(u^{k-1} - u^{k-2}v + \cdots + (-1)^{k-1}v^{k-1}\right)\\ &= (u+v)\left[(a^{k-1}+\delta_1) - (-a^{k-1}+\delta_2) + \cdots - (-a^{k-1}+\delta_k)\right]\\ &= (u+v)\left[ka^{k-1} + (\delta_1 - \delta_2 + \cdots - \delta_k)\right].\end{aligned}$$

Letting $\delta = \delta_1 - \delta_2 + \cdots - \delta_k$, one sees that $(u+v)(ka^{k-1}+\delta) = 0$ implies $(u+v) = 0$ as before. Hence, $u = -v$. $\square$

Given an invertible zeon u and even positive integer k, it now makes sense to define the *principal kth root* of u as the zeon w satisfying $w_\varnothing > 0$ and $w^k = u$. All roots of odd order can be considered principal.

Example 9.3.5. Set $w = 1 + 5\zeta_{\{1\}} - \zeta_{\{2\}} + 3\zeta_{\{1,3\}} - 6\zeta_{\{1,2,3\}}$ in $\mathcal{C}\ell_3{}^{\text{nil}}$. The principal square root of w is given by

$$w^{1/2} = 1 + \frac{5\zeta_{\{1\}}}{2} - \frac{\zeta_{\{2\}}}{2} + \frac{5}{4}\zeta_{\{1,2\}} + \frac{3}{2}\zeta_{\{1,3\}} - \frac{9}{4}\zeta_{\{1,2,3\}}.$$

The principal eighth root is

$$w^{1/8} = 1 + \frac{5\zeta_{\{1\}}}{8} - \frac{\zeta_{\{2\}}}{8} + \frac{35}{64}\zeta_{\{1,2\}} + \frac{3}{8}\zeta_{\{1,3\}} - \frac{27}{64}\zeta_{\{1,2,3\}}.$$

Example 9.3.6. Consider the following zeon element of $\mathcal{C}\ell_3{}^{\text{nil}}$:

$$u = 4 - 3\zeta_{\{1\}} + 3\zeta_{\{3\}} - 2\zeta_{\{1,2\}}.$$

Applying the result of Theorem 9.3.1, the principal fifth root of u is determined to be

$$u^{1/5} = 2^{2/5} - \frac{3\zeta_{\{1\}}}{10\ 2^{3/5}} + \frac{3\zeta_{\{3\}}}{10\ 2^{3/5}} - \frac{\zeta_{\{1,2\}}}{5\ 2^{3/5}} + \frac{9\zeta_{\{1,3\}}}{50\ 2^{3/5}} + \frac{3\zeta_{\{1,2,3\}}}{25\ 2^{3/5}}.$$

Explicit kth Root Formulas

While recursive constructions are convenient for proving existence of roots, they do not give a clear picture of the algebraic structure. The goal now is to give a more explicit formulation of zeon roots.

Given a polynomial function $f(x)$ and $a \in \mathbb{R}$, recall the *Taylor series expansion of f about a*:

$$f(x) = \sum_{j=0}^{\infty} \frac{f^{(j)}(a)}{j!}(x-a)^j.$$

Writing arbitrary $\alpha \in \mathcal{C}\ell_n{}^{\text{nil}}$ in the form $\alpha = a_\varnothing + \beta$, where $\beta \in \mathcal{C}\ell_n{}^{\text{nil}_\circ}$, the formal Taylor series of $f(\alpha)$ about $a_\varnothing$ is defined by

$$f(\alpha) = \sum_{j=0}^{\infty} \frac{f^{(j)}(a_\varnothing)}{j!}(\alpha - a_\varnothing)^j$$
$$= \sum_{j=0}^{\infty} \frac{f^{(j)}(a_\varnothing)}{j!}\beta^j.$$

With the formal series in hand, an explicit formula for the principal kth root is within reach.

Theorem 9.3.7. *Let $\alpha \in \mathcal{C}\ell_n{}^{\text{nil}_\star}$, where $n \geq 1$, and let $k \geq 2$ be a positive integer. The principal kth root of α is given by*

$$u^{1/k} = a_\varnothing{}^{1/k} + \sum_{I \neq \varnothing} \left(\sum_{j=1}^{|I|} a_\varnothing{}^{-j+\frac{1}{k}} \sum_{\ell=0}^{j} \frac{S_1(j,\ell)}{k^\ell} \sum_{\substack{\pi \in \mathcal{P}(I) \\ |\pi|=j}} a_\pi \right) \zeta_I.$$

Proof. Let $u \in \mathcal{C}\ell_n{}^{\text{nil}}$ be written in the form $a_\varnothing + \beta$, where $a_\varnothing \in \mathbb{R}$ is nonzero and β is nilpotent of index $m+1$. For fixed $k \in \mathbb{N}$, let $f(x) = x^{1/k}$ and consider the Taylor series expansion of $f(u)$ expanded about $a_\varnothing$. In particular,

$$u^{1/k} = \sum_{j=0}^{\infty} \frac{f^{(j)}(a_\varnothing)}{j!}\beta^j$$
$$= \sum_{j=0}^{m} \frac{f^{(j)}(a_\varnothing)}{j!}\beta^j.$$

Let $S_1(j,\ell)$ denote the Stirling number of the first kind defined by the following property: $(-1)^{(j-\ell)}S_1(j,\ell)$ is the number of permutations of j elements that contain exactly ℓ cycles. It is well known that Stirling numbers of the first kind are generated by the falling factorial $(x)_n$. In light of this result, it is not difficult to show that

$$\frac{d^j}{dx^j}(x^{1/k}) = \sum_{\ell=0}^{j} \frac{S_1(j,\ell)}{k^\ell x^{j-1/k}}.$$

Hence,

$$\frac{f^{(j)}(a_\varnothing)}{j!} = \frac{1}{j!}\sum_{\ell=0}^{j} \frac{S_1(j,\ell)}{k^\ell a_\varnothing{}^{j-1/k}} = \frac{a_\varnothing{}^{-j+\frac{1}{k}}}{j!}\sum_{\ell=0}^{j} \frac{S_1(j,\ell)}{k^\ell}.$$

Writing $\beta = \sum_{I \neq \varnothing} a_I \zeta_I$, Lemma 9.1.7 gives $\beta^j = j! \sum_{|I| \geq j} \zeta_I \sum_{\substack{\pi \in \mathcal{P}(I) \\ |\pi| = j}} a_\pi$, so that

$$\begin{aligned} u^{1/k} &= \sum_{j=0}^{m} \frac{f^{(j)}(a_\varnothing)}{j!} \beta^j \\ &= a_\varnothing{}^{1/k} + \sum_{j=1}^{m} \frac{a_\varnothing{}^{-j+\frac{1}{k}}}{j!} \sum_{\ell=0}^{j} \frac{S_1(j,\ell)}{k^\ell} \beta^j \\ &= a_\varnothing{}^{1/k} + \sum_{j=1}^{m} \frac{a_\varnothing{}^{-j+\frac{1}{k}}}{j!} \sum_{\ell=0}^{j} \frac{S_1(j,\ell)}{k^\ell} j! \sum_{|I| \geq j} \sum_{\substack{\pi \in \mathcal{P}(I) \\ |\pi| = j}} a_\pi \zeta_I \\ &= a_\varnothing{}^{1/k} + \sum_{j=1}^{m} a_\varnothing{}^{-j+\frac{1}{k}} \sum_{\ell=0}^{j} \frac{S_1(j,\ell)}{k^\ell} \sum_{|I| \geq j} \sum_{\substack{\pi \in \mathcal{P}(I) \\ |\pi| = j}} a_\pi \zeta_I. \end{aligned}$$

For fixed nontrivial multi-index I, the coefficient of ζ_I in $u^{1/k}$ is now given by rearranging the summation. Namely,

$$\langle u^{1/k}, \zeta_I \rangle = \sum_{j=1}^{|I|} a_\varnothing{}^{-j+\frac{1}{k}} \sum_{\ell=0}^{j} \frac{S_1(j,\ell)}{k^\ell} \sum_{\substack{\pi \in \mathcal{P}(I) \\ |\pi| = j}} a_\pi.$$

Expanding $u^{1/k}$ in terms of basis blades then reveals the desired result:

$$u^{1/k} = a_\varnothing{}^{1/k} + \sum_{I \neq \varnothing} \left(\sum_{j=1}^{|I|} a_\varnothing{}^{-j+\frac{1}{k}} \sum_{\ell=0}^{j} \frac{S_1(j,\ell)}{k^\ell} \sum_{\substack{\pi \in \mathcal{P}(I) \\ |\pi| = j}} a_\pi \right) \zeta_I.$$

□

Example 9.3.8. A closed formula for the principal 4th root of $\sum_{I \in 2^{[2]}} a_I \zeta_I \in \mathcal{C}\ell_2{}^{\mathrm{nil}}$, as computed by *Mathematica*, is

$$\frac{5a_\varnothing \left(a_{\{1,2\}}\zeta_{\{1,2\}} + a_{\{1\}}\zeta_{\{1\}} + a_{\{2\}}\zeta_{\{2\}}\right) - 4a_{\{1\}}a_{\{2\}}\zeta_{\{1,2\}} + 25a_\varnothing{}^2}{25a_\varnothing{}^{9/5}}.$$

Collecting terms by basis blade, this is seen to be

$$\sqrt[4]{a_\varnothing} + \frac{a_{\{1\}}\zeta_{\{1\}}}{4a_\varnothing{}^{3/4}} + \frac{a_{\{2\}}\zeta_{\{2\}}}{4a_\varnothing{}^{3/4}} + \frac{\left(4a_\varnothing a_{\{1,2\}} - 3a_{\{1\}}a_{\{2\}}\right)\zeta_{\{1,2\}}}{16a_\varnothing{}^{7/4}}.$$

Closed formulas for $n = 1, 2, 3$.

This approach leads to a number of dimension-dependent special formulas. For example, in $\mathcal{C}\ell_1{}^{\mathrm{nil}_\star}$, the principal kth root of $\alpha = a_\varnothing + a_{\{1\}}\zeta_{\{1\}}$ is

$$\alpha^{1/k} = a_\varnothing{}^{1/k} + \frac{a_\varnothing^{\frac{1}{k}-1} a_{\{1\}}\zeta_{\{1\}}}{k}.$$

In $\mathcal{C}\ell_2{}^{\mathrm{nil}_\star}$, the principal kth root of $\alpha = a_\varnothing + a_{\{1\}}\zeta_{\{1\}} + a_{\{2\}}\zeta_{\{2\}} + a_{\{1,2\}}\zeta_{\{1,2\}}$ is

$$\begin{aligned}\alpha^{1/k} = a_\varnothing{}^{1/k} &+ \frac{a_\varnothing{}^{\frac{1}{k}-1}a_{\{1\}}\zeta_{\{1\}}}{k} + \frac{a_\varnothing{}^{\frac{1}{k}-1}a_{\{2\}}\zeta_{\{2\}}}{k}\\ &+ \left[\frac{a_\varnothing{}^{\frac{1}{k}-1}a_{\{1,2\}}}{k} + \frac{a_\varnothing{}^{\frac{1}{k}-2}a_{\{1\}}a_{\{2\}}}{k^2} - \frac{a_\varnothing{}^{\frac{1}{k}-2}a_{\{1\}}a_{\{2\}}}{k}\right]\zeta_{\{1,2\}}.\end{aligned}$$

In $\mathcal{C}\ell_3{}^{\mathrm{nil}_\star}$, the principal kth root of $\alpha = \sum_{I\in 2^{[3]}} a_I\zeta_I$ is given by

$$\begin{aligned}\alpha^{1/k} = a_\varnothing{}^{1/k} &+ \frac{a_\varnothing{}^{\frac{1}{k}-1}a_{\{1\}}}{k}\zeta_{\{1\}} + \frac{a_\varnothing{}^{\frac{1}{k}-1}a_{\{2\}}}{k}\zeta_{\{2\}} + \frac{a_\varnothing{}^{\frac{1}{k}-1}a_{\{3\}}}{k}\zeta_{\{3\}}\\
&+ \left[\frac{a_\varnothing{}^{\frac{1}{k}-1}a_{\{1,2\}}}{k} + \frac{a_\varnothing{}^{\frac{1}{k}-2}a_{\{1\}}a_{\{2\}}}{k^2} - \frac{a_\varnothing{}^{\frac{1}{k}-2}a_{\{1\}}a_{\{2\}}}{k}\right]\zeta_{\{1,2\}}\\
&+ \left[\frac{a_\varnothing{}^{\frac{1}{k}-1}a_{\{1,3\}}}{k} + \frac{a_\varnothing{}^{\frac{1}{k}-2}a_{\{1\}}a_{\{3\}}}{k^2} - \frac{a_\varnothing{}^{\frac{1}{k}-2}a_{\{1\}}a_{\{3\}}}{k}\right]\zeta_{\{1,3\}}\\
&+ \left[\frac{a_\varnothing{}^{\frac{1}{k}-1}a_{\{2,3\}}}{k} + \frac{a_\varnothing{}^{\frac{1}{k}-2}a_{\{2\}}a_{\{3\}}}{k^2} - \frac{a_\varnothing{}^{\frac{1}{k}-2}a_{\{2\}}a_{\{3\}}}{k}\right]\zeta_{\{2,3\}}\\
&+ \left[\frac{a_\varnothing{}^{\frac{1}{k}-3}a_{\{1\}}a_{\{2\}}a_{\{3\}}}{k^3} - \frac{3a_\varnothing{}^{\frac{1}{k}-3}a_{\{1\}}a_{\{2\}}a_{\{3\}}}{k^2} + \frac{2a_\varnothing{}^{\frac{1}{k}-3}a_{\{1\}}a_{\{2\}}a_{\{3\}}}{k}\right]\zeta_{\{1,2,3\}}\\
&+ \left[\frac{a_\varnothing{}^{\frac{1}{k}-2}a_{\{3\}}a_{\{1,2\}}}{k^2} + \frac{a_\varnothing{}^{\frac{1}{k}-2}a_{\{2\}}a_{\{1,3\}}}{k^2} - \frac{a_\varnothing{}^{\frac{1}{k}-2}a_{\{3\}}a_{\{1,2\}}}{k}\right]\zeta_{\{1,2,3\}}\\
&+ \left[\frac{a_\varnothing{}^{\frac{1}{k}-2}a_{\{1\}}a_{\{2,3\}}}{k^2} - \frac{a_\varnothing{}^{\frac{1}{k}-2}a_{\{2\}}a_{\{1,3\}}}{k} - \frac{a_\varnothing{}^{\frac{1}{k}-2}a_{\{1\}}a_{\{2,3\}}}{k}\right]\zeta_{\{1,2,3\}}\\
&+ \frac{a_\varnothing{}^{\frac{1}{k}-1}a_{\{1,2,3\}}}{k}\zeta_{\{1,2,3\}}.\end{aligned}$$

Exercises

Exercise 9.1: Find elements $u, v \in \mathcal{C}\ell_3{}^{\text{nil}}$ such that $\kappa(u) = 3$ and $\kappa(v) = 4$.

Exercise 9.2: Prove Proposition 9.2.2.

Exercise 9.3: Prove Proposition 9.2.3.

Exercise 9.4: Compute the principal square root:

$$1 + 4\zeta_{\{1\}} - 10\zeta_{\{2\}} - 20\zeta_{\{1,2\}}.$$

Exercise 9.5: Compute the principal fourth root:

$$256 - 768\zeta_{\{1\}} + 512\zeta_{\{2\}} - 1664\zeta_{\{1,2\}}.$$

Exercise 9.6: Compute the principal fourth root:

$$256 - 512\zeta_{\{2\}} - 768\zeta_{\{3\}} - 512\zeta_{\{1,2\}} - 256\zeta_{\{1,3\}} + 896\zeta_{\{2,3\}} + 2304\zeta_{\{1,2,3\}}.$$

Exercise 9.7: Compute the principal cube root:

$$64 + 240\zeta_{\{1\}} - 96\zeta_{\{2\}} + 144\zeta_{\{3\}} - 432\zeta_{\{1,2\}} + 408\zeta_{\{1,3\}} + 96\zeta_{\{2,3\}} + 84\zeta_{\{1,2,3\}}.$$

Exercise 9.8: Compute the principal cube root:

$$27 - 27\zeta_{\{1\}} - 81\zeta_{\{2\}} + 27\zeta_{\{3\}} + 27\zeta_{\{1,2\}} + 90\zeta_{\{1,3\}} + 27\zeta_{\{2,3\}} - 297\zeta_{\{1,2,3\}}.$$

Exercise 9.9: Compute the principal fifth root:

$$243 + 1620\zeta_{\{1\}} - 405\zeta_{\{2\}} - 810\zeta_{\{3\}} - 2970\zeta_{\{1,2\}} - 3915\zeta_{\{1,3\}} + 1485\zeta_{\{2,3\}} + 8100\zeta_{\{1,2,3\}}.$$

Exercise 9.10: Compute the principal fourth root:

$$a + b\zeta_{\{1,2\}} + c\zeta_{\{1,4\}} + d\zeta_{\{2,3,4\}}.$$

Chapter 10

Zeon Polynomials

Analogous to real polynomial functions, zeon polynomial functions are defined as zeon-valued functions of a zeon variable. In this chapter, properties of zeon polynomials and their zeros are considered. Nilpotent and invertible zeon zeros of polynomials with real coefficients are characterized, and necessary conditions are established for the existence of zeros of polynomials with zeon coefficients. Quadratic polynomials with zeon coefficients are considered in detail, the "zeon quadratic formula" is developed, and solutions of $ax^2 + bx + c = 0$ are characterized with respect to the "zeon discriminant" of the equation. Zeros of zeon polynomials were first considered in [54], which serves as the basis for this chapter.

In this and subsequent chapters, the notion of "minimal grade" of a zeon element will be useful.

Definition 10.0.1. For a zeon $u \neq 0$, it is useful to define the *minimal grade* of u by

$$\natural u = \begin{cases} \min\{k \in \mathbb{N} : \langle \mathfrak{D}u \rangle_k \neq 0\} & \mathfrak{D}u \neq 0, \\ 0 & u = \mathfrak{R}u. \end{cases}$$

Note that $\natural u = 0$ if and only if u is trivial.

10.1 Zeon Polynomials with Real Coefficients

Let $f(x) = a_m x^m + \cdots + a_1 x + a_0$ ($a_m \neq 0$) be a polynomial function with real coefficients, and recall that a real number r such that $f(r) = 0$ is called a *zero* of $f(x)$. By the Fundamental Theorem of Algebra, $f(x)$ has at most m real zeros (and exactly m complex zeros).

If $f(x)$ can be written in the form $f(x) = (x-r)^\ell g(x)$, where $m \in \mathbb{N}$ and $g(r) \neq 0$, then r is said to be a *zero of multiplicity* ℓ of $f(x)$. Equivalently, one sees that r is a zero of multiplicity ℓ of $f(x)$ if and only if $f(r) = f'(r) = \cdots = f^{(\ell-1)}(r) = 0$ and $f^{(\ell)}(r) \neq 0$. Here $f^{(\ell)}(r)$ denotes the ℓth derivative of $f(x)$ evaluated at r. For convenience, $\mu_f(r)$ will denote the multiplicity of r as a zero of $f(x)$.

Letting $\varphi : \mathcal{C}\ell_n{}^{\text{nil}} \to \mathcal{C}\ell_n{}^{\text{nil}}$ denote the zeon extension of f, one is led to wonder about the number of zeros this polynomial may have in $\mathcal{C}\ell_n{}^{\text{nil}}$.

The easiest case to consider is the existence of nilpotent zeros of extended polynomials $\varphi(u)$ in $\mathcal{C}\ell_n{}^{\text{nil}}$. As illustrated in the following lemma, a nonzero polynomial either has no nilpotent zeros or infinitely many of them.

Lemma 10.1.1. *Let $f(x)$ be a nonzero polynomial, and let d be the least nonnegative integer such that $f^{(d)}(0) \neq 0$. Let φ denote the zeon extension of $f(x)$ over $\mathcal{C}\ell_n{}^{\text{nil}}$, where $n \geq 1$. If $d \in \{0, 1\}$, then φ has no nilpotent zeros. If $d \geq 2$, then φ has infinitely many nilpotent zeros in $\mathcal{C}\ell_n{}^{\text{nil}}$.*

Proof. Assuming d is the least nonnegative integer such that $f^{(d)}(0) \neq 0$, it is clear that $f(x)$ can be written as $f(x) = a_m x^m + a_{m-1}x^{m-1} + \cdots + a_d x^d$.

If $d = 0$ or $d = 1$, let u be any nilpotent element of $\mathcal{C}\ell_n{}^{\text{nil}}$. For any positive integer ℓ, the minimum grade of u^ℓ is either zero or at least $\ell \natural u$. Hence, no cancellation of the nonzero scalar term can occur in the $d = 0$ case, and the grade-$\natural u$ part of u is preserved by the degree-1 terms of the polynomial in the $d = 1$ case. Thus, $\natural\varphi(u) \geq \natural u > 0$.

When $d \geq 2$, let $u = a\zeta_I$ for nonzero scalar a and multi-index $I \neq \varnothing$. Then, $\varphi(u) = a_m u^m \cdots + a_d u^d = 0$. □

While nilpotent zeros provide an easy special case, the discussion now turns to more general zeros of zeon polynomials.

Proposition 10.1.2. *Let f be a nonzero polynomial with real coefficients, and let φ be its zeon extension. If f has no real zeros, then φ has no zeros. If r is a real zero of f, then any zeon element w satisfying $\Re w = r$ and $\kappa(\mathfrak{D}w) \leq \mu_f(r)$ is a zero of φ. Moreover, the zeros of φ are*

$$\varphi^{-1}(0) = \{w \in \mathcal{C}\ell_n{}^{\text{nil}\star} : f(\Re w) = 0, \kappa(\mathfrak{D}w) \leq \mu_f(\Re w)\}.$$

Proof. First, if f has no real zero, then $\Re(\varphi(u)) = f(\Re u)$ guarantees that φ has no zero. Otherwise, suppose f has distinct real zeros $r_1, \ldots, r_k$. It follows that f can be written multiplicatively as $f(x) = \prod_{i=1}^{k}(x - r_i)^{\mu_f(r_i)} g(x)$, where $g(x)$ is a constant or polynomial over $\mathbb{R}$ having no real zeros.

Next, let r be a real zero of f and label any remaining real zeros as $r_1, \ldots, r_{k-1}$. Let $u \in \mathcal{C}\ell_n{}^{\text{nil}}$ satisfy $\Re u = r$. Writing $\rho_i = \Re u - r_i$ for each i, and letting γ be the zeon extension of g, the multiplicative form of the zeon extension φ of f evaluated at u can now be written as

$$\begin{aligned}
\varphi(u) &= (u-r)^{\mu_f(r)} \prod_{i=1}^{k-1} (u - r_i)^{\mu_f(r_i)} \gamma(u) \\
&= (\Re u - r + \mathfrak{D}u)^{\mu_f(r)} \prod_{i=1}^{k-1} (\Re u - r_i + \mathfrak{D}u)^{\mu_f(r_i)} \gamma(u) \\
&= (\mathfrak{D}u)^{\mu_f(r)} \prod_{i=1}^{k-1} (\rho_i + \mathfrak{D}u)^{\mu_f(r_i)} \gamma(u).
\end{aligned}$$

Note that as an element of $\mathcal{C}\ell_n{}^{\text{nil}}$, $\gamma(u)$ has a multiplicative inverse; otherwise, $\gamma(\Re u) = g(\Re u) = 0$, contradicting the assumption that g has no real zeros. For each i, $(\rho_i + \mathfrak{D}u)$ is similarly invertible, so the product $\prod_{i=1}^{k-1} (\rho_i + \mathfrak{D}u)^{\mu_f(r_i)} g(u)$ is invertible. Consequently, $\varphi(u) = 0$ if and only if $(\mathfrak{D}u)^{\mu_f(r)} = 0$. In other words, $\varphi(u) = 0$ if and only if both $f(\Re u) = 0$ and $\kappa(\mathfrak{D}u) \leq \mu_f(r)$ are true. □

Remark 10.1.3. If r is a simple zero of f, then the only element $u \in \mathcal{C}\ell_n{}^{\text{nil}}$ satisfying $\Re u = r$ and $\varphi(u) = 0$ is $u = r$.

Example 10.1.4. The roots of $f(x) = x^2 + x$ are $x = 0$ and $x = -1$. Since both are simple, Remark 10.1.3 states that the only zeros of $\varphi(u) = 0$ are $u = 0$ and $u = -1$. Clearly, these are solutions, and according to Proposition 10.1.2, any other zero should be of the form $u = \mathfrak{D}u$ or $u = -1 + \mathfrak{D}u$ where $\kappa(\mathfrak{D}u) \leq 1$. This holds only for $\mathfrak{D}u = 0$.

An immediate corollary is obtained when f has a zero of multiplicity two or greater.

Corollary 10.1.5. *Let f be a nonzero polynomial function with real coefficients and let φ be its zeon extension. If $f(r) = f'(r) = 0$ for some $r \in \mathbb{R}$, then φ has infinitely many zeros in $\mathcal{C}\ell_n{}^{\text{nil}}$ $(n \geq 1)$.*

Proof. If $f(r) = f'(r) = 0$, then $\mu_f(r) \geq 2$. Letting $w = r + a\zeta_{\{1\}}$ for any nonzero $a \in \mathbb{R}$, Proposition 10.1.2 implies $\varphi(w) = 0$. □

Example 10.1.6. Consider the polynomial $f(x) = x^2 + x + 1/4$. The nontrivial zeon solutions of $\varphi(u) = 0$ in ${\mathcal{C}\ell_1}^{\rm nil}$ are seen to be $\{a\zeta_{\{1\}} - 1/2 : a \neq 0\}$.

Proposition 10.1.2 also implies that nonzero polynomials having real zeros of multiplicity exceeding n have infinitely many zeros in ${\mathcal{C}\ell_n}^{\rm nil}$, since the condition $\kappa(\mathfrak{D}w) \leq \mu_f(\Re w)$ is satisfied automatically when $\mu_f(\Re w) > n$. More formally, we have the following corollary.

Corollary 10.1.7. *Let f be a nonzero polynomial function with real coefficients and let φ be its zeon extension. If r is a zero of multiplicity m of f and $n < m$, then $\varphi(w) = 0$ for any $w \in {\mathcal{C}\ell_n}^{\rm nil}$ such that $\Re w = r$.*

Example 10.1.8. Consider the polynomial $f(x) = (x-1)^4$. Writing $w = 1 + \mathfrak{D}w \in {\mathcal{C}\ell_2}^{\rm nil}$, then $\kappa(\mathfrak{D}w) \leq 3 < 4 = \mu_f(1)$ and $\varphi(w) = 0$.

10.2 Polynomials with Zeon Coefficients

In this section, polynomials with zeon coefficents are considered. In particular, for $n \in \mathbb{N}$, the *algebra of zeon polynomials in the indeterminate x* is defined as

$${\mathcal{C}\ell_n}^{\rm nil}[x] = \{\alpha_m x^m + \cdots + \alpha_1 x + \alpha_0 : \alpha_0, \ldots, \alpha_m \in {\mathcal{C}\ell_n}^{\rm nil}, m \in \mathbb{N} \cup \{0\}\}.$$

The task at hand is to characterize the nilpotent and invertible zeros of such polynomials over the zeons.

In working with polynomials, it is useful to know which algebraic operations leave a polynomial's zeros invariant. To that end, consider the following.

Lemma 10.2.1. *Let $\varphi(u) = \sum_{i=0}^{m} \alpha_i u^i \in {\mathcal{C}\ell_n}^{\rm nil}[u]$. If w is an invertible zeon, then $\varphi(u) = 0 \Leftrightarrow w\varphi(u) = 0$. Hence, the solution set of $\varphi(u) = 0$ is invariant under multiplication of the equation by w.*

Proof. Let $u, w \in {\mathcal{C}\ell_n}^{\rm nil}$ be fixed. Clearly, $\varphi(u) = 0 \Rightarrow w\varphi(u) = 0$. Conversely, if $\Re w \neq 0$, then $w\varphi(u) = 0$ implies

$$\begin{aligned}\varphi(u) &= (w^{-1}w)\varphi(u)\\ &= w^{-1}(w\varphi(u))\\ &= 0.\end{aligned}$$

□

Corollary 10.2.2. *Let* $\varphi(u) = \sum_{i=0}^{m} \alpha_i u^i \in \mathcal{C}\ell_n{}^{\text{nil}}[u]$. *If* α_m *is invertible, then* $\varphi(u)$ *can be replaced by a monic zeon polynomial having the same zeros.*

We can now establish our first result characterizing the zeros of a zeon polynomial with zeon coefficients.

Proposition 10.2.3. *Let* $\varphi(u) = \sum_{i=0}^{m} \alpha_i u^i \in \mathcal{C}\ell_n{}^{\text{nil}}[u]$.

(i.) If $\varphi(0) = 0$ *and* α_1 *is nilpotent, then* φ *has infinitely many nilpotent zeros.*

(ii.) If $\varphi(0)$ *is invertible and* α_i *is nilpotent for* $i = 1, 2, \ldots, m$, *then* φ *has no zeros in* $\mathcal{C}\ell_n{}^{\text{nil}}$.

Proof. Proof of (i.) Suppose $\varphi(0) = 0$ and write $\varphi(u) = \alpha_m u^m + \cdots + \alpha_1 u + \alpha_0$. Suppose κ is the index of nilpotency for α_1, and note that $\kappa \geq 2$. Then, $j(\kappa - 1) - \kappa = (j-1)\kappa - j > j - 2 \geq 0$ provided $j \geq 2$. Thus, $j(\kappa - 1) \geq \kappa$ for $j \geq 2$. It follows that for arbitrary real scalar $r \neq 0$,

$$\begin{aligned}\varphi(r{\alpha_1}^{\kappa-1}) &= \alpha_m (r{\alpha_1}^{\kappa-1})^m + \cdots + \alpha_2 (r{\alpha_1}^{\kappa-1})^2 + r\alpha_1^{\kappa} \\ &= \alpha_m r^m ({\alpha_1}^{m(\kappa-1)})^m + \cdots + \alpha_2 r^2 ({\alpha_1}^{2(\kappa-1)}) \\ &= 0.\end{aligned}$$

Proof of (ii.) Suppose $\varphi(u) = 0$ and note that $\varphi(0) = \alpha_0$, then

$$\begin{aligned}0 &= \alpha_0 + \sum_{\ell=1}^{m} \alpha_\ell u^\ell \\ &= \Re\alpha_0 + \beta,\end{aligned}$$

where $\beta = \mathfrak{D}\alpha_0 + \sum_{\ell=1}^{m} \alpha_\ell u^\ell$, which is nilpotent. Thus, $\varphi(u) = 0$ implies $\Re(\varphi(0)) = 0$. □

Recalling the zeon extension φ of a real polynomial function f, the next lemma establishes the notion of a real function f obtained from a zeon function φ. The function f obtained in this way will be helpful for considering zeros of φ.

Lemma 10.2.4. *Let* $\varphi(u) = \sum_{i=0}^{m} \alpha_i u^i \in \mathcal{C}\ell_n{}^{\text{nil}}[u]$. *Then, defining the real polynomial* $f(x) = \sum_{i=0}^{m} \Re(\alpha_i) x^i \in \mathbb{R}[x]$, *one finds that*

$$\Re(\varphi(u)) = f(\Re u).$$

Proof. Write $\varphi(u) = \alpha_m u^m + \cdots + \alpha_1 u + \alpha_0$, where each α_i is an element of ${\mathcal{C}\ell_n}^{\text{nil}}$. Then,

$$\begin{aligned}\Re(\varphi(u)) &= \Re(\alpha_m u^m + \cdots + \alpha_1 u + \alpha_0)\\ &= \sum_{\ell=0}^{m} \Re(\alpha_\ell)\Re(u^\ell)\\ &= \sum_{\ell=0}^{m} \Re(\alpha_\ell)(\Re u)^\ell\\ &= f(\Re u).\end{aligned}$$

□

Define $\varphi \searrow f$ when f is obtained from φ as in Lemma 10.2.4. This leads to the following necessary condition for $\varphi(u) = 0$: if $\varphi \searrow f$ and $\varphi(u) = 0$, then $f(\Re u) = 0$. Further, it is not difficult to see that $\varphi(u) \in {\mathcal{C}\ell_n}^{\text{nil}}[u]$ for a particular positive integer n implies $\varphi(u) \in {\mathcal{C}\ell_{n'}}^{\text{nil}}[u]$ for all $n' \geq n$. Hence, the following necessary condition for the existence of "zeon zeros" is established.

Corollary 10.2.5. *Let* $\varphi(u) = \sum_{i=0}^{m} \alpha_i u^i \in {\mathcal{C}\ell_n}^{\text{nil}}[u]$ *and suppose that* $\varphi \searrow f \in \mathbb{R}[x]$ *as described in Lemma 10.2.4. Then* $\varphi(u) = 0$ *only if* $f(\Re u) = 0$. *In particular, if* f *has no zeros in* $\mathbb{R}$, *then* φ *has no zeros in* ${\mathcal{C}\ell_{n'}}^{\text{nil}}$ *for any* $n' \geq n$.

Example 10.2.6. Consider the zeon polynomial $\varphi(u) \in {\mathcal{C}\ell_5}^{\text{nil}}[u]$ defined as follows:

$$\begin{aligned}\varphi(u) = (3 + \zeta_{\{1,2\}} - 4\zeta_{\{5\}})u^6 + (5 - 4\zeta_{\{1,4,5\}} + \zeta_{\{2,3,5\}})u^4\\ -(2 + 7\zeta_{\{1\}} - \zeta_{\{3\}})u^2 + (1 - 5\zeta_{\{3,5\}})u + 1 + \zeta_{\{1,3,5\}}.\end{aligned}$$

The real polynomial $f(x) \in \mathbb{R}[x]$ such that $\varphi \searrow f$ is seen to be

$$f(x) = 3x^6 + 5x^4 - 2x^2 + x + 1,$$

which has no real zeros. Hence, $\varphi(u)$ has no zeros in ${\mathcal{C}\ell_n}^{\text{nil}}$ for any $n \geq 5$.

The Zeon Quadratic Formula

Motivated by the usual quadratic formula for polynomials with real or complex coefficients, the goal is to obtain a similar characterization of solutions for polynomials with zeon coefficients. In particular, consider solutions of the equation $\alpha u^2 + \beta u + \gamma = 0$, where $\alpha, \beta, \gamma \in {\mathcal{C}\ell_n}^{\text{nil}}$ and $\alpha \in {\mathcal{C}\ell_n}^{\text{nil}\star}$.

As a special case of Theorem 9.3.1, a recursive formulation of the principal square root of $\alpha \in \mathcal{C}\ell_n{}^{\text{nil}\star}$ is obtained. In particular, when $\Re\alpha > 0$, one can easily verify that

$$v^{1/2} = \alpha^{1/2} + \zeta_{\{n\}} \frac{\alpha^{-1/2}}{2} \beta. \tag{10.1}$$

It is worth noting that if $\Re\alpha < 0$ and k is even, there exists no element u such that $u^{2j} = \alpha$, since $\Re(u^{2j}) = (\Re u)^{2j}$ by Lemma 10.2.4. In particular, α has no square roots when $\Re\alpha < 0$.

With principal square roots of invertible zeons available, a nice result is obtained for quadratic zeon polynomials having invertible leading coefficient.

Theorem 10.2.7 (Quadratic Formula). *Let $\varphi(u) = \alpha u^2 + \beta u + \gamma$ be a quadratic function with zeon coefficients from $\mathcal{C}\ell_n{}^{\text{nil}}$, where $\Re\alpha \neq 0$. Let $\Delta_\varphi = \beta^2 - 4\alpha\gamma$ denote the zeon discriminant of φ. The zeros of φ are given by*

$$\varphi^{-1}(0) = \left\{ \frac{\alpha^{-1}}{2}(w - \beta) : w^2 = \beta^2 - 4\alpha\gamma \right\}.$$

In particular,

i. If $\Delta_\varphi = 0$, then the zeros of φ are given by $u = -\alpha^{-1}\beta/2 + \eta$ for any $\eta \in \mathcal{C}\ell_n{}^{\text{nil}}$ satisfying $\eta^2 = 0$.
ii. If $\Re\Delta_\varphi > 0$, then $\varphi(u) = 0$ has two distinct solutions.
iii. If $\Re\Delta_\varphi < 0$, then $\varphi(u) = 0$ has no solution.
iv. If $\Delta_\varphi \neq 0$ is nilpotent and $\varphi(u) = 0$ has a solution, then it has infinitely many solutions.

Proof. Writing

$$\alpha u^2 + \beta u + \gamma = \frac{\alpha^{-1}}{4}((2\alpha u + \beta)^2 - (\beta^2 - 4\alpha\gamma)),$$

it follows that the zeros of the polynomial are precisely the elements

$$\varphi^{-1}(0) = \left\{ \frac{\alpha^{-1}}{2}(w - \beta) : w^2 = \beta^2 - 4\alpha\gamma \right\}.$$

Thus, the problem of finding zeros of the polynomial is essentially reduced to finding square roots of the discriminant. We now proceed by cases.

i.) If $\Delta_\varphi = 0$, set $u_0 = -\alpha^{-1}\beta/2$ and let $\eta \in \mathcal{C}\ell_n{}^{\text{nil}}$ such that $\eta^2 = 0$. We show that $u_0 + \eta$ is in $\varphi^{-1}(0)$. Letting $w = 2\alpha\eta$, it follows that $w^2 = 0$,

and

$$\begin{aligned}u_0+\eta &= u_0+\frac{\alpha^{-1}}{2}w\\ &= -\frac{\alpha^{-1}}{2}\beta+\frac{\alpha^{-1}}{2}w\\ &= \frac{\alpha^{-1}}{2}(w-\beta)\in\varphi^{-1}(0).\end{aligned}$$

ii.) When $\Re\Delta_\varphi > 0$, the discriminant has two square roots by Theorem 9.3.4. Letting w be the principal square root of Δ_φ, computed as in (10.1), the zeros of $\varphi(u)$ are

$$u=\frac{\alpha^{-1}}{2}\left(-\beta\pm w\right).$$

iii.) When $\Re\Delta_\varphi < 0$, the discriminant has no square roots as a consequence of Lemma 10.2.4, noted previously. Hence, $\varphi(u) = 0$ has no solutions.

iv.) For the nilpotent case, suppose $\beta^2 - 4\alpha\gamma$ is nilpotent and note that by Lemma 9.3.2, existence of a square root of Δ_φ implies the existence of infinitely many square roots. Hence, if Δ_φ is nilpotent and $\varphi(u) = 0$ has a solution in $\mathcal{C}\ell_n{}^{\text{nil}}$, then $\varphi(u) = 0$ has infinitely many solutions. □

Example 10.2.8. Suppose $\beta^2 - 4\alpha\gamma = a\zeta_I$ for some positive scalar a and multi-index I of cardinality at least 2. Partitioning $I = (I_1|I_2)$ into two nonempty subsets, choose scalars a_1, a_2 such that $a_1a_2 = a/2$ and consider the square of $a_1\zeta_{I_1} + a_2\zeta_{I_2}$:

$$\begin{aligned}(a_1\zeta_{I_1}+a_2\zeta_{I_2})^2 &= 2a_1a_2\zeta_{I_1\cup I_2}\\ &= \beta^2-4\alpha\gamma.\end{aligned}$$

In this case, there are clearly infinitely many choices of "square root" for $\beta^2 - 4\alpha\gamma$, and thus infinitely many solutions to $\varphi(u) = 0$.

On the other hand, if $\beta^2 - 4\alpha\gamma = a\zeta_{\{j\}}$ for nonzero scalar a and $j \in [n]$, it is not difficult to see that the monomial $a\zeta_{\{j\}}$ has no square roots.

Example 10.2.9. Consider the following quadratic equation with coefficients in $\mathcal{C}\ell_3{}^{\text{nil}}$:

$$\varphi(u) = (1+5\zeta_{\{1,2\}}-3\zeta_{\{2,3\}})u^2+(5-\zeta_{\{1\}})u+4\zeta_{\{1,3\}}=0.$$

The zeon discriminant of the polynomial is $\Delta_\varphi = 25 - 10\zeta_{\{1\}} - 16\zeta_{\{1,3\}}$. Since $\Re\Delta_\varphi = 25 > 0$, the equation has two distinct solutions. By the Zeon Quadratic Formula, these solutions are

$$\begin{aligned}u_1 &= -\frac{4}{5}\zeta_{\{1,3\}}\\ u_2 &= -5+\zeta_{\{1\}}+25\zeta_{\{1,2\}}+\frac{4}{5}\zeta_{\{1,3\}}-15\zeta_{\{2,3\}}+3\zeta_{\{1,2,3\}}.\end{aligned}$$

Theorem 10.2.7 lends itself nicely to a convenient corollary. First, consider the expansion of Δ_φ when $\varphi(u) = \alpha u^2 + \beta u + \gamma$, and note that $\Re\Delta_\varphi = (\Re\beta)^2 - 4\Re\alpha\Re\gamma$, as follows:

$$\begin{aligned}\Delta_\varphi &= \beta^2 - 4\alpha\gamma \\ &= (\Re\beta)^2 + 2\Re\beta\mathfrak{D}\beta + (\mathfrak{D}\beta)^2 - 4(\Re\alpha\Re\gamma + \Re\alpha\mathfrak{D}\gamma + \mathfrak{D}\alpha\Re\gamma + \mathfrak{D}\alpha\mathfrak{D}\gamma) \\ &= (\Re\beta)^2 - 4\Re\alpha\Re\gamma + \mathfrak{D}\beta\,(2\Re\beta + (\mathfrak{D}\beta)) - 4(\Re\alpha\mathfrak{D}\gamma + \Re\gamma\mathfrak{D}\alpha + \mathfrak{D}\alpha\mathfrak{D}\gamma).\end{aligned} \tag{10.2}$$

Corollary 10.2.10. *Let $\varphi(u) = \alpha u^2 + \beta u + \gamma$ be a quadratic function with zeon coefficients, where $\Re\alpha \neq 0$ and γ is nilpotent. Then,*

(1) φ has two zeros if β is invertible;
(2) φ has no zeros or infinitely many zeros if β is nilpotent.

Proof. By (10.2), $\Re\Delta_\varphi = (\Re\beta)^2 - 4\Re\alpha\Re\gamma$. The rest follows from Theorem 10.2.7. □

Example 10.2.11. Consider the following quadratic polynomial with coefficients in $\mathcal{C}\ell_3{}^{\text{nil}}$:

$$\begin{aligned}\varphi(u) = u^2 &\left(-3\zeta_{\{1,2\}} + 2\zeta_{\{2,3\}} + 2\zeta_{\{1,2,3\}} - 4\zeta_{\{1\}} - 3\zeta_{\{2\}} - 3\zeta_{\{3\}} - 4\right) \\ &+ u\left(4\zeta_{\{1,2\}} + 2\zeta_{\{1,3\}} + 2\zeta_{\{2,3\}} - 3\zeta_{\{1,2,3\}} - \zeta_{\{1\}} - 3\zeta_{\{2\}} + \zeta_{\{3\}} + 4\right) \\ &- \zeta_{\{1\}} - 2\zeta_{\{3\}} - \zeta_{\{1,2\}} + 3\zeta_{\{1,3\}} + 4\zeta_{\{1,2,3\}}.\end{aligned}$$

Writing $\varphi(u) = \alpha u^2 + \beta u + \gamma$, it is clear that β is invertible and γ is nilpotent. In light of Corollary 10.2.10, this polynomial has two distinct zeros. By the Zeon Quadratic Formula, these zeros are

$$\begin{aligned}u_1 &= \frac{7}{16}\zeta_{\{1,2\}} - \frac{7}{16}\zeta_{\{1,3\}} + \frac{3}{8}\zeta_{\{2,3\}} - \frac{37}{32}\zeta_{\{1,2,3\}} + \frac{\zeta_{\{1\}}}{4} + \frac{\zeta_{\{3\}}}{2} \\ u_2 &= \frac{9}{4}\zeta_{\{1,2\}} + \frac{19}{8}\zeta_{\{1,3\}} + \frac{17}{8}\zeta_{\{2,3\}} - \frac{85}{16}\zeta_{\{1,2,3\}} - \frac{1}{2}3\zeta_{\{1\}} - \zeta_{\{3\}} - \frac{3\zeta_{\{2\}}}{2} + 1.\end{aligned}$$

Example 10.2.12. Figure 10.1 shows how *Mathematica* can be used for finding and verifying solutions of $\varphi(x) = \alpha x^2 + \beta x + \gamma = 0$ when $\Re\Delta_\varphi > 0$.

Example 10.2.13. Consider the following quadratic polynomial with coefficients in $\mathcal{C}\ell_3{}^{\text{nil}}$:

$$\begin{aligned}\varphi(u) = u^2 &\left(\zeta_{\{1,2\}} + \zeta_{\{1,3\}} - 2\zeta_{\{2,3\}} + \zeta_{\{1,2,3\}} - \zeta_{\{1\}} - 3\zeta_{\{2\}} - \zeta_{\{3\}} - 3\right) \\ &+ u\left(-\zeta_{\{1,3\}} - 2\zeta_{\{2,3\}} - \zeta_{\{1\}} + 3\zeta_{\{3\}} + 1\right) \\ &- 1 - \zeta_{\{1\}} - 5\zeta_{\{2\}} - 3\zeta_{\{3\}} + 3\zeta_{\{1,2\}} - 5\zeta_{\{1,3\}} - 3\zeta_{\{2,3\}} - \zeta_{\{1,2,3\}}.\end{aligned}$$

```
        α = 3 + 7 ζ{1,2} - 4 ζ{1,3};
        β = 4 - ζ{2,3};
        γ = 1 + 4 ζ{1,2,3};

In[∘]:= Δ = zeonDiscriminant[α x^2 + β x + γ, x]

Out[∘]= 4 - 28 ζ{1,2} + 16 ζ{1,3} - 8 ζ{2,3} - 48 ζ{1,2,3}

        (* Two solutions *)

        sols = zeonQuadraticFormula[α x^2 + β x + γ, x];
        Print["Sol. 1: "];
        Print[sols[[1]]];
        Print["---"];
        Print["Sol. 2: "];
        Print[sols[[2]]];

        Sol. 1:
```

$$-\frac{1}{3}-\frac{7}{18}\zeta_{\{1,2\}}+\frac{2}{9}\zeta_{\{1,3\}}-\frac{1}{6}\zeta_{\{2,3\}}-2\zeta_{\{1,2,3\}}$$

```
        ---
        Sol. 2:
```

$$-1+\frac{7}{2}\zeta_{\{1,2\}}-2\zeta_{\{1,3\}}+\frac{1}{2}\zeta_{\{2,3\}}+2\zeta_{\{1,2,3\}}$$

```
        (* Check the solutions by evaluation. *)

In[∘]:= zeonQuadraticEval[α x^2 + β x + γ, x, sols[[1]]]

Out[∘]= 0

In[∘]:= zeonQuadraticEval[α x^2 + β x + γ, x, sols[[2]]]

Out[∘]= 0
```

Figure 10.1 A *Mathematica* example.

The zeon discriminant of the polynomial is
$\Delta_\varphi = -11+8\zeta_{\{1,2\}}-80\zeta_{\{1,3\}}-104\zeta_{\{2,3\}}-40\zeta_{\{1,2,3\}}-18\zeta_{\{1\}}-72\zeta_{\{2\}}-34\zeta_{\{3\}}$.
Since $\Re\Delta_\varphi = -11 < 0$, the polynomial has no zeros.

The "indeterminate" case occurs when Δ_φ is nilpotent. In this case, φ has either no zeros or infinitely many zeros. The next corollary provides conditions for the existence of solutions.

Corollary 10.2.14. *Let $\varphi(u) = \alpha u^2 + \beta u + \gamma$ be a quadratic function with zeon coefficients such that $\Re\alpha \neq 0$. Suppose that the zeon discriminant Δ_φ is nonzero and nilpotent. If there exist scalars $\{a_I : I \in 2^{[n]} \setminus \varnothing\}$ such that*

$$\Delta_\varphi = \sum_{I\neq\varnothing} \sum_{\substack{J\subsetneq I \\ J\neq\varnothing}} a_J a_{I\setminus J}\zeta_I, \tag{10.3}$$

then the equation $\varphi(u) = 0$ has infinitely many zeros. Otherwise, $\varphi(u) = 0$ has no solutions.

Proof. If there exists $u \in \mathcal{C}\ell_n{}^{\text{nil}}$ such that $\Delta_\varphi = u^2$ is nilpotent, then u must also be nilpotent. Writing $u = \sum_{I\neq\varnothing} a_I\zeta_I$, it is not difficult to see that the square of a nilpotent zeon element must be of the form

$$u^2 = \sum_{I\neq\varnothing} \sum_{\substack{J\subsetneq I \\ J\neq\varnothing}} a_J a_{I\setminus J}\zeta_I. \tag{10.4}$$

Moreover, $\Delta_\varphi = u^2$ implies $\Delta_\varphi = (u + a\zeta_{[n]})^2$ for arbitrary $a \in \mathbb{R}$. □

Since u is assumed to be nilpotent in (10.4), the minimal grade of u^2 is at least 2. An immediate consequence is that $\varphi(u) = 0$ has no solutions if Δ_φ is nilpotent and $\natural\Delta_\varphi = 1$.

Example 10.2.15. Consider $\varphi(x) = 1 + x\left(\zeta_{\{1,2\}} - 2\right) + x^2\left(\zeta_{\{1\}} + 1\right) \in \mathcal{C}\ell_2{}^{\text{nil}}[x]$. The discriminant is $\Delta_\varphi = -4\zeta_{\{1,2\}} - 4\zeta_{\{1\}}$, which is clearly nilpotent. However, the discriminant is not an element of the form (10.4), so $\varphi(x)$ has no zeros. In particular, the coefficient of $\zeta_{\{1\}}$ in Δ_φ would have to satisfy $2a_\varnothing a_{\{1\}} = 1$, but $a_\varnothing = 0$ by hypothesis.

Example 10.2.16. Consider $\varphi(x) = \alpha x^2 + \beta x + \gamma \in \mathcal{C}\ell_4{}^{\text{nil}}$, where

$$\begin{aligned}
\alpha &= \zeta_{\{1,2\}} + \zeta_{\{3,4\}} + \zeta_{\{1,2,4\}} - \zeta_{\{1,3,4\}} + \zeta_{\{3\}} + \zeta_{\{4\}} - 1 \\
\beta &= \zeta_{\{1,2\}} + \zeta_{\{1,4\}} - 2\zeta_{\{2,3\}} + \zeta_{\{1,2,3\}} + \zeta_{\{1,2,4\}} + \zeta_{\{1,3,4\}} - \zeta_{\{2,3,4\}} \\
&\quad - 2\zeta_{\{1\}} + \zeta_{\{2\}} + \zeta_{\{3\}} - \zeta_{\{4\}} - 2 \\
\gamma &= -\zeta_{\{1,2\}} - \zeta_{\{1,4\}} + \zeta_{\{3,4\}} + \zeta_{\{1,2,3\}} + \zeta_{\{1,2,4\}} + \zeta_{\{1,3,4\}} + \zeta_{\{1,2,3,4\}} \\
&\quad - \zeta_{\{2\}} - \zeta_{\{3\}} + \zeta_{\{4\}} - 1.
\end{aligned}$$

The discriminant of φ is nilpotent:

$$\begin{aligned}\Delta_\varphi &= 8\zeta_{\{1\}} - 8\zeta_{\{2\}} - 4\zeta_{\{3\}} + 12\zeta_{\{4\}} \\ &\quad - 8\zeta_{\{1,2\}} - 4\zeta_{\{1,3\}} - 4\zeta_{\{1,4\}} + 14\zeta_{\{2,3\}} + 2\zeta_{\{2,4\}} + 6\zeta_{\{3,4\}} \\ &\quad + 18\zeta_{\{1,2,3\}} + 4\zeta_{\{1,2,4\}} + 2\zeta_{\{1,3,4\}} + 12\zeta_{\{2,3,4\}} - 2\zeta_{\{1,2,3,4\}}.\end{aligned}$$

Since $\natural\Delta_\varphi = 1$, it follows from Corollary 10.2.14 that $\varphi(x)$ has no zeros.

While invertibility of the leading coefficient is required for the quadratic formula, solutions of $\alpha u^2 + \beta u + \gamma = 0$ can also be obtained when α is a nonzero nilpotent. For example, if $\varphi(u) = \alpha u^2 + \beta u$ is a quadratic function with zeon coefficients from ${\mathcal{C}\ell_n}^{\text{nil}}$, where $\Re\alpha = \Re\beta = 0$. It follows immediately that $\varphi(t\zeta_{[n]}) = 0$ for every real number t.

10.3 Lagrange Interpolation of Zeon Polynomials

The goal of this section is to extend the method of Lagrange interpolation to polynomials with zeon coefficients. We begin by recalling that for any arbitrary finite collection $\{x_1, \ldots, x_k\}$ of distinct real numbers, one can define a family of polynomial functions $\{f_i(x) : 1 \le i \le k\}$ having the property that for $1 \le \ell, j \le k$,

$$f_j(x_\ell) = \begin{cases} 1 & \text{if } j = \ell, \\ 0 & \text{otherwise.} \end{cases}$$

It is not difficult to see that the polynomials $f_j(x)$ are defined by

$$f_j(x) = \prod_{\ell \neq j} \frac{x - x_\ell}{x_j - x_\ell}.$$

It now follows that for any collection of real numbers $\{y_1, \ldots, y_k\}$, the polynomial $g(x)$ defined by

$$g(x) = \sum_{j=1}^{k} y_j f_j(x)$$

has the property that $g(x_j) = y_j$ for $j = 1, \ldots, k$. The polynomials $\{f_i(x) : 1 \le i \le k\}$ are called *Lagrange basis polynomials*, and the function $g(x)$ is the *Lagrange interpolation polynomial* of the collection $\{(x_j, y_j) : 1 \le j \le k\}$.

Along these lines, we consider the problem of constructing a Lagrange interpolation polynomial for a collection $\{(u_j, z_j) : 1 \le j \le k\}$ of ordered pairs of zeons. Key to this discussion is that distinct values of u_j are not sufficient to define the basis polynomials. More specifically, the collection $\{\Re u_j : 1 \le j \le k\}$ must be distinct. With this restriction in mind, one defines the *zeon Lagrange basis polynomials* as follows.

Definition 10.3.1. Given a collection $\{u_j : j = 1, \ldots, k\} \subset {\mathcal{C}\ell_n}^{\text{nil}}$ satisfying $\Re u_\ell \ne \Re u_j$ whenever $\ell \ne j$, the *zeon Lagrange basis polynomials* are polynomial functions $\phi_j : {\mathcal{C}\ell_n}^{\text{nil}} \to \{0, 1\}$ defined by

$$\phi_j(u) = \prod_{\ell \ne j} (u - u_\ell)(u_j - u_\ell)^{-1}$$

for $1 \le j \le k$. These polynomials have the property that

$$\phi_j(u_\ell) = \begin{cases} 1 & j = \ell, \\ 0 & j \ne \ell. \end{cases} \tag{10.5}$$

The zeon Lagrange interpolation of a collection $\{(u_j, z_j) : 1 \le j \le k\}$ can now be expressed as a linear combination of zeon Lagrange basis polynomials.

Definition 10.3.2. Given a collection $Z = \{(u_j, z_j) : 1 \le j \le k\}$ satisfying $\Re u_\ell \ne \Re u_j$ whenever $\ell \ne j$, the *zeon Lagrange interpolation polynomial* of Z is the zeon polynomial $\gamma : {\mathcal{C}\ell_n}^{\text{nil}} \to {\mathcal{C}\ell_n}^{\text{nil}}$ defined by

$$\gamma(u) = \sum_{j=1}^{k} z_j \phi_j(u).$$

This polynomial satisfies $\gamma(u_j) = z_j$ for $j = 1, \ldots, k$.

Example 10.3.3. Consider the set $Z = \{(u_i, z_i) : 1 \le i \le 3\} \subset {\mathcal{C}\ell_3}^{\text{nil}} \times {\mathcal{C}\ell_3}^{\text{nil}}$, where

$$\begin{aligned}
(u_1, z_1) &= (\zeta_{\{1\}} + 3\zeta_{\{2\}}, 4\zeta_{\{1,2,3\}}) \\
(u_2, z_2) &= (3 - \zeta_{\{1,3\}}, 2\zeta_{\{1,2\}} - \zeta_{\{2,3\}}) \\
(u_3, z_3) &= (5 + 2\zeta_{\{2\}}, 3 - \zeta_{\{2\}} + \zeta_{\{3\}}).
\end{aligned}$$

The Lagrange basis polynomials are as follows:

$$\begin{aligned}
\phi_1(u) = 1 &+ \frac{8\zeta_{\{1\}}}{15} + \frac{8\zeta_{\{2\}}}{5} + \frac{92}{75}\zeta_{\{1,2\}} + \frac{1}{3}\zeta_{\{1,2,3\}} \\
&+ u\left(-\frac{1}{9}\zeta_{\{1,3\}} - \frac{16}{45}\zeta_{\{1,2,3\}} - \frac{688\zeta_{\{1,2\}}}{1125} - \frac{64\zeta_{\{1\}}}{225} - \frac{58\zeta_{\{2\}}}{75} - \frac{8}{15}\right) \\
&+ u^2\left(\frac{76\zeta_{\{1,2\}}}{1125} + \frac{1}{45}\zeta_{\{1,3\}} + \frac{11}{225}\zeta_{\{1,2,3\}} + \frac{8\zeta_{\{1\}}}{225} + \frac{2\zeta_{\{2\}}}{25} + \frac{1}{15}\right);
\end{aligned}$$

$$\begin{aligned}
\phi_2(u) &= -\frac{5\zeta_{\{2\}}}{2} - \frac{5\zeta_{\{1\}}}{6} - \frac{7}{6}\zeta_{\{1,2\}} + \frac{5}{12}\zeta_{\{1,2,3\}} \\
&\quad + u\left(\frac{5}{9}\zeta_{\{1,2\}} - \frac{5}{36}\zeta_{\{1,3\}} + \frac{5}{9}\zeta_{\{1,2,3\}} + \frac{4\zeta_{\{1\}}}{9} + \frac{5\zeta_{\{2\}}}{6} + \frac{5}{6}\right) \\
&\quad + u^2\left(-\frac{1}{18}\zeta_{\{1,2\}} + \frac{1}{36}\zeta_{\{1,3\}} - \frac{5}{36}\zeta_{\{1,2,3\}} - \frac{\zeta_{\{1\}}}{18} - \frac{1}{6}\right); \\
\phi_3(u) &= \frac{3\zeta_{\{1\}}}{10} + \frac{9\zeta_{\{2\}}}{10} - \frac{3}{50}\zeta_{\{1,2\}} - \frac{3}{4}\zeta_{\{1,2,3\}} \\
&\quad + u\left(\frac{7}{125}\zeta_{\{1,2\}} + \frac{1}{4}\zeta_{\{1,3\}} - \frac{1}{5}\zeta_{\{1,2,3\}} - \frac{1}{25}4\zeta_{\{1\}} - \frac{3\zeta_{\{2\}}}{50} - \frac{3}{10}\right) \\
&\quad + u^2\left(-\frac{3}{250}\zeta_{\{1,2\}} - \frac{1}{20}\zeta_{\{1,3\}} + \frac{9}{100}\zeta_{\{1,2,3\}} + \frac{\zeta_{\{1\}}}{50} - \frac{2\zeta_{\{2\}}}{25} + \frac{1}{10}\right).
\end{aligned}$$

The zeon Lagrange interpolation polynomial $\gamma(u)$ is then given by

$$\begin{aligned}
\gamma(u) &= \frac{3u^2}{10} - \frac{9u}{10} + \left(\frac{3u^2}{50} - \frac{12u}{25} + \frac{9}{10}\right)\zeta_{\{1\}} + \left(-\frac{17u^2}{50} + \frac{3u}{25} + \frac{27}{10}\right)\zeta_{\{2\}} \\
&\quad + \left(\frac{u^2}{10} - \frac{3u}{10}\right)\zeta_{\{3\}} + \left(-\frac{146u^2}{375} + \frac{748u}{375} - \frac{12}{25}\right)\zeta_{\{1,2\}} \\
&\quad + \left(-\frac{13u^2}{100} + \frac{59u}{100} + \frac{3}{10}\right)\zeta_{\{1,3\}} + \left(\frac{13u^2}{150} - \frac{67u}{75} + \frac{9}{10}\right)\zeta_{\{2,3\}} \\
&\quad + \left(\frac{709u^2}{1125} - \frac{15173u}{4500} + \frac{757}{300}\right)\zeta_{\{1,2,3\}}.
\end{aligned}$$

It is left as an exercise for the reader to verify the interpolation.

Exercises

Exercise 10.1: Evaluate the polynomial at the given point of ${\mathcal{C}\ell_3}^{\mathrm{nil}}$.

$$\varphi_1(x) = -5x - 5x^2.$$

$$u_1 = -5\zeta_{\{1,2\}} - 4\zeta_{\{2,3\}} - 2\zeta_{\{1,2,3\}} + 5\zeta_{\{3\}} + 1.$$

Exercise 10.2: Evaluate the polynomial at the given point of ${\mathcal{C}\ell_3}^{\mathrm{nil}}$.

$$\varphi_2(x) = 3x^3 - 5x^2 - x.$$

$$u_2 = \zeta_{\{1,2,3\}} + \zeta_{\{1\}} - \zeta_{\{2\}}.$$

Exercise 10.3: Evaluate the polynomial at the given point of ${\mathcal{C}\ell_3}^{\text{nil}}$.

$$\varphi_3(x) = -4x^6 - 2x^5 + 3x^4 + x^3 + 5x^2.$$

$$u_3 = 4\zeta_{\{2,3\}} + 3\zeta_{\{1,2,3\}} + 2.$$

Exercise 10.4: Evaluate the polynomial at the given point of ${\mathcal{C}\ell_4}^{\text{nil}}$.

$$\varphi_4(x) = 5x^8 + 2x^7 - 2x^6 + x^5 - 3x^3 + x^2 - 5x.$$

$$u_4 = -3\zeta_{\{1,3\}} - \zeta_{\{2,3\}} + 4\zeta_{\{2,4\}} - 2\zeta_{\{1,2,4\}} + 4\zeta_{\{2,3,4\}} + 3\zeta_{\{1,2,3,4\}} - 3\zeta_{\{2\}} - \zeta_{\{3\}}.$$

Exercise 10.5: Evaluate the polynomial at the given point of ${\mathcal{C}\ell_4}^{\text{nil}}$.

$$\varphi_5(x) = 4x - 5x^2.$$

$$u_5 = -\zeta_{\{1,4\}} - \zeta_{\{2,3\}} - \zeta_{\{2,4\}} + \zeta_{\{3,4\}} - 3\zeta_{\{1,3,4\}} - 5\zeta_{\{2,3,4\}} + 4\zeta_{\{1\}}.$$

Exercise 10.6: Solve the zeon quadratic equation $\alpha u^2 + \beta u + \gamma = 0$ for solutions in ${\mathcal{C}\ell_3}^{\text{nil}}$, where α, β, and γ are given as follows:

$$\begin{aligned}
\alpha &= -3 + +\zeta_{\{1\}} + 4\zeta_{\{2\}} + 3\zeta_{\{3\}}4\zeta_{\{1,2\}} + 3\zeta_{\{1,3\}} + 4\zeta_{\{2,3\}} - 4\zeta_{\{1,2,3\}} \\
\beta &= -3 - 4\zeta_{\{1\}} - 4\zeta_{\{2\}} + 3\zeta_{\{3\}} + \zeta_{\{1,2\}} - 2\zeta_{\{1,3\}} + 2\zeta_{\{2,3\}} + 5\zeta_{\{1,2,3\}} \\
\gamma &= -\zeta_{\{1,2\}} - 3\zeta_{\{1,3\}} - 5\zeta_{\{2,3\}} + 4\zeta_{\{1,2,3\}} - 5\zeta_{\{1\}} + \zeta_{\{2\}} - 2\zeta_{\{3\}}.
\end{aligned}$$

Exercise 10.7: Solve the zeon quadratic equation $\alpha u^2 + \beta u + \gamma = 0$ for solutions in ${\mathcal{C}\ell_2}^{\text{nil}}$, where α, β, and γ are given as follows:

$$\begin{aligned}
\alpha &= -3 + 2\zeta_{\{1\}} - 2\zeta_{\{2\}} + 3\zeta_{\{1,2\}} \\
\beta &= -4 + 5\zeta_{\{1\}} - \zeta_{\{1,2\}} \\
\gamma &= 4 - 5\zeta_{\{1\}} + \zeta_{\{2\}} - \zeta_{\{1,2\}}.
\end{aligned}$$

Exercise 10.8: Prove the assertion of (10.5); i.e., prove that the zeon Lagrange basis polynomials have the property that

$$\varphi_j(u_\ell) = \begin{cases} 1 & j = \ell, \\ 0 & j \neq \ell. \end{cases}$$

Exercise 10.9: Construct the zeon Lagrange basis polynomials and interpolation polynomial of the collection

$$Z = \left\{(3\zeta_{\{1,2\}}, 6 - 4\zeta_{\{1,2\}}), (5 - \zeta_{\{1\}}, 3 + 2\zeta_{\{1\}})\right\} \subset {\mathcal{C}\ell_2}^{\text{nil}}.$$

Exercise 10.10: Verify the interpolation of Example 10.3.3 by showing that the zeon polynomial $\gamma(u)$ defined by

$$\begin{aligned}\gamma(u) = {} & \frac{3u^2}{10} - \frac{9u}{10} + \left(\frac{3u^2}{50} - \frac{12u}{25} + \frac{9}{10}\right)\zeta_{\{1\}} + \left(-\frac{17u^2}{50} + \frac{3u}{25} + \frac{27}{10}\right)\zeta_{\{2\}} \\ & + \left(\frac{u^2}{10} - \frac{3u}{10}\right)\zeta_{\{3\}} + \left(-\frac{146u^2}{375} + \frac{748u}{375} - \frac{12}{25}\right)\zeta_{\{1,2\}} \\ & + \left(-\frac{13u^2}{100} + \frac{59u}{100} + \frac{3}{10}\right)\zeta_{\{1,3\}} + \left(\frac{13u^2}{150} - \frac{67u}{75} + \frac{9}{10}\right)\zeta_{\{2,3\}} \\ & + \left(\frac{709u^2}{1125} - \frac{15173u}{4500} + \frac{757}{300}\right)\zeta_{\{1,2,3\}}\end{aligned}$$

satisfies the following equations:

$$\begin{aligned}\gamma(\zeta_{\{1\}} + 3\zeta_{\{2\}}) &= 4\zeta_{\{1,2,3\}}, \\ \gamma(3 - \zeta_{\{1,3\}}) &= 2\zeta_{\{1,2\}} - \zeta_{\{2,3\}}, \\ \gamma(5 + 2\zeta_{\{2\}}) &= 3 - \zeta_{\{2\}} + \zeta_{\{3\}}.\end{aligned}$$

Chapter 11

Norms and Inequalities in Zeon Algebras

In this chapter, based on [65], norms are defined on ${\mathcal{C}\ell_n}^{\text{nil}}$, multiplicative inequalities are established for those norms, and the norm inequalities are applied to establish a convergence theorem for geometric series in zeon algebras.

The zeon p-norms and infinity norms are introduced in Section 11.1. Of the norms considered, the infinity norm is the most straightforward. Inequalities involving the infinity norm, which is not sub-multiplicative, are developed in Section 11.1. In Section 11.1, the 1-norm is shown to be the only sub-multiplicative p-norm on ${\mathcal{C}\ell_n}^{\text{nil}}$. The inner product norm ($p = 2$) is considered in some detail in Section 11.1, where multiplicative inequalities are again established.

In Section 11.2, equivalence of norms in ${\mathcal{C}\ell_n}^{\text{nil}}$ is discussed and used to establish a number of multiplicative inequalities between p-norms and the infinity norm. Tight upper bounds on p-norms of products are established with respect to products of infinity norms.

The zeon geometric series is considered in Section 11.4. After defining the spectral seminorm on ${\mathcal{C}\ell_n}^{\text{nil}}$, necessary and sufficient conditions for convergence are established, and the series limit is expressed as a finite sum.

For convenience, the following combinatorial identities are recalled from Chapter 2. The *Stirling numbers of the second kind*, written $\left\{ {n \atop k} \right\}$, count the number of ways a set of n labelled objects can be partitioned into k nonempty subsets. Observing that the null-square properties of zeon generators guarantees that the product $\zeta_I \zeta_J$ is nonzero only if $I \cap J = \varnothing$ (that is, $\{I|J\}$ is a partition of $I \cup J$), it should not be surprising that Stirling numbers of the second kind often appear when working with zeons.

By definition, $\left\{ {n \atop 0} \right\} = 0$ for all positive integers n. Similarly, $\left\{ {n \atop k} \right\}$ is defined to be zero whenever $k > n$. Recall the following closed formula for $\left\{ {m \atop k} \right\}$, presented in Chapter 2 as (2.1).:

$$\left\{ {m \atop k} \right\} = \frac{1}{k!}\sum_{j=0}^{k}(-1)^j\binom{k}{j}(k-j)^m.$$

Recall the following result of Lemma 2.1.2: for $n, m \in \mathbb{N}$,

$$\sum_{k=1}^{n}\frac{m!}{(m-k)!}\left\{ {n \atop k} \right\} = m^n.$$

Further, letting $2^{[n]m}$ denote the m-fold Cartesian product of the power set of $[n]$, let $I \in 2^{[n]}$ be a fixed multi-index. Lemma 2.1.3 reveals that the number of independent ordered m-tuples whose union is I is given by

$$\left|\left\{(I_1|\cdots|I_m) \in 2^{[n]m} : \bigcup_{j=1}^{m} I_j = I\right\}\right| = m^{|I|}.$$

These combinatorial results will be useful for obtaining upper bounds on multiplicative norm inequalities.

11.1 Zeon Norms

A *norm* on ${\mathcal{C}\ell_n}^{\text{nil}}$ is a mapping $\|\cdot\| : {\mathcal{C}\ell_n}^{\text{nil}} \to \mathbb{R}$ satisfying the following properties for $u, w \in {\mathcal{C}\ell_n}^{\text{nil}}$ and $a \in \mathbb{R}$:

$$\|u\| = 0 \Leftrightarrow u = 0; \tag{11.1}$$
$$\|u\| \geq 0; \tag{11.2}$$
$$\|au\| = |a|\|u\|; \tag{11.3}$$
$$\|u + w\| \leq \|u\| + \|w\|. \tag{11.4}$$

A mapping satisfying (11.2) through (11.4), is said to be a *seminorm*. A norm satisfying $\|uw\| \leq \|u\|\|w\|$ for all $u, w \in {\mathcal{C}\ell_n}^{\text{nil}}$ is said to be *submultiplicative*.

A number of norms can be defined on ${\mathcal{C}\ell_n}^{\text{nil}}$ as natural extensions of norms on the 2^n-dimensional vector space $\text{span}(\{\zeta_I : I \in 2^{[n]}\})$. Norms of particular interest in the current work are p-norms and the infinity norm.

The Zeon Infinity Norm

Following the standard definition of the infinity norm in finite-dimensional vector spaces, the zeon infinity norm is defined as follows: given $u = \sum_{I\in 2^{[n]}} u_I\zeta_I$,

$$\|u\|_\infty = \max\{|u_I| : I \in 2^{[n]}\}.$$

The first goal is to establish upper bounds on coefficients appearing in products of zeon elements.

Lemma 11.1.1. *Let $u_1, \ldots, u_m \in {\mathcal{C}\ell_n}^{\mathrm{nil}}$. Expanding $\prod_{\ell=1}^m u_\ell = \sum_{I\in 2^{[n]}} c_I\zeta_I$, let $J \in 2^{[n]}$ be an arbitrary nonempty multi-index. Then,*

$$|c_J| \le \sum_{k=1}^{\min\{|J|,m\}} \frac{m!}{(m-k)!} \left\{ {|J| \atop k} \right\} \prod_{\ell=1}^m \|u_\ell\|_\infty.$$

Proof. For general products of invertible zeons, consider the number of k-block partitions and the positions in which the $m-k$ scalars can appear. Choosing k positions for blocks, the remaining $m-k$ will be occupied by scalars. □

When the factors are all nilpotent, the following special case is easily established.

Corollary 11.1.2. *Let $u_1, \ldots, u_m \in {\mathcal{C}\ell_n}^{\mathrm{nil}_\circ}$. Expanding $\prod_{\ell=1}^m u_\ell = \sum_{I\in 2^{[n]}} c_I\zeta_I$, let $J \in 2^{[n]}$ be an arbitrary nonempty multi-index. If $m > |J|$, then $c_J = 0$. More generally,*

$$|c_J| \le m! \left\{ {|J| \atop m} \right\} \prod_{\ell=1}^m \|u_\ell\|_\infty.$$

Observing that each ℓ-subset of $[k]$ simultaneously determines a unique $(k-\ell)$-subset of $[k]$, the 2-partitions of $[k]$ are counted by summing (essentially) half of the binomial coefficients. It is therefore not surprising that when $k = 2$ and $m \geq 2$, the closed formula (2.1) yields the following:

$$\begin{aligned} \left\{ {m \atop 2} \right\} &= \frac{1}{2}\sum_{j=0}^{2}(-1)^j\binom{2}{j}(2-j)^m \\ &= \frac{1}{2}\left(2^m - 2\right) \\ &= 2^{m-1} - 1. \end{aligned} \tag{11.5}$$

Corollary 11.1.3. *Let $u, v \in \mathcal{C}\ell_n{}^{\text{nil}}$ and write $uv = \sum_{I\in 2^{[n]}} c_I \zeta_I$. Then, for given multi-index $I \in 2^{[n]}$, the infinity norm of the product uv satisfies*

$$\|uv\|_\infty \leq 2^n \|u\|_\infty \|v\|_\infty.$$

Proof. Write $uv = \sum_{I\in 2^{[n]}} c_I \zeta_I$. Noting that as a function of $\kappa \geq m$, $\varsigma(\kappa) = \left\{ {\kappa \atop m} \right\}$ is strictly increasing, it is sufficient to consider the coefficient $c_{[n]}$. Applying Lemma 11.1.1 directly,

$$\begin{aligned} |c_{[n]}| &\leq 2 + 2\left\{ {n \atop 2} \right\} \|u\|_\infty \|v\|_\infty \\ &= (2 + 2^n - 2)\|u\|_\infty \|v\|_\infty \\ &= 2^n \|u\|_\infty \|v\|_\infty. \end{aligned}$$

□

The extension to products of more than two zeon elements can be naturally established by induction.

Lemma 11.1.4. *For $m \geq 2$, let $u_1, u_2, \ldots, u_m \in \mathcal{C}\ell_n{}^{\text{nil}}$. Then,*

$$\left\| \prod_{\ell=1}^{m} u_\ell \right\|_\infty \leq 2^{n(m-1)} \prod_{\ell=1}^{m} \|u_\ell\|_\infty.$$

Alternatively, Lemma 11.1.1 leads to a tighter bound on the infinity norm of general products of nilpotent elements than Lemma 11.1.4, as seen in the next lemma.

Lemma 11.1.5. *Let $u_1, u_2, \ldots, u_m \in \mathcal{C}\ell_n{}^{\text{nil}_\circ}$. Then,*

$$\|u_1 u_2 \cdots u_m\|_\infty \leq m! \left\{ {n \atop m} \right\} \prod_{\ell=1}^{m} \|u_\ell\|_\infty.$$

Proof. Fixing m, the result follows from Corollary 11.1.2 by noting that as a function of $\kappa \geq m$, $\varsigma(\kappa) = \left\{ {\kappa \atop m} \right\}$ is strictly increasing. □

The most general product inequality for the infinity norm is given in Proposition 11.1.6. A tighter inequality can be obtained for products when the number of nilpotent factors is known.

Proposition 11.1.6. *Let* $u_1, u_2, \ldots, u_m \in \mathcal{C}\ell_n{}^{\text{nil}}$. *Then,*

$$\|u_1 u_2 \cdots u_m\|_\infty \le m^n \|u_1\|_\infty \cdots \|u_m\|_\infty. \tag{11.6}$$

Moreover, if $\exists \kappa > 0$ *such that* $\|u_1 u_2 \cdots u_m\|_\infty \le \kappa \|u_1\|_\infty \cdots \|u_m\|_\infty$ *for every choice of* $u_1, \ldots, u_m$, *then* $\kappa \ge m^n$.

Proof. For fixed m, the function $\varsigma(\kappa) = \left\{ {\kappa \atop m} \right\}$ is strictly increasing for $\kappa \ge m$. Applying Lemma 11.1.1, it follows that

$$\|u_1 u_2 \cdots u_m\|_\infty \le \sum_{k=0}^{n} \frac{m!}{(m-k)!} \left\{ {n \atop k} \right\} \|u_1\|_\infty \cdots \|u_m\|_\infty.$$

The inequality (11.6) now follows from Lemma 2.1.2. To see that this upper bound is tight, let $u = \sum_{I \in 2^{[n]}} \zeta_I$ and note that $\|u\|_\infty = 1$. By the multiplicative properties of zeons (Lemma 9.1.1), it follows that

$$\begin{aligned}\|u^m\|_\infty &\le |\langle u^m, \zeta_{[n]} \rangle| \\ &= \left| \sum_{k=0}^{n} \frac{m!}{(m-k)!} \left\{ {n \atop k} \right\} \right| \\ &= m^n \|u\|_\infty{}^m.\end{aligned}$$

□

Given a collection $\{u_1, \ldots, u_m\}$ in which the number of invertible elements is known, a tighter upper bound can be obtained.

Theorem 11.1.7. *Given* $u_1, u_2, \ldots, u_m \in \mathcal{C}\ell_n{}^{\text{nil}}$, *writing* $\prod_{\ell=1}^{m} u_\ell = \sum_{I \in 2^{[n]}} c_I \zeta_I$, *and assuming the number of invertible elements in the list is* $v \le m$, *the following inequality holds:*

$$|c_I| \le \sum_{j=0}^{v} \binom{v}{j} (m-j)! \left\{ {|I| \atop m-j} \right\} \|u_1\|_\infty \cdots \|u_m\|_\infty. \tag{11.7}$$

Proof. Again, Lemma 9.1.1 provides the starting point for this proof.

$$|c_I| = \sum_{\substack{(I_1|\cdots|I_m) \\ I_1 \cup \cdots \cup I_m = I}} \prod_{j=1}^{m} a_{j,I_j} \le \sum_{\substack{(I_1|\cdots|I_m) \\ I_1 \cup \cdots \cup I_m = I}} \prod_{j=1}^{m} \|u_j\|_\infty \le \prod_{j=1}^{m} \|u_j\|_\infty \sum_{\substack{(I_1|\cdots|I_m) \\ I_1 \cup \cdots \cup I_m = I}} 1.$$

With v invertible factors in the product, it is possible (but not necessary) to have v empty subsets in the list $(I_1, \ldots, I_m)$. First, consider the case where j of the m-tuples are empty subsets. In this case, $|I|$ must be partitioned among the subsets that are nonempty, namely $m-j$ of them. By definition, this is $\left\{ {|I| \atop m-j} \right\}$. All the permutations of the nonempty subsets must be accounted for, ignoring the empty subsets. The permutations of the nonempty sets are accounted for by multiplying by $(m-j)!$. The j empty subsets are chosen from the v available possible spaces using the combination $\binom{v}{j}$. Hence, when using j empty subsets, the greatest number of terms in the summation can be given by

$$\binom{v}{j}(m-j)!\left\{ {|I| \atop m-j} \right\}.$$

Summing over j completes the proof. □

With Theorem 11.1.7 in hand, a tight multiplicative inequality can now be established for the infinity norm.

Theorem 11.1.8. *Given $u_1, u_2, \ldots, u_m \in {\mathcal{C}\ell_n}^{\mathrm{nil}}$, and assuming the number of invertible elements in the list is $k \leq m$, the following inequality holds:*

$$\|u_1 \cdots u_m\|_\infty \leq \beta_n(m,k)\|u_1\|_\infty \cdots \|u_m\|_\infty,$$

where the numbers $\beta_n(m,k)$ satisfy the following recurrence relations:

$$\beta_n(m,0) = m!\left\{ {n \atop m} \right\}, \quad \beta_n(m,m) = m!\left\{ {n \atop m} \right\},$$

$$\beta_n(m,k) = \beta_n(m,k-1) + \beta_n(m-1,k-1), \ \ (0 < k < m).$$

Moreover, this upper bound is tight in the sense that if $\exists \lambda > 0$ such that $\|u_1 u_2 \cdots u_m\|_\infty \leq \lambda \|u_1\|_\infty \cdots \|u_m\|_\infty$ for every choice of $u_1, \ldots, u_m$ having a fixed k-subset of invertible elements, then $\lambda \geq \beta_n(m,k)$.

Proof. First, note that $\left\{ {|I| \atop m-j} \right\} \leq \left\{ {n \atop m-j} \right\}$ for all $I \subseteq [n]$. Setting

$$\beta_n(m,k) = \sum_{j=0}^{k} \binom{k}{j}(m-j)!\left\{ {n \atop m-j} \right\},$$

it is sufficient to show that the recurrence is satisfied. When $k=0$, the sum is easily reduced to the single term $\beta_n(m,0) = m!\left\{ {n \atop m} \right\}$, in accordance with Corollary 11.1.2. When $k=m$, the result follows from arguments used in Lemma 2.1.3 and is equivalent to Proposition 11.1.6.

For $k = 1, \ldots, m-1$, recall that $\beta_n(m,k)$ represents the number of ways that the n-set can be partitioned among m "boxes" while leaving at most a (fixed) k-subset of the boxes empty. Beginning with a collection of $m-1$ boxes, consider adding an mth box. This box is either allowed to be empty (if the mth factor it represents is invertible) or forbidden from being empty (mth factor is nilpotent). If the mth box is allowed to be empty, the number of valid assignments allowing up to k empty boxes is equal to the number of assignments allowing up to $k-1$ empty boxes among the previous $m-1$ boxes; i.e., $\beta_n(m-1,k-1)$. If the box is forbidden from being empty, then at most $k-1$ of the m boxes are allowed to be empty; that is, $\beta_n(m,k-1)$. Since emptiness being allowed or being forbidden are mutually exclusive, the number $\beta_n(m,k)$ of valid assignments is given by the sum $\beta_n(m,k-1) + \beta_n(m-1,k-1)$.

As in the proof of Proposition 11.1.6, tightness of the upper bound is established by construction. Letting $a, b > 0$, setting $u = a\sum_I \zeta_I$, and setting $v = b\sum_{I\neq\varnothing} \zeta_I$, direct computation shows that

$$\begin{aligned}\|u^k v^{m-k}\|_\infty &= \left|\langle u^k v^{m-k}, \zeta_{[n]}\rangle\right| \\ &= \beta_n(m,k) a^k b^{m-k} \\ &= \beta_n(m,k)(\|u\|_\infty)^k (\|v\|_\infty)^{m-k}.\end{aligned}$$

□

Zeon p-Norms

Given $u = \sum_{I\in 2^{[n]}} u_I \zeta_I \in \mathcal{C}\ell_n{}^{\text{nil}}$ and $p \geq 1$, the *p-norm of u* is defined by

$$\|u\|_p = \left(\sum_{I\in 2^{[n]}} |u_I|^p\right)^{1/p}.$$

The first result on p-norms may be somewhat surprising. Unlike the infinity norm, the 1-norm is sub-multiplicative on $\mathcal{C}\ell_n{}^{\text{nil}}$.

Lemma 11.1.9. *Let $u, v \in \mathcal{C}\ell_n{}^{\text{nil}}$ be arbitrary. Then,*

$$\|uv\|_1 \leq \|u\|_1 \|v\|_1.$$

In other words, the 1-norm is sub-multiplicative on $\mathcal{C}\ell_n{}^{\text{nil}}$.

Proof. Let $u, v \in \mathcal{C}\ell_n{}^{\text{nil}}$ be arbitrary, and write $u = \sum_I u_I\zeta_I$, $v = \sum_I v_I\zeta_I$. Then, by Lemma 9.1.1,

$$\begin{aligned}
\|uv\|_1 &= \sum_{I\in 2^{[n]}} \left| \sum_{K\subseteq I} u_K v_{I\setminus K} \right| \\
&\le \sum_{I\in 2^{[n]}} \sum_{K\subseteq I} |u_K||v_{I\setminus K}| \\
&\le \sum_{I\in 2^{[n]}} \sum_{K\in 2^{[n]}} |u_K||v_{I\setminus K}| \\
&\le \sum_{I\in 2^{[n]}} \sum_{K\in 2^{[n]}} |u_K||v_I| \\
&= \|u\|_1\|v\|_1.
\end{aligned}$$

□

Example 11.1.10. Setting $u = 3 - 2\zeta_{\{1,2\}} - 2\zeta_{\{1,3\}} - 2\zeta_{\{2,3\}} + 3\zeta_{\{1,2,3\}} + 4\zeta_{\{2\}} - \zeta_{\{3\}}$ and $v = 4 + 2\zeta_{\{1,2\}} + 3\zeta_{\{1,3\}} + 5\zeta_{\{2,3\}} - \zeta_{\{1\}} + 3\zeta_{\{2\}} - 3\zeta_{\{3\}}$ in $\mathcal{C}\ell_n{}^{\text{nil}}$ for any $n \ge 3$, we find that

$$uv = 12 - 6\zeta_{\{1,2\}} + 2\zeta_{\{1,3\}} - 8\zeta_{\{2,3\}} + 24\zeta_{\{1,2,3\}} - 3\zeta_{\{1\}} + 25\zeta_{\{2\}} - 13\zeta_{\{3\}}.$$

Thus, $\|uv\|_1 = 93$, while $\|u\|_1\|v\|_1 = (17)(21) = 357$.

On the other hand, it should not be surprising that the zeon p-norms are not generally sub-multiplicative. Three examples motivate the next result.

Example 11.1.11. Let $u = 1 + \zeta_{\{1\}} + \zeta_{\{2\}} + \zeta_{\{3\}} \in \mathcal{C}\ell_n{}^{\text{nil}}$ for any $n \ge 3$. Let $p = 1.4$. Then, $\|u\|_p = 2.6918$, $(\|u\|_p)^2 = 7.24579$ and $\|u^2\|_p = 7.51363$. Hence, $\|u^2\|_p \ge (\|u\|_p)^2$, showing that the p-norm is not sub-multiplicative for $p = 1.4$.

Example 11.1.12. Let $u = 1 + \zeta_{\{2\}} + \zeta_{\{3\}} + \zeta_{\{4\}} + \zeta_{\{5\}} + \zeta_{\{6\}} \in \mathcal{C}\ell_n{}^{\text{nil}}$ for any $n \ge 6$, and let $p = 1.25$. Then, $\|u\|_p = 4.19296$, $(\|u\|_p)^2 = 17.5809$, and $\|u^2\|_p = 17.8446$. Hence, $\|u^2\|_p \ge (\|u\|_p)^2$, showing that the p-norm is not sub-multiplicative for $p = 1.25$.

Example 11.1.13. Let $u = 1000000 + \zeta_{\{1\}}$, $v = 5000 + \zeta_{\{1\}}$ in $\mathcal{C}\ell_n{}^{\text{nil}}$ for any $n \ge 1$. Let $p = 1.0001$. Then, $\|u\|_p = 1. \times 10^6$, $\|v\|_p = 5001$, and $\|uv\|_p = 5001 \times 10^9$. More specifically, $\|uv\|_p - \|u\|_p\|v\|_p = 2.14904$. This shows that the p-norm is not sub-multiplicative for $p = 1.0001$.

Based on the examples above, we conjecture that for $p > 1$, the p-norm is *not* sub-multiplicative on ${\mathcal{C}\ell_n}^{\rm nil}$. The next lemma verifies that the 1-norm is indeed the exceptional case.

Lemma 11.1.14. *For $p > 1$, the p-norm on ${\mathcal{C}\ell_n}^{\rm nil}$ $(n \geq 1)$ is* not *sub-multiplicative.*

Proof. The result is established by construction. To that end, set $z = a + \zeta_{\{1\}} \in {\mathcal{C}\ell_n}^{\rm nil}$ for $a \in \mathbb{R}$, and observe that $\|z\|_p = (|a|^p + 1)^{1/p}$. Also note that $\|z^2\|_p = (|a^2|^p + |2a|^p)^{1/p}$. Consider the difference ${\|z^2\|_p}^p - ({\|z\|_p}^p)^2$:

$$\begin{aligned}
{\|z^2\|_p}^p - ({\|z\|_p}^2)^p &= {\|z^2\|_p}^p - ({\|z\|_p}^p)^2 \\
&= |a|^{2p} + 2^p|a|^p - (|a|^p + 1)^2 \\
&= |a|^{2p} + 2^p|a|^p - (|a|^{2p} + 2|a|^p + 1) \\
&= 2^p|a|^p - 2|a|^p - 1 \\
&= (2^p - 2)|a|^p - 1.
\end{aligned}$$

Choosing $a > \left(\dfrac{1}{2^p - 2}\right)^{1/p}$, the difference is positive, which implies $\|z^2\|_p > (\|z\|_p)^2$. Hence, the result. □

Formulating multiplicative inequalities for p-norms is more daunting than was the case for the infinity norm. Generally, if $z_1, \ldots, z_m \in {\mathcal{C}\ell_n}^{\rm nil}$ with $z_j = \sum_I a_{j,I}\zeta_I$ for $1 \leq j \leq m$, then

$$(\|z_1 \cdots z_m\|_p)^p = \sum_{I \in 2^{[n]}} \left| \sum_{\substack{(I_1|\cdots|I_m) \\ I_1 \cup \cdots \cup I_m = I}} \prod_{j=1}^m a_{j,I_j} \right|^p .$$

Establishing general multiplicative inequalities for other p-norms will be facilitated by utilizing norm equivalence in Section 11.2.

The special case $p = 2$ provides some additional tools that simplify the process. Multiplicative inequalities for the inner product norm will be established directly in Section 11.1 below.

Zeon Inner Product Norm

Knowing that the inner product norm is not sub-multiplicative, an inequality is now sought. The first step is recalling the Cauchy-Schwarz inequality for the inner product on ${\mathcal{C}\ell_n}^{\rm nil}$.

Lemma 11.1.15. *Let* $u, v \in \mathcal{C}\ell_n{}^{\text{nil}}$. *Then,*

$$|\langle u, v\rangle| \leq \|u\|_2 \|v\|_2.$$

Proof. Given arbitrary elements $u, v \in \mathcal{C}\ell_n{}^{\text{nil}}$ and $a \in \mathbb{R}$, it is clear that

$$\begin{aligned}\|u + av\|_2{}^2 &= \langle u + av, u + av\rangle \\ &= \langle u, u\rangle + \langle av, av\rangle + 2\langle u, av\rangle \\ &= \|u\|_2{}^2 + a^2\|v\|_2{}^2 + 2a\langle u, v\rangle.\end{aligned}$$

Similarly,

$$(\|u\|_2 + \|av\|_2)^2 = \|u\|_2{}^2 + a^2\|v\|_2{}^2 + 2|a|\|u\|\|v\|.$$

Applying the triangle inequality, the result is then obtained by subtraction:

$$\begin{aligned}0 &\leq (\|u\|_2 + \|av\|_2)^2 - \|u + av\|_2{}^2 \\ &= 2|a|\|u\|\|v\| - 2a\langle u, v\rangle.\end{aligned}$$

Thus $\frac{a}{|a|}\langle u, v\rangle \leq \|u\|_2\|v\|_2$ for all nonzero $a \in \mathbb{R}$; hence, the result. □

With the Cauchy-Schwarz inequality in hand, we can now show that the inner product norm satisfies the following inequality.

Lemma 11.1.16. *Let* $u, v \in \mathcal{C}\ell_n{}^{\text{nil}}$. *Then,*

$$\|uv\|_2 \leq 2^{n/2}\|u\|_2\|v\|_2.$$

Proof. Write $u = \sum_I u_I\zeta_I$ and $v = \sum_I v_I\zeta_I$. Applying Lemmas 9.1.1 and 11.1.15, we obtain

$$\begin{aligned}\|uv\|_2{}^2 &= \langle uv, uv\rangle \\ &= \sum_{I\in 2^{[n]}} \sum_{K\subseteq I} u_K v_{I\setminus K} \sum_{J\subseteq I} u_J v_{I\setminus J} \\ &\leq \sum_{I\in 2^{[n]}} \sum_{K\subseteq I} |u_K v_{I\setminus K}| \sum_{J\subseteq I} |u_J v_{I\setminus J}| \\ &< \sum_{I\in 2^{[n]}} |\langle u, v\rangle|^2 \\ &= 2^n|\langle u, v\rangle|^2 \\ &\leq 2^n(\|u\|_2\|v\|_2)^2.\end{aligned}$$

□

The following inductive extension of Lemma 11.1.16 is a natural corollary.

Corollary 11.1.17. *Let* $u_1, \ldots, u_m \in \mathcal{C}\ell_n{}^{\text{nil}}$. *Then,*

$$\left\| \prod_{j=1}^{m} u_j \right\|_2 \leq 2^{n(m-1)/2} \prod_{j=1}^{m} \|u_j\|_2.$$

11.2 Norm Equivalence and Multiplicative Inequalities

Given a vector space X, recall that two norms $\|\cdot\|_\dagger$ and $\|\cdot\|_\ddagger$ are said to be *equivalent* if there exist positive real numbers c and d such that

$$c\|x\|_\dagger \leq \|x\|_\ddagger \leq d\|x\|_\dagger$$

for all $x \in X$. A well-known result of linear algebra states that all norms are equivalent when X is finite-dimensional. The goal of the current section is to establish some specific multiplicative inequalities based on norm equivalence.

Lemma 11.2.1. *Given* $u \in \mathcal{C}\ell_n{}^{\text{nil}}$ *and* $p \geq 1$,

$$\|u\|_\infty \leq \|u\|_p \leq 2^{n/p}\|u\|_\infty.$$

In particular, the zeon infinity norm is equivalent to the zeon p-norm for all $p \geq 1$.

Proof. Given $u \in \mathcal{C}\ell_n{}^{\text{nil}}$ and $p \geq 1$, it is not difficult to see that

$$\|u\|_\infty \leq \|u\|_p.$$

Further,

$$\|u\|_p{}^p = \sum_{I \in 2^{[n]}} |u_I|^p \leq \sum_{I \in 2^{[n]}} (\|u\|_\infty)^p = 2^n(\|u\|_\infty)^p,$$

so that $\|u\|_p \leq 2^{n/p}\|u\|_\infty$. □

The inductive extension is easily obtained as seen in Corollary 11.2.2. However, a tighter bound can be achieved as in Proposition 11.2.3.

Corollary 11.2.2. *Let* $p \geq 1$, *and let* $u_1, \ldots, u_m \in \mathcal{C}\ell_n{}^{\text{nil}}$ *be arbitrary. Then,*

$$\|u_1 \cdots u_m\|_p \leq (2^{1/p}m)^n \|u_1\|_\infty \cdots \|u_m\|_\infty.$$

Proof. By Lemma 11.2.1 and Corollary 11.2.4, one obtains

$$\begin{aligned}\|u_1 \cdots u_m\|_p &\leq 2^{n/p}\|u_1 \cdots u_m\|_\infty \\ &\leq 2^{n/p} \sum_{k=0}^{n} \frac{m!}{(m-k)!} \left\{ {n \atop k} \right\} \|u_1\|_\infty \cdots \|u_m\|_\infty.\end{aligned}$$

Applying the identity of Lemma 2.1.2 completes the proof. □

Proposition 11.2.3. *Let $p \geq 1$, and let $u_1, \ldots, u_m \in \mathcal{C}\ell_n{}^{\text{nil}}$ be arbitrary. Then,*

$$\|u_1 \cdots u_m\|_p \leq (1 + m^p)^{n/p} \|u_1\|_\infty \cdots \|u_m\|_\infty.$$

Moreover, if $\exists \kappa > 0$ such that $\|u_1 u_2 \cdots u_m\|_p \leq \kappa \|u_1\|_\infty \cdots \|u_m\|_\infty$ for every choice of $u_1, \ldots, u_m$, then $\kappa \geq (1 + m^p)^{n/p}$.

Proof. For each $j = 1, \ldots, m$, write $u_j = \sum_I u_{j\,I} \zeta_I$. Applying Lemmas 9.1.1 and 2.1.3, we find that

$$\begin{aligned}\left\| \prod_{j=1}^{m} u_j \right\|_p^p &= \sum_{I \in 2^{[n]}} |\langle u_1 \cdots u_j, \zeta_I \rangle|^p \\ &= \sum_{I \in 2^{[n]}} \left| \sum_{\substack{(I_1, \ldots, I_m) \\ I_1 \cup \cdots \cup I_m = I}} \prod_{j=1}^{m} u_{j\,I_j} \right|^p \\ &\leq \sum_{I \in 2^{[n]}} \left| \sum_{\substack{(I_1, \ldots, I_m) \\ I_1 \cup \cdots \cup I_m = I}} \|u_1\|_\infty \cdots \|u_m\|_\infty \right|^p \\ &= (\|u_1\|_\infty \cdots \|u_m\|_\infty)^p \sum_{I \in 2^{[n]}} m^{|I|p} \\ &= (\|u_1\|_\infty \cdots \|u_m\|_\infty)^p \sum_{k=0}^{n} \binom{n}{k} m^{pk} \\ &= (\|u_1\|_\infty \cdots \|u_m\|_\infty)^p (1 + m^p)^n.\end{aligned}$$

ng pth roots establishes the upper bound. Tightness of the upper d is shown by construction. To see that this upper bound is tight, $= a \sum_I \zeta_I$ for arbitrary $a > 0$, and note that $\|w\|_\infty = a$. By the multiplicative properties of zeons (Lemma 9.1.1), it follows that

$$\begin{aligned}
\|w^m\|_p{}^p &= \sum_I \left| \sum_{\substack{(I_1,\ldots,I_m)\\ I_1\cup\cdots\cup I_m=I}} \prod_{j=1}^m w_{I_j} \right|^p \\
&= \sum_I \left| \sum_{\substack{(I_1,\ldots,I_m)\\ I_1\cup\cdots\cup I_m=I}} a^m \right|^p \\
&= \sum_I \left| m^{|I|} a^m \right|^p \\
&= \sum_{k=0}^n \binom{n}{k} \left| m^k a^m \right|^p \\
&= a^{mp} \sum_{k=0}^n \binom{n}{k} m^{kp} \\
&= (\|w\|_\infty)^{mp} (1+m^p)^n.
\end{aligned}$$

Hence, there exists $w \in \mathcal{C}\ell_n{}^{\mathrm{nil}}$ such that $\|w^m\|_p = (1+m^p)^{p/n}\|w\|_\infty{}^m$. □

Since every p-norm is equivalent to the infinity norm, Lemma 11.2.1 shows that all zeon p-norms are equivalent. This is made more explicit in the following corollary.

Corollary 11.2.4. *Let $p, q \geq 1$ be distinct real numbers, and let $u \in \mathcal{C}\ell_n{}^{\mathrm{nil}}$ be arbitrary. Then,*

$$2^{-n/q}\|u\|_q \leq \|u\|_p \leq 2^{n/p}\|u\|_q.$$

Hence, all zeon p-norms are equivalent.

Proof. By Lemma 11.2.1, $\|u\|_\infty \leq \|u\|_p$ and $2^{-n/p}\|u\|_p \leq \|u\|_\infty$. Hence,

$$2^{-n/p}\|u\|_p \leq \|u\|_\infty \leq \|u\|_p.$$

Similarly,

$$2^{-n/q}\|u\|_q \leq \|u\|_\infty \leq \|u\|_q.$$

Combining these and again appealing to Lemma 11.2.1,

$$2^{-n/q}\|u\|_q \leq \|u\|_p \leq 2^{n/p}\|u\|_q.$$

□

11.3 Sequences in Zeon Algebras

As in the case of real sequences, a *zeon sequence* is a function $\psi : \mathbb{N} \to {\mathcal{C}\ell_n}^{\text{nil}}$, commonly denoted by (ξ_k), such that for each $k = 1, 2, \ldots$, $\xi_k = \psi(k)$. Fixing a norm on ${\mathcal{C}\ell_n}^{\text{nil}}$, convergence is defined in the familiar way.

Definition 11.3.1. A zeon sequence (ξ_k) is said to *converge* to $\xi \in {\mathcal{C}\ell_n}^{\text{nil}}$ if for each $\varepsilon > 0$, there exists $n^* \in \mathbb{N}$ such that $\forall k \geq n^*$,

$$\|\xi_k - \xi\| \leq \varepsilon.$$

When (ξ_k) converges to ξ, we say that ξ is the *limit* of the sequence (ξ_k) and write $\xi = \lim_{k\to\infty} \xi_k$.

Definition 11.3.2. A zeon sequence (ξ_k) is said to be *Cauchy* if, for each $\varepsilon > 0$, there exists $k^* \in \mathbb{N}$ such that $\|\xi_\ell - \xi_m\| < \varepsilon$ whenever $\ell, m \geq k^*$.

Proposition 11.3.3. *Let (ξ_k), (η_k) be convergent zeon sequences having limits ξ and η, respectively. Then,*

i. the sequence $(\xi_k + \eta_k)$ converges to $\xi + \eta$, and
ii. the sequence $(\xi_k \eta_k)$ converges to $\xi\eta$.

Proof. Proof of *i.*: Let $\varepsilon > 0$ be given. By convergence of (ξ_k), there exists $k_1 \in \mathbb{N}$ such that $\forall k \geq k_1$, $\|\xi_k - \xi\| \leq \varepsilon/2$. Similarly, $\exists k_2 \in \mathbb{N}$ such that $\forall k \geq k_2$, $\|\eta_k - \eta\| < \varepsilon/2$. Letting $k^* = \max\{k_1, k_2\}$, it follows that for all $k \geq k^*$, the following inequality holds:

$$\begin{aligned}\|(\xi_k + \eta_k) - (\xi + \eta)\| &= \|\xi_k - \xi + \eta_k - \eta\| \\ &\leq \|\xi_k - \xi\| + \|\eta_k - \eta\| \\ &< \frac{\varepsilon}{2} + \frac{\varepsilon}{2} = \varepsilon.\end{aligned}$$

The proof of the second part is left as an exercise. □

Proposition 11.3.4. *Let $(\xi_k) \subset {\mathcal{C}\ell_n}^{\text{nil}\star}$ be a convergent zeon sequence of invertible elements having invertible limit ξ. Then,*

$$\lim_{k\to\infty} {\xi_k}^{-1} = \xi^{-1}.$$

Proof. The proof is left as an exercise. □

Given the importance of monotonic sequences in real analysis, the zeon analogue follows naturally from the vector space isomorphism $\mathcal{C}\ell_n{}^{\text{nil}} \cong \mathbb{R}^{2^n}$.

Definition 11.3.5. Let (ξ_k) be a sequence in $\mathcal{C}\ell_n{}^{\text{nil}}$, and let I be a fixed multi-index. If $k_1 \leq k_2$ implies

$$\langle \xi_{k_1}, \zeta_I \rangle - \langle \xi_{k_2}, \zeta_I \rangle \geq 0,$$

then (ξ_k) is said to be *monotonically decreasing* in ζ_I. If $k_1 \leq k_2$ implies

$$\langle \xi_{k_1}, \zeta_I \rangle - \langle \xi_{k_2}, \zeta_I \rangle \leq 0,$$

then (ξ_k) is said to be *monotonically increasing in* ζ_I. In either case, (ξ_k) is said to be *monotonic* in ζ_I. More generally, the sequence (ξ_k) is said to be *monotonic* if it is monotonic in ζ_I for each multi-index I.

In the usual way, a zeon sequence (ξ_k) is said to be *bounded above* if there exists a real number β such that $\langle \xi_k, \zeta_I \rangle \leq \beta$ for all $k \in \mathbb{N}$ and for all multi-indices I. Similarly, (ξ_k) is said to be *bounded below* if there exists a real number α such that $\langle \xi_k, \zeta_I \rangle \geq \alpha$ for all $k \in \mathbb{N}$ and for all multi-indices I.

A zeon sequence is *bounded* if there exist $\alpha, \beta \in \mathbb{R}$ such that $\alpha \leq \langle \xi_k, \zeta_I \rangle \leq \beta$ for all $k \in \mathbb{N}$ and all $I \in 2^{[n]}$. Equivalently, there exists $M > 0$ such that $\|\xi_k\| \leq M$ for all k.

Theorem 11.3.6 (Bounded Convergence Theorem). *A monotonic zeon sequence (ξ_k) is convergent if and only if it is bounded.*

Proof. The proof is analogous to the Bounded Convergence Theorem for real sequences. For each multi-index I, the sequence $(\langle \xi_k, \zeta_I \rangle)$ is bounded and monotone, and therefore convergent. Details are left as an exercise. □

Theorem 11.3.7 (Bolzano Weierstrass). *Every bounded zeon sequence (ξ_k) has a convergent subsequence.*

Proof. The proof is analogous to the Bolzano-Weierstrass Theorem for real sequences. For each multi-index I, the sequence $(\langle \xi_k, \zeta_I \rangle)$ is bounded and therefore has a convergent subsequence. Details are left as an exercise. □

Proposition 11.3.8. *Every zeon sequence (ξ_k) has a monotonic subsequence.*

Proof. The proof is left as an exercise. □

11.4 Geometric Series in Zeon Algebras

A *zeon geometric series* is a series of the form $\sum_{\ell=0}^{\infty} ru^\ell$, where $r \in \mathcal{C}\ell_n{}^{\text{nil}}$ is fixed and $u \in \mathcal{C}\ell^{\text{nil}\star}$. The series is said to *converge* to ξ (or to have *sum* ξ) if the sequence of partial sums (ξ_m) defined by $\xi_m = \sum_{\ell=0}^{m} ru^\ell$ for $m \geq 0$ converges to ξ.

Remark 11.4.1. It is clear that if $u \in \mathcal{C}\ell_n{}^{\text{nil}_\circ}$, then $\sum_{\ell=0}^{\infty} ru^\ell = \sum_{\ell=0}^{n} ru^\ell$. Thus, any zeon geometric series involving powers of a nilpotent zeon u would be trivially convergent. The more interesting questions concern convergence of geometric series of invertible zeons. Hence, we require u to be invertible in the definition of zeon geometric series.

The Spectral Seminorm

Recall that the spectrum of an element z in a unital algebra is the collection of all scalars λ for which $z-\lambda$ is not invertible. Given $u = \Re u + \mathfrak{D} u \in \mathcal{C}\ell_n{}^{\text{nil}}$, one sees immediately that the *spectrum* of u is $\Re u$.

Lemma 11.4.2. *The mapping* $\mathcal{C}\ell_n{}^{\text{nil}} \to \mathbb{R}$ *defined by* $z \mapsto |z|_* := \sqrt{(\Re z)^2} = |\Re z|$ *defines a seminorm on* $\mathcal{C}\ell_n{}^{\text{nil}}$. *That is,*

i. $|z|_* \geq 0$ *for all* $z \in \mathcal{C}\ell_n{}^{\text{nil}}$,
ii. $|z_1 + z_2|_* \leq |z_1|_* + |z_2|_*$ *for all* $z_1, z_2 \in \mathcal{C}\ell_n{}^{\text{nil}}$.

Proof. Proof is by direct calculation. The first part is obvious. For the second part, observe that $\Re(z_1 + z_2) = \Re(z_1) + \Re(z_2)$. Hence, the triangle inequality for absolute values of real numbers gives

$$\begin{aligned} |z_1 + z_2|_* &= |\Re z_1 + \Re z_2| \\ &\leq |\Re z_1| + |\Re z_2| \\ &= |z_1|_* + |z_2|_*. \end{aligned}$$

□

For positive integers m, n with $m \geq n$, the *falling factorial* $(m)_n$ is defined by $(m)_n := \prod_{\ell=0}^{n}(m-\ell)$. This notation will be useful in the proof of the next theorem.

Theorem 11.4.3 (Spectral Criterion for Zeon Geometric Series). *Let* $r \in \mathcal{C}\ell_n{}^{\text{nil}}$ *and let* $u \in \mathcal{C}\ell_n{}^{\text{nil}\star}$. *The zeon geometric series* $\sum_\ell ru^\ell$ *converges* *if and only if* $|u|_* < 1$. *Moreover, the sum of the series is given by the following finite sum:*

$$\sum_{\ell=0}^{\infty} ru^\ell = \frac{r}{1-\Re u}\sum_{j=0}^{n}(1-\Re u)^{-j}(\mathfrak{D}u)^j.$$

Proof. First note that $u \in \mathcal{C}\ell_n{}^{\mathrm{nil}\star}$ implies $|u|_* > 0$. Since r is fixed, it is sufficient to consider convergence of $\sum_\ell u^\ell$. Observing that $\Re\left(\sum_{\ell=0}^{m} u^\ell\right) = \sum_{\ell=0}^{m}(\Re u)^\ell$ is a real geometric series, we sees that $|u|_* = |\Re u| < 1$ is required for convergence. Thus, necessity of the criterion is established.

To establish sufficiency of the criterion, suppose $0 < |u|_* < 1$ and set $\omega = \max\{\|(\mathfrak{D}u)^\ell\|_\infty : 0 \le \ell \le n\}$. Breaking u into its real and dual parts and assuming $m \ge n$, it follows that

$$\begin{aligned}
\|u^m\|_\infty &= \left\|\sum_{\ell=0}^{m}\binom{m}{\ell}(\Re u)^{m-\ell}(\mathfrak{D}u)^\ell\right\|_\infty \\
&\le \sum_{\ell=0}^{m}\binom{m}{\ell}|\Re u|^{m-\ell}\left\|\mathfrak{D}u^\ell\right\|_\infty \\
&= \sum_{\ell=0}^{n}\binom{m}{\ell}|\Re u|^{m-\ell}\left\|\mathfrak{D}u^\ell\right\|_\infty \\
&\le \omega\sum_{\ell=0}^{n}\binom{m}{\ell}|\Re u|^{m-\ell} \\
&= \omega\left[|\Re u|^m + m|\Re u|^{m-1} + \cdots + \frac{(m)_n}{n!}|\Re u|^{m-n}\right] \\
&\le \frac{\omega(m)_n}{|\Re u|^n(n-1)!}|\Re u|^m.
\end{aligned}$$

Observing that the falling factorial $(m)_n := \prod_{\ell=0}^{n}(m-\ell)$ satisfies $(m)_n < m^n$, we see that

$$\|u^m\|_\infty \le \frac{\omega m^n}{|\Re u|^n(n-1)!}|\Re u|^m,$$

where n and ω are constants. It follows that convergence of the zeon geometric series depends only on the spectrum of u.

For fixed $\gamma > 0$, define the continuous function $f(\lambda) = \dfrac{\lambda^n}{(1+\gamma)^\lambda}$. By repeated application of L'Hôpital's Rule, we obtain

$$\begin{aligned}\lim_{\lambda\to\infty} f(\lambda) &= \lim_{\lambda\to\infty} \frac{n!}{(1+\gamma)^\lambda(\log(1+\gamma))^n}\\ &= 0.\end{aligned}$$

Substituting ℓ for λ, it follows that for any $\kappa > 1$, there exists $N \in \mathbb{N}$ such that $\ell^n < \kappa^\ell$ for all $\ell \geq N$. Choosing $\kappa < 1/\alpha$, one can assume that $\ell^n\alpha^\ell < (\alpha\kappa)^\ell$ for all $\ell \geq N$, where $\alpha\kappa < 1$.

The sequence (t_m) of partial sums defined by $t_m = \sum_{\ell=1}^m (\alpha\kappa)^\ell$ is now clearly convergent and therefore Cauchy. Thus, for arbitrary $\varepsilon > 0$, there exists $\eta \in \mathbb{N}$ such that $M > N \geq \eta$ implies $|t_M - t_N| < \dfrac{|\Re u|^n(n-1)!}{\omega}\varepsilon$.

Hence, for integers $M > N \geq \eta$, the partial sums $s_N = \sum_{j=0}^N u^j$ satisfy

$$\begin{aligned}\|s_M - s_N\|_\infty &= \left\|\sum_{\ell=M+1}^{N} u^\ell\right\|_\infty\\ &\leq \sum_{\ell=M+1}^{N} \|u^\ell\|_\infty\\ &\leq \frac{\omega}{|\Re u|^n(n-1)!}\sum_{\ell=M+1}^{N} \ell^n|\Re u|^\ell\\ &= \frac{\omega}{|\Re u|^n(n-1)!}|t_M - t_N|\\ &< \varepsilon.\end{aligned}$$

It then follows that for each multi-index I, the sequence $(\langle s_m, \zeta_I\rangle)$ is a Cauchy sequence of real numbers converging to some $\varsigma_I \in \mathbb{R}$ such that

$$\sum_{\ell=0}^{\infty} u^\ell = \sum_{I\in 2^{[n]}} \varsigma_I\zeta_I.$$

Considering the mth partial sum of the zeon geometric series, one sees that

$$(1-u)\sum_{\ell=0}^{m} u^\ell = 1 - u^{m+1},$$

where $u^{m+1} \to 0$ as $m \to \infty$, by the assumed convergence of the series. Consequently, recalling Proposition 9.1.3, the sum of the zeon geometric series is now obtained from a finite sum as follows:

$$\begin{aligned}\sum_{\ell=0}^{\infty} u^\ell &= (1-u)^{-1}\\ &= ((1-\Re u)-\mathfrak{D}u)^{-1}\\ &= (1-\Re u)^{-1}\sum_{j=0}^{n}(-1)^j(1-\Re u)^{-j}(-\mathfrak{D}u)^j\\ &= \frac{1}{1-\Re u}\sum_{j=0}^{n}(1-\Re u)^{-j}(\mathfrak{D}u)^j.\end{aligned}$$

Multiplying both sides by the constant $r \in \mathcal{C}\ell_n{}^{\text{nil}}$ completes the proof. □

Example 11.4.4. In $\mathcal{C}\ell_3{}^{\text{nil}}$, let $u = \frac{3}{4} - \frac{1}{2}\zeta_{\{1,2\}} - \frac{1}{2}\zeta_{\{1,3\}} - \frac{1}{2}\zeta_{\{2,3\}} + \frac{3}{4}\zeta_{\{1,2,3\}} + \zeta_{\{2\}} - \frac{1}{4}\zeta_{\{3\}}$, and let $r = 4 + 3\zeta_{\{1\}}$. By Theorem 11.4.3, the series converges to

$$\begin{aligned}\sum_{\ell=0}^{\infty} ru^\ell &= 16 + 16\zeta_{\{1,2\}} - 44\zeta_{\{1,3\}} - 160\zeta_{\{2,3\}} - 264\zeta_{\{1,2,3\}} + 12\zeta_{\{1\}}\\ &\quad + 64\zeta_{\{2\}} - 16\zeta_{\{3\}}.\end{aligned}$$

Some terms of the sequence of partial sums (s_N) obtained with *Mathematica* reveal the following:

$$\begin{aligned}s_{10} &= 15.3242 + 12.8464\zeta_{\{1,2\}} - 35.3277\zeta_{\{1,3\}} - 95.4272\zeta_{\{2,3\}}\\ &\quad -137.632\zeta_{\{1,2,3\}} + 11.4932\zeta_{\{1\}} + 51.3858\zeta_{\{2\}} - 12.8464\zeta_{\{3\}}\\ s_{30} &= 15.9979 + 15.9757\zeta_{\{1,2\}} - 43.9332\zeta_{\{1,3\}} - 158.871\zeta_{\{2,3\}}\\ &\quad -261.606\zeta_{\{1,2,3\}} + 11.9984\zeta_{\{1\}} + 63.9029\zeta_{\{2\}} - 15.9757\zeta_{\{3\}}\\ s_{60} &= 16. + 16.\zeta_{\{1,2\}} - 44.\zeta_{\{1,3\}} - 159.999\zeta_{\{2,3\}} - 263.998\zeta_{\{1,2,3\}}\\ &\quad +12.\zeta_{\{1\}} + 64.\zeta_{\{2\}} - 16.\zeta_{\{3\}}.\end{aligned}$$

Exercises

Exercise 11.1: Show that the limit of a convergent zeon sequence is unique.

Exercise 11.2: Suppose (ξ_k) and (η_k) are zeon sequences such that $\lim(\xi_k) = 0$, and (η_k) satisfies $3 < \|\eta_n\| < 10$ for all $k \in \mathbb{N}$. Give a formal ε-k^* proof that $\lim(\xi_k/\eta_k) = 0$.

Exercise 11.3: Show that every zeon Cauchy sequence converges to a limit in ${\mathcal{C}\ell_n}^{\mathrm{nil}}$.

Exercise 11.4: Suppose (ξ_k) is a convergent zeon sequence and that $\lim_{k\to\infty} \xi_k = \xi$. Show that $(\Re\xi_k)$ converges to $\Re\xi$.

Exercise 11.5: Prove Proposition 11.3.3.

Exercise 11.6: Prove Proposition 11.3.4 and deduce the limit of zeon quotients; i.e., if $(\xi_k) \subset {\mathcal{C}\ell_n}^{\mathrm{nil}}$ and $(\psi_k) \subset {\mathcal{C}\ell_n}^{\mathrm{nil}\star}$ such that $\lim_{k\to\infty} \xi_k = \xi$ and $\lim_{k\to\infty} \psi_k = \psi$, where $\Re\psi \neq 0$, then

$$\lim_{k\to\infty} \frac{\xi_k}{\psi_k} = \frac{\xi}{\psi}.$$

Exercise 11.7: Prove Theorem 11.3.6.

Exercise 11.8: Prove Proposition 11.3.8.

Exercise 11.9: Prove Theorem 11.3.7.

Exercise 11.10: Give an example of two non-convergent zeon sequences (ξ_k) and (ψ_k) whose product $(\xi_k\psi_k)$ is convergent.

Exercise 11.11: Give an example of two nonconstant monotone zeon sequences (ξ_k) and (ψ_k) whose product $(\xi_k\psi_k)$ is monotone.

Exercise 11.12: If possible, give an example of two nonconstant, nilpotent zeon sequences $(\xi_k), (\psi_k) \subset {\mathcal{C}\ell_n}^{\mathrm{nil}_\circ}$ whose product $(\xi_k\psi_k)$ is nonconstant and monotone.

Exercise 11.13: Give an example of two monotone zeon sequences (ξ_k) and (ψ_k) whose product $(\xi_k\psi_k)$ is not monotone.

Exercise 11.14: Show directly from the definitions that if (ξ_k) and (η_k) are zeon Cauchy sequences, then $(\xi_k + \eta_k)$ and $(\xi_k\, \eta_n)$ are zeon Cauchy sequences.

Exercise 11.15: In ${\mathcal{C}\ell_3}^{\rm nil}$, let

$$\begin{aligned} u &= -2\zeta_{\{1,2\}} - 3\zeta_{\{1\}} + 3\zeta_{\{3\}} + 2 \\ v &= -7\zeta_{\{2,3\}} + \zeta_{\{1\}} + 5\zeta_{\{2\}} + 1 \\ w &= 3\zeta_{\{1,3\}} - 6\zeta_{\{1,2,3\}} + 5\zeta_{\{1\}} - \zeta_{\{2\}}. \end{aligned}$$

Compute the following norms:

i. $\|u\|_\infty$
ii. $\|v\|_1$
iii. $\|w\|_2$
iv. $\|uv\|_\infty$.

Exercise 11.16: In ${\mathcal{C}\ell_3}^{\rm nil}$, let $u = \frac{3}{4} - \frac{1}{2}\zeta_{\{1,2\}} - \frac{1}{2}\zeta_{\{1,3\}} - \frac{1}{2}\zeta_{\{2,3\}} + \frac{3}{4}\zeta_{\{1,2,3\}} + \zeta_{\{2\}} - \frac{1}{4}\zeta_{\{3\}}$, and let $r = 4 + 3\zeta_{\{1\}}$. Compute the limit of the geometric series $\sum_\ell ru^\ell$.

Exercise 11.17: Show that the series $\displaystyle\sum_{k=1}^{\infty}\left(\frac{-1}{k} + \frac{1}{k}\zeta_{\{1\}}\right)^2$ converges and compute its sum.

Exercise 11.18: Show that the sequence $\left(\left(1 + \frac{1}{k}\zeta_{\{1\}}\right)^k\right)$ converges and determine its limit.

Exercise 11.19: Show that the series $\displaystyle\sum_{k=0}^{\infty}\left(\frac{1}{2} + \zeta_{\{1\}}\right)^k$ converges and determine its sum.

Exercise 11.20: Show that the series $\displaystyle\sum_{k=0}^{\infty}\left(\frac{1}{2} + \zeta_{\{1\}} + \zeta_{\{2\}}\right)^k$ converges and determine its sum.

Exercise 11.21: Show that the series $\displaystyle\sum_{k=0}^{\infty}\left(\frac{1}{2} + \zeta_{\{1\}} + \zeta_{\{2\}} + \zeta_{\{3\}}\right)^k$ converges and determine its sum.

Exercise 11.22: Show that the series $\displaystyle\sum_{k=0}^{\infty}\left(\frac{1}{2} + \zeta_{\{1\}} + \cdots + \zeta_{\{n\}}\right)^k$ converges in ${\mathcal{C}\ell_n}^{\rm nil}$ and determine a general formula for its sum.

Chapter 12

Zeon Matrices

Constructing matrices with entries from ${\mathcal{C}\ell_n}^{\rm nil}$ has proven useful in numerous applications related to graph theory. In this chapter we consider broader theoretical aspects of zeon matrices.

12.1 Invertibility of Zeon Matrices

Recall that a zeon element u is invertible if and only if $\Re u \neq 0$, in which case

$$u^{-1} = \frac{1}{\Re u} \sum_{\ell=0}^{\kappa(\mathfrak{D}u)-1} (-1)^\ell (\Re u)^{-\ell} (\mathfrak{D}u)^\ell.$$

With this in mind, it is not difficult to see that if A is a square matrix having zeon entries, then A can be written in the form $\Re A + \mathfrak{D}A$, where $\Re A = (\Re a_{ij})$ and $\mathfrak{D}A = A - \Re A$. Thus, $\mathfrak{D}A$ is clearly nilpotent, and the notation $\kappa(\mathfrak{D}A)$ can be used to denote the index of nilpotency of $\mathfrak{D}A$.

Lemma 12.1.1. *Let X be an $r \times c$ matrix having entries in ${\mathcal{C}\ell_n}^{\rm nil_o}$, and let U be a $c \times t$ matrix having entries in ${\mathcal{C}\ell_n}^{\rm nil}$. Then, the entries of XU are in ${\mathcal{C}\ell_n}^{\rm nil_o}$, and XU is therefore nilpotent.*

Proof. Write $X = (x_{ij})$, $U = (u_{ij})$, and $XU = (y_{ij})$. Fixing indices i, j such that $1 \leq i \leq r$ and $1 \leq j \leq t$, one has $y_{ij} = \sum_{\ell=1}^{c} x_{i\ell} u_{\ell j}$. Since ${\mathcal{C}\ell_n}^{\rm nil_o}$ is an ideal, it is clear from the properties of zeon multiplication that $x_{i\ell} u_{\ell j} \in {\mathcal{C}\ell_n}^{\rm nil_o}$ for all i, j, ℓ. Hence, the result. □

With Lemma 12.1.1 established, conditions for invertibility of zeon matrices can be discussed.

Proposition 12.1.2. *Let $A = (a_{ij})$ be a square matrix having entries from ${\mathcal{C}\ell_n}^{\text{nil}}$, and write $A = \Re A + \mathfrak{D}A$, where $\Re A = (\Re a_{ij})$. It follows that A is invertible if and only if $\Re A$ is invertible. In this case, the inverse is given by*

$$A^{-1} = (\Re A)^{-1} \sum_{\ell=0}^{\kappa(\mathfrak{D}A(\Re A)^{-1})-1} (-1)^\ell (\mathfrak{D}A(\Re A)^{-1})^\ell.$$

Proof. Suppose A, U are $m \times m$ matrices having entries in ${\mathcal{C}\ell_n}^{\text{nil}}$. Writing $AU = \Re(AU) + \mathfrak{D}(AU)$ and observing that

$$\begin{aligned} AU &= (\Re A + \mathfrak{D}A)(\Re U + \mathfrak{D}U) \\ &= \Re A \Re U + \Re A + \mathfrak{D}U + \mathfrak{D}A \Re U + \mathfrak{D}A\mathfrak{D}U \\ &= \Re(AU) + \mathfrak{D}(AU), \end{aligned}$$

it is not difficult to see that $\Re(AU) = \Re A \Re U$. Hence, $U = A^{-1}$ if and only if $\Re A$ is invertible and $\Re U = (\Re A)^{-1}$.

Now, assume $\Re A$ is invertible, and note that by Lemma 12.1.1, $\mathfrak{D}A(\Re A)^{-1}$ is nilpotent. Letting $k = \kappa(\mathfrak{D}A(\Re A)^{-1})$ denote the index of nilpotency of $\mathfrak{D}A(\Re A)^{-1}$, one easily verifies that

$$\begin{aligned} A\left[(\Re A)^{-1}\sum_{\ell=0}^{k-1}(-1)^\ell(\mathfrak{D}A(\Re A)^{-1})^\ell\right] &= I + \sum_{\ell=1}^{k-1}(-1)^\ell(\mathfrak{D}A(\Re A)^{-1})^\ell \\ &\quad + \mathfrak{D}A(\Re A)^{-1}\sum_{\ell=0}^{k-1}(-1)^\ell(\mathfrak{D}A(\Re A)^{-1})^\ell \\ &= I + \sum_{\ell=1}^{k-1}(-1)^\ell(\mathfrak{D}A(\Re A)^{-1})^\ell \\ &\quad + \sum_{\ell=1}^{k-1}(-1)^{\ell-1}(\mathfrak{D}A(\Re A)^{-1})^\ell \\ &= I. \end{aligned}$$

□

Example 12.1.3. Consider $A \in \text{Mat}_2({\mathcal{C}\ell_3}^{\text{nil}})$:

$$A = \begin{pmatrix} 1 + \zeta_{\{1,2\}} - 2\zeta_{\{3\}} & 4 + \zeta_{\{1\}} - \zeta_{\{2,3\}} \\ -2 + \zeta_{\{2\}} + \zeta_{\{1,2,3\}} & 5 + 3\zeta_{\{1\}} \end{pmatrix}$$

Then,

$$\Re A = \begin{pmatrix} 1 & 4 \\ -2 & 5 \end{pmatrix}, \qquad \mathfrak{D}A = \begin{pmatrix} \zeta_{\{1,2\}} - 2\zeta_{\{3\}} & 4 + \zeta_{\{1\}} - \zeta_{\{2,3\}} \\ \zeta_{\{2\}} + \zeta_{\{1,2,3\}} & 3\zeta_{\{1\}} \end{pmatrix}.$$

Since $|\Re A| = 13$, A is invertible; in particular,

$$(\Re A)^{-1} = \begin{pmatrix} \frac{5}{13} & \frac{-4}{13} \\ \frac{2}{13} & \frac{1}{13} \end{pmatrix},$$

and

$$\mathfrak{D}A(\Re A)^{-1}$$
$$= \begin{pmatrix} \frac{2\zeta_{\{1\}}}{13} + \frac{5\zeta_{\{1,2\}}}{13} - \frac{2\zeta_{\{2,3\}}}{13} - \frac{10\zeta_{\{3\}}}{13} & \frac{\zeta_{\{1\}}}{13} + \frac{8\zeta_{\{3\}}}{13} - \frac{4\zeta_{\{1,2\}}}{13} - \frac{1}{13}\zeta_{\{2,3\}} \\ \frac{6\zeta_{\{1\}}}{13} + \frac{5\zeta_{\{2\}}}{13} + \frac{5\zeta_{\{1,2,3\}}}{13} & \frac{3\zeta_{\{1\}}}{13} - \frac{4\zeta_{\{1,2,3\}}}{13} - \frac{4\zeta_{\{2\}}}{13} \end{pmatrix}.$$

A *Mathematica* calculation verifies that $A^{-1} = (\Re A)^{-1} \sum_{\ell=0}^{3} (-\mathfrak{D}A(\Re A)^{-1})^{\ell}$.

12.2 The Zeon-Frobenius Norm

Given two complex $r \times c$ matrices A and B, the *Frobenius inner product* of A and B is defined by

$$\langle A, B \rangle_F = \operatorname{tr}\left(B^\dagger A\right),$$

where $B^\dagger$ is the conjugate transpose of the matrix B. A familiar norm for matrices with real or complex entries is the Frobenius norm (or Hilbert-Schmidt norm). Given an $r \times c$ matrix $A = (a_{ij})$ with entries in $\mathbb{C}$, the *Frobenius norm* of A is defined by

$$\|A\|_{\mathrm{F}}{}^2 = \operatorname{tr}(A^\dagger A).$$

This has the equivalent expression

$$\|A\|_{\mathrm{F}}{}^2 = \sum_{i=1}^{r} \sum_{j=1}^{c} a_{ij}\overline{a_{ij}} = \sum_{\substack{1 \le i \le r \\ 1 \le j \le c}} |a_{ij}|^2.$$

However, if the elements of A are in $\mathcal{C}\ell_n{}^{\mathrm{nil}}$, then $\operatorname{tr}(A^\dagger A) \in \mathcal{C}\ell_n{}^{\mathrm{nil}}$. Consequently, the positive-definite requirement of a norm may not be satisfied. In order to obtain a norm on zeon matrices, the zeon-Hodge dual is defined.

Definition 12.2.1. Let $u = \sum_{I \in 2^{[n]}} u_I \zeta_I \in \mathcal{C}\ell_n{}^{\mathrm{nil}}$. The *zeon-Hodge dual* (or, more simply, dual) of u, denoted $u^\star$, is defined by

$$u^\star = \sum_{I \in 2^{[n]}} u_I \zeta_{[n]\setminus I}.$$

In other words, the coefficient of a basis blade ζ_I in u is the coefficient of the blade $\zeta_{I'}$ in $u^\star$, where $I' = [n] \setminus I$ is the relative complement of I in $[n]$.

It is not difficult to see that $(u^\star)^\star = u$ and that the mapping is clearly linear: $(\alpha u + v)^\star \mapsto \alpha u^\star + v^\star$ for $\alpha \in \mathbb{R}$ and $u, v \in \mathcal{C}\ell_n{}^{\text{nil}}$. However, one sees that $u \mapsto u^\star$ is not an involution because $(uv)^\star \neq u^\star v^\star$ for general u, v.

Lemma 12.2.2. *An inner product is defined on* $\mathcal{C}\ell_n{}^{\text{nil}}$ *by*

$$\langle u, v\rangle = \Re\left((u^\star v)^\star\right).$$

Proof. The definition clearly implies that $\langle\cdot,\cdot\rangle$ is a well-defined function $\mathcal{C}\ell_n{}^{\text{nil}} \times \mathcal{C}\ell_n{}^{\text{nil}} \to \mathbb{R}$. Further, this function is bilinear. Linearity in the first component is established for $c \in \mathbb{R}$ as follows:

$$\begin{aligned}\langle cu + v, w\rangle &= \Re\left(((cu+v)^\star w)^\star\right)\\ &= \Re\left((cu)^\star w)^\star + ((v^\star w)^\star\right)\\ &= \Re\left(((cu)^\star w)^\star\right) + \Re\left((v^\star w)^\star\right)\\ &= c\Re\left((u^\star w)^\star\right) + \Re\left((v^\star w)^\star\right)\\ &= c\langle u, w\rangle + \langle v, w\rangle.\end{aligned}$$

Linearity in the second component follows similarly. □

As stated in the next lemma, the mapping $u \mapsto (\Re(u^\star u)^\star)^{1/2}$ is equivalent to the inner product norm $u \mapsto \|u\|_2$ on $\mathcal{C}\ell_n{}^{\text{nil}}$.

Lemma 12.2.3. *A norm is defined on* $\mathcal{C}\ell_n{}^{\text{nil}}$ *by*

$$\|u\| = \langle u, u\rangle^{1/2} = (\Re(u^\star u)^\star)^{1/2}.$$

Proof. Given $u = \sum_{I\in 2^{[n]}} u_I\zeta_I \in \mathcal{C}\ell_n{}^{\text{nil}}$, straightforward calculation shows that

$$\langle u, u\rangle = \Re((u^\star u)^\star) = \sum_{I\in 2^{[n]}} {u_I}^2.$$

In other words, $(\langle u, u\rangle)^{1/2}$ is the Euclidean norm of u in the vector space spanned by the basis blades of $\mathcal{C}\ell_n{}^{\text{nil}}$. □

The zeon-Hodge dual extends naturally (entry-wise) to matrices with zeon coefficients; i.e., $A = (a_{ij})$ implies $A^\star = ({a_{ij}}^\star)$.

Definition 12.2.4. Let $A = (a_{ij})$ be an $r \times c$ matrix with entries in $\mathcal{C}\ell_n{}^{\text{nil}}$, and let $A^\dagger = A^{\star\top}$ denote the dual-transpose of A. The *zeon-Frobenius* *norm* of A is defined by

$$\begin{aligned}\|A\|_{\mathrm{zF}} &= \sqrt{\Re\left(\operatorname{tr}(A^\dagger A)^\star\right)}\\ &= \left(\Re \sum_{\substack{1\le i\le r\\ 1\le j\le c}} (a_{ij}{a_{ij}}^\star)^\star\right)^{1/2}\\ &= \left(\sum_{\substack{1\le i\le r\\ 1\le j\le c}} < a_{ij}, a_{ij} >\right)^{1/2}.\end{aligned}$$

Example 12.2.5. The zeon-Frobenius norm of matrix

$$A = \begin{pmatrix} 1+\zeta_{\{1\}} & 2-3\zeta_{\{1,3\}} & 4\zeta_{\{2\}}\\ 1-\zeta_{\{1,2,3\}} & 2\zeta_{\{2,3\}} & -5\zeta_{\{1,2\}}\end{pmatrix}$$

having entries in ${\mathcal{C}\ell_3}^{\mathrm{nil}}$ is computed and verified using *Mathematica* in Figure 12.1.

Proposition 12.2.6. *Suppose A is $r_1 \times c_1$ and B is $c_1 \times c_2$, both having entries from ${\mathcal{C}\ell_n}^{\mathrm{nil}}$. Then,*

$$\|AB\|_{\mathrm{zF}} \le 2^{n/2} c_1 \sqrt{r_1 c_2}\, \|A\|_{\mathrm{zF}} \|B\|_{\mathrm{zF}}.$$

Proof. By the definition of matrix multiplication, the $r_1 \times c_2$ zeon matrix $M = (m_{ij}) = AB$ has entries determined by

$$m_{ij} = \sum_{\ell=1}^{c_1} a_{i\ell} b_{\ell j}$$

for $1 \le i \le r_1$ and $1 \le j \le c_2$. It follows from the triangle inequality that

$$\begin{aligned}\|m_{ij}\|_2 &\le \sum_{\ell=1}^{c_1} \|a_{i\ell} b_{\ell j}\|_2\\ &\le \sum_{\ell=1}^{c_1} 2^{n/2} \|a_{i\ell}\|_2 \|b_{\ell j}\|_2\\ &\le 2^{n/2} \sum_{\ell=1}^{c_1} \|a_{i\ell}\|_2 \|B\|_{\mathrm{zF}}\\ &\le 2^{n/2} c_1 \|A\|_{\mathrm{zF}} \|B\|_{\mathrm{zF}}.\end{aligned}$$

```
Print["A = ",
  MatrixForm[A = {{1 + ζ{1}, 2 - 3 ζ{1,3}, 4 ζ{2}}, {1 - ζ{1,2,3}, 2 ζ{2,3}, -5 ζ{1,2}}}]];
```

$$A = \begin{pmatrix} 1+\zeta_{\{1\}} & 2-3\zeta_{\{1,3\}} & 4\zeta_{\{2\}} \\ 1-\zeta_{\{1,2,3\}} & 2\zeta_{\{2,3\}} & -5\zeta_{\{1,2\}} \end{pmatrix}$$

The dual transpose of A:

```
Print["A† = ", MatrixForm[ad = Transpose[zeonDual[A]]]];
```

$$A^\dagger = \begin{pmatrix} \zeta_{\{2,3\}}+\zeta_{\{1,2,3\}} & -1+\zeta_{\{1,2,3\}} \\ -3\zeta_{\{2\}}+2\zeta_{\{1,2,3\}} & 2\zeta_{\{1\}} \\ 4\zeta_{\{1,3\}} & -5\zeta_{\{3\}} \end{pmatrix}$$

```
Print["A†A = ",
 MatrixForm[adA = clExpand[cliffordMatrixProduct[A, Transpose[zeonDual[A]]]]]]
```

$$A^\dagger A = \begin{pmatrix} -6\zeta_{\{2\}}+\zeta_{\{2,3\}}+31\zeta_{\{1,2,3\}} & -1+3\zeta_{\{1\}}-20\zeta_{\{2,3\}}+\zeta_{\{1,2,3\}} \\ \zeta_{\{2,3\}}+\zeta_{\{1,2,3\}} & -1+31\zeta_{\{1,2,3\}} \end{pmatrix}$$

The trace of $(A^\dagger A)^*$:

```
Print["tr[A†A]* = ",
 x = Tr[zeonDual[clExpand[cliffordMatrixProduct[A, Transpose[zeonDual[A]]]]]]]
```

$\mathrm{tr}[A^\dagger A]^* = 62+\zeta_{\{1\}}-6\zeta_{\{1,3\}}-\zeta_{\{1,2,3\}}$

The zeon-Frobenius norm:

```
Print["√R(tr[A† A]*) = ", zf = Sqrt[zRe[x]]];
```

$\sqrt{\mathfrak{R}\left(\mathrm{tr}[A^\dagger A]^*\right)} = \sqrt{62}$

Verify using sum of squares of inner product norms.

```
Print["√(∑ ‖aij‖²) = ",
 Sqrt[Sum[zeonInnerProduct[A[[i, j]], A[[i, j]]], {i, 2}, {j, 3}]]]
```

$\sqrt{\sum \|a_{ij}\|^2} = \sqrt{62}$

Figure 12.1 Computation of zeon-Frobenius norm.

Squaring and summing over all entries, one obtains

$$\begin{aligned}(\|AB\|_{\mathrm{zF}})^2 &\le \sum_{\substack{1\le i\le r_1\\1\le j\le c_2}} \|m_{ij}\|_2{}^2\\ &\le \sum_{\substack{1\le i\le r_1\\1\le j\le c_2}} 2^n c_1{}^2(\|A\|_{\mathrm{zF}}\|B\|_{\mathrm{zF}})^2\\ &= 2^n r_1 c_2 c_1{}^2(\|A\|_{\mathrm{zF}}\|B\|_{\mathrm{zF}})^2.\end{aligned}$$

Hence, the result. □

Exercises

Exercise 12.1: Compute the inverse of

$$A = \begin{pmatrix} 3\zeta_{\{2\}} + 2 & 4\zeta_{\{2\}} \\ 4\zeta_{\{1,2\}} + 1 & 2 - 4\zeta_{\{2\}} \end{pmatrix}$$

over ${\mathcal{C}\ell_2}^{\text{nil}}$.

Exercise 12.2: Compute the inverse of

$$A = \begin{pmatrix} -3 & 5 - 3\zeta_{\{2\}} \\ 5 & 2\zeta_{\{2\}} - 3 \end{pmatrix}$$

over ${\mathcal{C}\ell_2}^{\text{nil}}$.

Exercise 12.3: Compute the inverse of

$$A = \begin{pmatrix} -2 & -4 & 1 - 2\zeta_{\{1\}} \\ \zeta_{\{1\}} + 3 & 0 & -2 \\ 0 & -2 & 1 \end{pmatrix}$$

over ${\mathcal{C}\ell_2}^{\text{nil}}$.

Exercise 12.4: Compute the inverse of

$$A = \begin{pmatrix} -3 & 2\zeta_{\{1\}} + 5 \\ 4\zeta_{\{1,2,3\}} + 3 & -3 \end{pmatrix}$$

over ${\mathcal{C}\ell_3}^{\text{nil}}$.

Exercise 12.5: Compute the zeon-Frobenius norm of the following matrices in ${\mathcal{C}\ell_3}^{\text{nil}}$:

$$A_1 = \begin{pmatrix} 0 & -1 - \zeta_{\{1\}} + 2\zeta_{\{2\}} - \zeta_{\{2,3\}} & 2 + \zeta_{\{1\}} + 2\zeta_{\{2\}} + \zeta_{\{2,3\}} \end{pmatrix}$$

$$A_2 = \begin{pmatrix} -\zeta_{\{2\}} + \zeta_{\{3\}} + 2\zeta_{\{1,2\}} + 2\zeta_{\{2,3\}} \\ -\zeta_{\{1\}} - \zeta_{\{2,3\}} \\ 0 \end{pmatrix}$$

$$A_3 = \begin{pmatrix} 2\zeta_{\{2\}} - \zeta_{\{1,2,3\}} & \zeta_{\{1,2\}} \\ 2 - \zeta_{\{1,2\}} & -1 - \zeta_{\{1,3\}} + 2\zeta_{\{1,2,3\}} \end{pmatrix}$$

$$A_4 = \begin{pmatrix} -\zeta_{\{2\}} + \zeta_{\{2,3\}} \\ 2 + 2\zeta_{\{3\}} \end{pmatrix}$$

$$A_5 = \begin{pmatrix} \zeta_{\{2,3\}} & 2\zeta_{\{1,2,3\}} \\ -\zeta_{\{3\}} & 2\zeta_{\{2\}} + 2\zeta_{\{3\}} + \zeta_{\{1,2\}} + 2\zeta_{\{1,3\}} \end{pmatrix}.$$

Exercise 12.6: Show that the zeon-Frobenius norm is not submultiplicative.

Exercise 12.7: Using results from this chapter, determine a constant $k > 0$ such that

$$\|AB\|_{\mathrm{zF}} \leq k\|A\|_{\mathrm{zF}}\|B\|_{\mathrm{zF}}$$

for all 3×3 matrices A, B having entries in ${\mathcal{C}\ell_2}^{\mathrm{nil}}$.

Exercise 12.8: Suppose A is 3×2 and B is 2×1, both having entries from ${\mathcal{C}\ell_5}^{\mathrm{nil}}$. Determine a constant $k > 0$ such that

$$\|AB\|_{\mathrm{zF}} \leq k\|A\|_{\mathrm{zF}}\|B\|_{\mathrm{zF}}.$$

Exercise 12.9: Using previous results from this chapter, determine a constant $k > 0$ such that

$$\|AB\|_{\mathrm{zF}} \leq k\|A\|_{\mathrm{zF}}\|B\|_{\mathrm{zF}}$$

for all $m \times m$ matrices A, B having entries in ${\mathcal{C}\ell_n}^{\mathrm{nil}}$. Note that k can depend on m, n.

Chapter 13

Zeon Functions and Factorizations

In this chapter, algebraic properties of zeons are considered, including the existence of elementary factorizations and homogeneous factorizations of invertible zeons. A "zeon division algorithm" is established, showing that every nontrivial invertible zeon can be written as a sum of homogeneously decomposable zeons. Elementary functions (exponential, logarithmic, hyperbolic, and trigonometric) are extended to zeons, and a number of properties and identities are revealed. Finally, fast computation of logarithms is discussed for homogeneously decomposable zeons.

13.1 Elementary Factorization of Zeons

Definition 13.1.1. Let $u \in {\mathcal{C}\ell_n}^{\mathrm{nil}\star}$. If u can be expressed in the form $u = (1 + a\zeta_I)\mathfrak{z}$ for some nonzero scalar a, some nonempty multi-index I, and some invertible zeon $\mathfrak{z}$ such that $\natural\mathfrak{z} > 0$, then $(1 + a\zeta_I)$ is said to be an *elementary factor* of u.

An *elementary factorization* of $u \in {\mathcal{C}\ell_n}^{\mathrm{nil}\star}$ is a factorization of the form $u = a \prod_{I \in \mathfrak{J} \subseteq 2^{[n]}} (1 + a_I\zeta_I)$. As will be shown in Lemma 13.1.2, this factorization is unique and always exists. When all multi-indices in the set $\mathfrak{J}$ are of the same cardinality, the elementary factorization is said to be *homogeneous.* It is not difficult to see that if an invertible nontrivial zeon u has a homogeneous elementary factorization, then the cardinality of the multi-indices in $\mathfrak{J}$ must be $\natural u$.

Lemma 13.1.2. *Every nontrivial element $u \in {\mathcal{C}\ell_n}^{\mathrm{nil}\star}$ has a unique elementary factorization.*

Proof. First, let $u \in \mathcal{C}\ell_n{}^{\mathrm{nil}\star}$ have normalized[1] additive representation $u = \sum_{I\in 2^{[n]}} a_I\zeta_I$, and write

$$u = 1 + \sum_{\substack{I\in 2^{[n]}\\ I\neq\varnothing}} a_I\zeta_I.$$

Let I be any nonempty multi-index of minimal cardinality $\natural u$ appearing in the canonical expansion of u; viz., the coefficient a_I is nonzero, and note that $(1+a_I\zeta_I)(1-a_I\zeta_I) = 1$, so that $(1+a_I\zeta_I)^{-1} = (1-a_I\zeta_I)$. Consequently, writing

$$\begin{aligned}(1+a_I\zeta_I)^{-1}u &= (1-a_I\zeta_I)\left(1+a_I\zeta_I+\sum_{J\neq I} a_J\zeta_J\right)\\ &= 1+\sum_{J\neq I} a_J\zeta_J - a_I\zeta_I\sum_{J\neq I} a_J\zeta_J,\end{aligned} \tag{13.1}$$

one sees that the resulting canonical expansion contains no term indexed by I. Repeating the process with a multi-index appearing in (13.1), one eliminates all terms of grade $\natural u$ after at most $\binom{n}{\natural u}$ steps. Noting that there are no more than n nontrivial grades, the process terminates after some positive number of steps, say m. One thereby obtains the desired elementary factorization via $(1+a_{I_m}\zeta_{I_m})^{-1}\cdots(1+a_{I_1}\zeta_{I_1})^{-1}u = 1$, which implies $u = \prod_{\ell=1}^{m}(1+a_{I_\ell}\zeta_{I_\ell})$.

To establish uniqueness of the elementary factorization, suppose

$$\prod_{\varnothing\neq I\in 2^{[n]}}(1+a_I\zeta_I) = \prod_{\varnothing\neq I\in 2^{[n]}}(1+b_I\zeta_I).$$

Clearly, if $(1+a_I\zeta_I) = (1+b_I\zeta_I)$, then $a_I = b_I$. Proceeding by induction on the number of elementary factors, suppose that for $\mathfrak{J}\subseteq 2^{[n]}$ such that $|\mathfrak{J}| = k$, $\prod_{\varnothing\neq I\in\mathfrak{J}\subseteq 2^{[n]}}(1+a_I\zeta_I) = \prod_{\varnothing\neq I\in\mathfrak{J}\subseteq 2^{[n]}}(1+b_I\zeta_I)$ implies $a_I = b_I$ for all $I\in\mathfrak{J}$. Choose $J\in 2^{[n]}$ such that $J\notin\mathfrak{J}$ and suppose

$$\prod_{\varnothing\neq I\in(\mathfrak{J}\cup\{J\})\subseteq 2^{[n]}}(1+a_I\zeta_I) = \prod_{\varnothing\neq I\in(\mathfrak{J}\cup\{J\})\subseteq 2^{[n]}}(1+b_I\zeta_I).$$

[1]We can divide by $a_\varnothing$ if necessary.

Then,

$$
\begin{aligned}
1+a_J\zeta_J &= \prod_{\varnothing\neq I\in(\mathfrak{J}\cup\{J\})\subseteq 2^{[n]}}(1+b_I\zeta_I)\prod_{\varnothing\neq I\in\mathfrak{J}\subseteq 2^{[n]}}(1+a_I\zeta_I)^{-1}\\
&= \prod_{\varnothing\neq I\in(\mathfrak{J}\cup\{J\})\subseteq 2^{[n]}}(1+b_I\zeta_I)\prod_{\varnothing\neq I\in\mathfrak{J}\subseteq 2^{[n]}}(1+b_I\zeta_I)^{-1}\\
&= (1+b_J\zeta_J)\prod_{\varnothing\neq I\in\mathfrak{J}\subseteq 2^{[n]}}(1+b_I\zeta_I)\prod_{\varnothing\neq I\in\mathfrak{J}\subseteq 2^{[n]}}(1+b_I\zeta_I)^{-1}\\
&= 1+b_J\zeta_J.
\end{aligned}
$$

Thus $a_J = b_J$, and the result holds by induction. □

When an invertible nontrivial zeon u has a homogeneous elementary factorization, the grades of all nonzero terms in the canonical expansion of u must be congruent modulo $\natural u$. The converse, however, is not generally true.

Theorem 13.1.3. *An invertible zeon* $u = \sum_{I\in 2^{[n]}} a_I\zeta_I \in \mathcal{C}\ell_n{}^{\text{nil}\star}$ *has a homogeneous elementary factorization if and only if*

$$u = a_\varnothing \prod_{\{I:|I|=\natural u\}}\left(1+\frac{a_I}{a_\varnothing}\zeta_I\right).$$

Proof. If $u = a_\varnothing \prod_{\{I:|I|=\natural u\}}\left(1+\frac{a_I}{a_\varnothing}\zeta_I\right)$, then u has a homogeneous elementary factorization by definition.

Otherwise, suppose $u = \sum_{I\in 2^{[n]}} a_I\zeta_I \in \mathcal{C}\ell_n{}^{\text{nil}\star}$ has a homogeneous elementary factorization $u = b_\varnothing \prod_{\{I:|I|=\natural u\}}\left(1+\frac{b_I}{b_\varnothing}\zeta_I\right)$. Then,

$$
\begin{aligned}
u &= b_\varnothing \prod_{\{I:|I|=\natural u\}}\left(1+\frac{b_I}{b_\varnothing}\zeta_I\right)\\
&= b_\varnothing\left(1+\sum_{|I|=\natural u}\frac{b_I}{b_\varnothing}\zeta_I\right) + \text{higher-grade terms}\\
&= b_\varnothing + \sum_{|I|=\natural u} b_I\zeta_I + \text{higher-grade terms}.
\end{aligned}
$$

Hence, $a_\varnothing = b_\varnothing$ and $a_I = b_I$ whenever $|I| = \natural u$. Observing that u is determined uniquely by these $\binom{n}{\natural u} + 1$ scalars, equality of higher-grade coefficients follows. □

Example 13.1.4. Consider $u \in \mathcal{C}\ell_6{}^{\text{nil}\star}$ given by

$$\begin{aligned} u = 1 &- \zeta_{\{1,2\}} - 2\zeta_{\{1,3\}} - 3\zeta_{\{1,5\}} - 2\zeta_{\{1,6\}} - 2\zeta_{\{2,3\}} + 2\zeta_{\{2,4\}} + 3\zeta_{\{2,5\}} \\ &+\zeta_{\{2,6\}} + 3\zeta_{\{3,4\}} + 3\zeta_{\{3,5\}} - 3\zeta_{\{3,6\}} - 2\zeta_{\{4,5\}} - 2\zeta_{\{4,6\}} + 2\zeta_{\{5,6\}} \\ &-7\zeta_{\{1,2,3,4\}} - 3\zeta_{\{1,2,3,5\}} + 5\zeta_{\{1,2,3,6\}} - 4\zeta_{\{1,2,4,5\}} - 2\zeta_{\{1,2,4,6\}} \\ &-11\zeta_{\{1,2,5,6\}} - 5\zeta_{\{1,3,4,5\}} - 2\zeta_{\{1,3,4,6\}} - \zeta_{\{1,3,5,6\}} + 10\zeta_{\{1,4,5,6\}} \\ &+19\zeta_{\{2,3,4,5\}} + \zeta_{\{2,3,4,6\}} - 10\zeta_{\{2,3,5,6\}} - 4\zeta_{\{2,4,5,6\}} + 6\zeta_{\{3,4,5,6\}} \\ &-39\zeta_{\{1,2,3,4,5,6\}}. \end{aligned} \tag{13.2}$$

The elementary 2-homogeneous factorization of u is given by

$$\begin{aligned} u = &(1 - \zeta_{\{1,2\}})(1 - 2\zeta_{\{1,3\}})(1 - 3\zeta_{\{1,5\}})(1 - 2\zeta_{\{1,6\}})(1 - 2\zeta_{\{2,3\}}) \\ &(1 + 2\zeta_{\{2,4\}})(1 + 3\zeta_{\{2,5\}})(1 + \zeta_{\{2,6\}})(1 + 3\zeta_{\{3,4\}})(1 + 3\zeta_{\{3,5\}}) \\ &(1 - 3\zeta_{\{3,6\}})(1 - 2\zeta_{\{4,5\}})(1 - 2\zeta_{\{4,6\}})(1 + 2\zeta_{\{5,6\}}). \end{aligned} \tag{13.3}$$

An element $u \in \mathcal{C}\ell_n{}^{\text{nil}\star}$ for which an elementary homogeneous factorization exists will be said to be *homogeneously decomposable.* While not every invertible nontrivial zeon is homogeneously decomposable, the next theorem shows that every invertible nontrivial zeon can be written as a sum of homogeneously decomposable zeons.

Theorem 13.1.5 (Zeon Division Algorithm)**.** *Let $u, v \in \mathcal{C}\ell_n{}^{\text{nil}\star}$ be nontrivial. Then, there exist $q, r \in \mathcal{C}\ell_n{}^{\text{nil}}$ such that $u = qv + r$ and the following hold:*

i. qv is homogeneously decomposable,
ii. $r = 0$ or $r \in \mathcal{C}\ell_n{}^{\text{nil}\star}$ with $\natural r > \natural u$.

Proof. Let u, v be as stated, and let u have canonical expansion $\sum_{I\in 2^{[n]}} a_I\zeta_I$.

Let $q = v^{-1}a_\varnothing \prod_{\{J:|J|=\natural u\}} \left(1 + \frac{a_J}{a_\varnothing}\zeta_J\right)$. If $u = qv$, then u is homogeneously decomposable and $r = 0$.

On the other hand, if $u - qv \neq 0$ for the preceding choice of q, instead set

$$q = v^{-1}\frac{a_\varnothing}{2} \prod_{\{J:|J|=\natural u\}} \left(1 + \frac{2a_J}{a_\varnothing}\zeta_J\right).$$

Direct computation shows that

$$\Re qv + \langle qv \rangle_{\natural qv} = \frac{1}{2}\Re u + \langle u \rangle_{\natural u},$$

where $\natural u = \natural qv$. Hence, $r = u - qv \in \mathcal{C}\ell_n{}^{\text{nil}\star}$ and $\langle u - qv \rangle_{\natural u} = 0$, so that $\natural r > \natural u$.

□

As a consequence of Theorem 13.1.5, every invertible zeon can be expressed as a sum of homogeneously decomposable zeons. In the best case (i.e., when an invertible zeon $u \in \mathcal{C}\ell_n{}^{\text{nil}\star}$ is homogeneously decomposable), it can be written as a product of $\binom{n}{\natural u}$ or fewer elementary factors. Otherwise, u can be written as a sum of products of elementary factors.

13.2 Elementary Functions on Zeons

Following [76], for a function $f : A \subset \mathbb{R} \to \mathbb{R}$ with a sufficient number of derivatives, one can extend the domain of f by defining

$$f(z) = \sum_{k=0}^{n} \frac{1}{k!} f^{(k)}(\Re z)(\mathfrak{D} z)^k$$

for $z \in \mathcal{C}\ell_n{}^{\text{nil}}$ with $\Re z \in A$. In this case, the coefficient of ζ_I in $f(z)$ can be calculated using Lemma 9.1.1.

Proposition 13.2.1. *Let $f : A \subset \mathbb{R} \to \mathbb{R}$ be n times differentiable on A. If $z = \sum_I a_I \zeta_I \in \mathcal{C}\ell_n{}^{\text{nil}}$ with $a_\varnothing \in A$ and $f(z) = \sum_I b_I \zeta_I$, then*

$$b_I = \sum_{k=0}^{|I|} f^{(k)}(a_\varnothing) \sum_{\substack{\pi \in P(I) \\ |\pi| = k}} \prod_{J \in \pi} a_J,$$

where $P(I)$ is the set of partitions of I.

The Zeon Exponential and Logarithm

We define the exponential function on zeon algebras $\mathfrak{exp} : \mathcal{C}\ell_n{}^{\text{nil}} \to \mathcal{C}\ell_n{}^{\text{nil}}$ by

$$\mathfrak{exp}(z) = e^{\Re z} \sum_{k=0}^{n} \frac{1}{k!} (\mathfrak{D} z)^k.$$

Since zeon algebras are commutative, the following useful identity can be obtained.

Proposition 13.2.2. *For* $z_1, z_2 \in \mathcal{C}\ell_n{}^{\text{nil}}$, $\mathfrak{exp}(z_1)\mathfrak{exp}(z_2) = \mathfrak{exp}(z_1 + z_2)$.

Proof. The calculations in this proof are similar to the real case:

$$\begin{aligned}
\mathfrak{exp}(z_1)\mathfrak{exp}(z_2) &= \left(e^{\Re z_1}\sum_{k=0}^{n}\frac{1}{k!}(\mathfrak{D}z_1)^k\right)\left(e^{\Re z_2}\sum_{k=0}^{n}\frac{1}{k!}(\mathfrak{D}z_2)^k\right)\\
&= e^{\Re(z_1+z_2)}\sum_{k=0}^{n}\sum_{\ell=0}^{k}\frac{1}{\ell!(k-\ell)!}(\mathfrak{D}z_1)^\ell(\mathfrak{D}z_2)^{k-\ell}\\
&= e^{\Re(z_1+z_2)}\sum_{k=0}^{n}\frac{1}{k!}\sum_{\ell=0}^{k}\binom{k}{\ell}(\mathfrak{D}z_1)^\ell(\mathfrak{D}z_2)^{k-\ell}\\
&= e^{\Re(z_1+z_2)}\sum_{k=0}^{n}\frac{1}{k!}(\mathfrak{D}(z_1+z_2))^k\\
&= \mathfrak{exp}(z_1+z_2).
\end{aligned}$$

Note that all terms of the form $(\mathfrak{D}z_1)^p(\mathfrak{D}z_2)^q$ vanish when $p + q > n$. Going forward, this will be done without mention. The binomial theorem was used in the third equation. □

The nilpotency associated with the zeon algebras, along with the commutativity relation shown above, gives us the following useful interpretation: if $z = \sum_I a_I\zeta_I$, then

$$\mathfrak{exp}(z) = e^{a_\varnothing}\prod_{I\neq\varnothing}(1 + a_I\zeta_I).$$

The classic logarithm can also be extended to zeon algebras. For $z = \sum_I a_I\zeta_I \in \mathcal{C}\ell_n{}^{\text{nil}}$ with $\Re z > 0$, we define

$$\mathfrak{log}\, z = \log(\Re z) + \sum_{k=1}^{n}\frac{(-1)^{k-1}}{k(\Re z)^k}(\mathfrak{D}z)^k.$$

The goal now is to show that $\mathfrak{exp}(\mathfrak{log}\, z) = z$ for all $z \in \mathcal{C}\ell_n{}^{\text{nil}}$ with $\Re z > 0$. To do this, we will need a result which extends the chain rule of elementary calculus.

Theorem 13.2.3 (Faà Di Bruno's Formula)**.** *If f and g are functions with a sufficient number of derivatives, then*

$$\frac{d^m}{dt^m}g(f(t)) = \sum\frac{m!}{b_1!\cdots b_m!}g^{(k)}(f(t))\left(\frac{f'(t)}{1!}\right)^{b_1}\cdots\left(\frac{f^{(m)}(t)}{m!}\right)^{b_m},$$

where $k = b_1 + \cdots + b_m$, and the sum is taken over all nonnegative integers $b_1, \ldots, b_m$ such that $b_1 + \cdots + mb_m = m$.

For a proof relying on zeon algebras, see [76]. Theorem 13.2.3 will be useful in establishing the next lemma.

Lemma 13.2.4. *For $m \geq 2$, the following hold:*

$$0 = \sum \frac{(-1)^k}{b_1!1^{b_1} \cdots b_m!m^{b_m}} = \sum \frac{(-1)^{k-1}(k-1)!}{b_1!(1!)^{b_1} \cdots b_m!(m!)^{b_m}}$$

where $k = b_1 + \cdots + b_m$, and the sum is taken over all nonnegative integers $b_1, \ldots, b_m$ such that $b_1 + \cdots + mb_m = m$.

Proof. To establish the result, Theorem 13.2.3 is applied twice, first using $f = \log$ and $g = \exp$ (considered as functions over $\mathbb{R}$), and then using $g = \log$ and $f = \exp$. □

Lemma 13.2.4 can now be used to verify the required inverse relationship of the zeon exponential and the zeon logarithm.

Theorem 13.2.5. *Let $z \in \mathcal{C}\ell_n{}^{\text{nil}}$. Then, $\mathfrak{log}\,(\mathfrak{exp}\, z) = z$. Further, $\mathfrak{exp}(\mathfrak{log}\, z) = z$, provided $\Re z > 0$.*

Proof. First, assume $\Re z = 1$. Then apply the multinomial theorem:

$$\begin{aligned}
\mathfrak{exp}(\mathfrak{log}(z)) &= \sum_{k=0}^{n} \frac{1}{k!} \left(\sum_{j=1}^{n} (-1)^{j-1} \frac{(\mathfrak{D} z)^j}{j} \right)^k \\
&= \sum_{k=0}^{n} \frac{1}{k!} \sum \binom{k}{b_1, \ldots, b_n} \prod_{j=1}^{n} \left((-1)^{j-1} \frac{(\mathfrak{D} z)^j}{j} \right)^{b_j} \\
&= \sum_{k=0}^{n} \sum \frac{(-1)^k}{b_1! \cdots b_m!} \prod_{j=1}^{n} (-1)^{jb_j} \frac{(\mathfrak{D} z)^{jb_j}}{j^{b_j}},
\end{aligned}$$

where the unlabeled sums are taken over all nonnegative integers $b_1, \ldots, b_n$ such that $b_1 + \cdots + b_n = k$. Next, we combine terms with common powers of z:

$$\begin{aligned}
\mathfrak{exp}(\mathfrak{log}\, z) = 1 + \mathfrak{D} z + \sum_{m=2}^{n} (-1)^m \left[\sum \frac{(-1)^k}{b_1!1^{b_1} \cdots b_m!m^{b_m}} \right] (\mathfrak{D} z)^m \\
+ \text{(higher order terms)},
\end{aligned}$$

where in this equation, the unlabeled sum is taken over all nonnegative integers $b_1, \ldots, b_m$ such that $b_1 + \cdots + mb_m = m$, and for convenience $k = b_1 + \cdots + b_m$ on the right side of the equation. By Lemma 13.2.4, all of the inner sums are zero, and since $\mathfrak{D} z$ is nilpotent of order at most $n + 1$, we see that $\mathfrak{exp}(\mathfrak{log}\, z) = z$.

To complete the result for $\mathfrak{exp}(\mathfrak{log}\, z)$, now assume that $\Re z > 0$. It follows that

$$\begin{aligned}\mathfrak{exp}(\mathfrak{log}\, z) &= \mathfrak{exp}\left(\mathfrak{log}\left(\Re z + \mathfrak{D} z\right)\right) = \mathfrak{exp}\left(\mathfrak{log}\left(\Re z\left(1+\frac{\mathfrak{D} z}{\Re z}\right)\right)\right)\\ &= \Re z\, \mathfrak{exp}\left(\mathfrak{log}\left(1+\frac{\mathfrak{D} z}{\Re z}\right)\right) = \Re z\left(1+\frac{\mathfrak{D} z}{\Re z}\right) = z.\end{aligned}$$

Next, to establish the result for $\mathfrak{log}(\mathfrak{exp}\, z)$, suppose $\Re z = 0$. Then,

$$\begin{aligned}\mathfrak{log}(\mathfrak{exp}(z)) &= \sum_{k=1}^{n} \frac{(-1)^{k-1}}{k}\left(\sum_{j=1}^{n}\frac{z^j}{j!}\right)^k\\ &= \sum_{k=1}^{n}\frac{(-1)^{k-1}}{k}\sum_{b_1+\cdots+b_n=k}\binom{k}{b_1,\ldots,b_n}\prod_{j=1}^{n}\left(\frac{z^j}{j!}\right)^{b_j}\\ &= \sum_{k=1}^{n}\sum_{b_1+\cdots+b_n=k}\left[\frac{(-1)^{k-1}(k-1)!}{b_1!(1!)^{b_1}\cdots b_m!(m!)^{b_m}}\right] z^{\sum_j j b_j}.\end{aligned}$$

Collecting terms in powers of z, we obtain

$$\begin{aligned}\mathfrak{log}(\mathfrak{exp}(z)) = z + \sum_{m=2}^{n}\left[\sum \frac{(-1)^{k-1}(k-1)!}{b_1!(1!)^{b_1}\cdots b_m!(m!)^{b_m}}\right] z^m\\ + \text{(higher order terms)},\end{aligned}$$

where in this equation, the sum is taken over all nonnegative integers $b_1, \ldots, b_m$ such that $b_1 + \cdots + m b_m = m$, and for convenience $k = b_1 + \cdots + b_m$. Thus, by Lemma 13.2.4, all of the inner sums are zero, and due to nilpotency the higher order terms are zero. Hence, $\mathfrak{log}(\mathfrak{exp}\, z) = z$ is obtained for the case that $\Re z = 0$, and the result for invertible zeons quickly follows. □

The previous results now imply that the familiar additive property $\mathfrak{log}(z_1 z_2) = \mathfrak{log}(z_1) + \mathfrak{log}(z_2)$ holds for $z_1, z_2 \in \{z \in \mathcal{C}\ell_n{}^{\text{nil}\star} : \Re z > 0\}$. As a consequence of the exponential and logarithm being inverses, a method for calculating rational powers of zeons is obtained.

Proposition 13.2.6. *Let $z \in \mathcal{C}\ell_n{}^{\text{nil}}$ with $\Re z > 0$ and let $m \in \mathbb{Q}$. Then,*

$$\mathfrak{exp}(m\,\mathfrak{log}\, z) = z^m.$$

Proof. Write $m = \frac{p}{q}$ for $p, q \in \mathbb{Z}$ with q nonzero. Then,

$$\mathfrak{exp}(m\,\mathfrak{log}\, z)^q = \mathfrak{exp}(q\frac{1}{q}\mathfrak{log}\, z)^p = z^p.$$

□

For invertible zeons, the following identities can be used in conjunction with elementary factorizations to greatly simplify computations involving elementary functions. The proof is by direct computation.

With the elementary factorization of an invertible zeon in hand, the logarithm is easily computed.

Theorem 13.2.7. *Let $u \in \mathcal{C}\ell_n{}^{\mathrm{nil}\star}$ have elementary factorization $u = (\Re u) \prod_{I \neq \varnothing} (1 + a_I \zeta_I)$, where $\Re u > 0$. Then, $\mathfrak{log}\, u = \log(\Re u) + \sum_{I \neq \varnothing} a_I \zeta_I$.*

Proof. First note that the logarithm of any nontrivial elementary zeon is given by $\mathfrak{log}(1 + a\zeta_I) = a\zeta_I$. Thus, if $u = (\Re u) \prod_{I \neq \varnothing} (1 + a_I \zeta_I)$, one finds

$$\mathfrak{log}\, u = \log(\Re u) + \sum_{I \neq \varnothing} \mathfrak{log}(1 + a_I \zeta_I) = \log(\Re u) + \sum_{I \neq \varnothing} a_I \zeta_I.$$

□

Example 13.2.8. Returning to Example 13.1.4, let u denote the invertible zeon defined in (13.2). In light of Theorem 13.2.7, one computes the logarithm of $u = \sum_I a_I \zeta_I$ by applying the factorization seen in (13.3):

$$\begin{aligned}
\mathfrak{log}\, u &= \mathfrak{log}\left(\prod_{|I| = \natural u} (1 + a_I \zeta_I) \right) \\
&= -\zeta_{\{1,2\}} - 2\zeta_{\{1,3\}} - 3\zeta_{\{1,5\}} - 2\zeta_{\{1,6\}} - 2\zeta_{\{2,3\}} + 2\zeta_{\{2,4\}} + 3\zeta_{\{2,5\}} \\
&\quad + \zeta_{\{2,6\}} + 3\zeta_{\{3,4\}} + 3\zeta_{\{3,5\}} - 3\zeta_{\{3,6\}} - 2\zeta_{\{4,5\}} - 2\zeta_{\{4,6\}} + 2\zeta_{\{5,6\}}.
\end{aligned}$$

As Example 13.2.8 shows, logarithms of homogeneously decomposable zeons are particularly easy to compute.

Corollary 13.2.9. *Let $u \in \mathcal{C}\ell_n{}^{\mathrm{nil}\star}$ be homogeneously decomposable with $\Re u > 0$. Then,*

$$\mathfrak{log}\, u = \log \Re u + (\Re u)^{-1} \langle u \rangle_{\natural u}.$$

Hyperbolic Functions of Zeons

Once the zeon exponential has been defined, the zeon hyperbolic functions follow quite naturally. In particular, we define $\mathfrak{sinh}\, z = \frac{1}{2}(\mathfrak{exp}(z) - \mathfrak{exp}(-z))$ and $\mathfrak{cosh}\, z = \frac{1}{2}(\mathfrak{exp}(z) + \mathfrak{exp}(-z))$. In light of the established properties of the exponential, one can see immediately that $\mathfrak{cosh}^2 z - \mathfrak{sinh}^2 z = 1$ and that $\mathfrak{exp}\, z = \mathfrak{sinh}\, z + \mathfrak{cosh}\, z$. Other properties follow easily from finite power series expansions.

Lemma 13.2.10. *Let $z \in \mathcal{C}\ell_n{}^{\text{nil}}$. Then,*

$$\mathfrak{sinh}\, z = \sinh \Re z \mathfrak{cosh} \mathfrak{D} z + \cosh \Re z \mathfrak{sinh} \mathfrak{D} z,$$
$$\mathfrak{cosh}\, z = \cosh \Re z \mathfrak{cosh} \mathfrak{D} z + \sinh \Re z \mathfrak{sinh} \mathfrak{D} z,$$

where

$$\mathfrak{sinh} \mathfrak{D} z = \sum_{\substack{0 \le k \le n \\ k \text{ odd}}} \frac{(\mathfrak{D} z)^k}{k!}, \qquad \mathfrak{cosh} \mathfrak{D} z = \sum_{\substack{0 \le k \le n \\ k \text{ even}}}^{n} \frac{(\mathfrak{D} z)^k}{k!}.$$

Proof. Let $z \in \mathcal{C}\ell_n{}^{\text{nil}}$. Then,

$$\begin{aligned}
\mathfrak{sinh}\, z &= \mathfrak{sinh}\,(\Re z + \mathfrak{D} z) \\
&= \frac{1}{2}\left(\exp(\Re z) \sum_{k=0}^{n} \frac{(\mathfrak{D} z)^k}{k!} - \exp(-\Re z) \sum_{k=0}^{n} \frac{(-1)^k (\mathfrak{D} z)^k}{k!}\right) \\
&= \sum_{k=0}^{n} \left[\exp(\Re z) - (-1)^k \exp(-\Re z)\right] \frac{(\mathfrak{D} z)^k}{k!}, \\
&= \sinh \Re z \sum_{\substack{0 \le k \le n \\ k \text{ even}}} \frac{(\mathfrak{D} z)^k}{k!} + \cosh \Re z \sum_{\substack{0 \le k \le n \\ k \text{ odd}}} \frac{(\mathfrak{D} z)^k}{k!} \\
&= \sinh \Re z \mathfrak{cosh} \mathfrak{D} z + \sinh \Re z \mathfrak{cosh} \mathfrak{D} z.
\end{aligned}$$

The formulation of $\mathfrak{cosh}\, z$ is established similarly, or by using $\mathfrak{cosh}\, z = \mathfrak{exp}\, z - \mathfrak{sinh}\, z$. □

It is evident from Lemma 13.2.10 that $\Re(\mathfrak{sinh}\, z) = \sinh \Re z$ and $\Re(\mathfrak{cosh}\, z) = \cosh \Re z$. Thus, for all $z \in \mathcal{C}\ell_n{}^{\text{nil}}$, $\Re(\mathfrak{cosh}\, z) \neq 0$ so that $(\mathfrak{cosh}\, z)^{-1} = \operatorname{sech} z$ is well-defined. It is therefore natural to define the hyperbolic tangent by

$$\mathfrak{tanh}\, z = \mathfrak{sinh}\, z (\mathfrak{cosh}\, z)^{-1} = \mathfrak{sinh}\, z \operatorname{sech} z, \quad (z \in \mathcal{C}\ell_n{}^{\text{nil}}).$$

Inverse Hyperbolic Functions

With logarithms, roots, and multiplicative inverses in hand, it is possible to define a number of elementary inverse functions on the invertible zeons. To begin, recall the following elementary inverse functions:

$$\sinh^{-1} u = \log\left(u + \sqrt{u^2 + 1}\right),$$
$$\cosh^{-1} u = \log\left(u + \sqrt{u^2 - 1}\right),$$
$$\tanh^{-1} u = \frac{1}{2} \log\left(\frac{1 + u}{1 - u}\right).$$

In light of previous results, these formulas hold for elementary functions on zeons, provided some restrictions on the domain are satisfied. These are summarized in the following lemma.

Lemma 13.2.11. *Let* $z \in \mathcal{C}\ell_n{}^{\text{nil}}$. *Then,*

$$\begin{aligned}
\mathfrak{sinh}^{-1} z &= \mathfrak{log}\left(z + \sqrt{z^2+1}\right),\\
\mathfrak{cosh}^{-1} z &= \mathfrak{log}\left(z + \sqrt{z^2-1}\right), \ (\Re z \geq 1),\\
\mathfrak{tanh}^{-1} z &= \frac{1}{2}\mathfrak{log}\left(\frac{1+z}{1-z}\right), \ (-1 < \Re z < 1).
\end{aligned}$$

Proof. To compute $\mathfrak{log} z$, it is necessary that $\Re z > 0$. Let $z \in \mathcal{C}\ell_n{}^{\text{nil}}$ and observe that

$$\begin{aligned}
\Re(z + \sqrt{z^2+1}) &= \Re z + \Re(\sqrt{((\Re z)^2+1) + \mathfrak{D}(z^2)})\\
&= \Re z + \sqrt{(\Re z)^2+1}\\
&> 0.
\end{aligned}$$

Setting $u = \mathfrak{log}\left(z + \sqrt{z^2+1}\right)$, which is defined for all zeons, it is easy to verify that

$$\begin{aligned}
\mathfrak{sinh}\left(\mathfrak{log}\left(z+\sqrt{z^2+1}\right)\right) &= \frac{\mathfrak{exp}(u) - \mathfrak{exp}(-u)}{2}\\
&= \frac{1}{2}\left(z + \sqrt{z^2+1} - \left(z+\sqrt{z^2+1}\right)^{-1}\right)\\
&= \frac{1}{2}\left(z + \sqrt{z^2+1} - \frac{z - \sqrt{z^2+1}}{-1}\right)\\
&= z.
\end{aligned}$$

Thus, $\mathfrak{sinh}^{-1} z = \mathfrak{log}\left(z + \sqrt{z^2+1}\right)$ is defined for all zeons. Similarly, one finds that $\mathfrak{log}\left(z + \sqrt{z^2-1}\right)$ is defined for invertible zeons whose real part is greater than 1; i.e.,

$$\begin{aligned}
\Re(z + \sqrt{z^2-1}) &= \Re z + \Re(\sqrt{((\Re z)^2-1) + \mathfrak{D}(z^2)})\\
&= \Re z + \sqrt{(\Re z)^2-1}\\
&> 0, \text{ provided } \Re z \geq 1.
\end{aligned}$$

The verification that $\mathfrak{cosh}\left(z+\sqrt{z^2-1}\right) = z$ then establishes the result; the details are left as an exercise. Finally,

$$\begin{aligned}\Re\left(\frac{1+z}{1-z}\right) &= \Re\left((1+z)(1-z)^{-1}\right)\\ &= \frac{1+\Re z}{1-\Re z}\\ &> 0, \text{ provided } -1 < \Re z < 1.\end{aligned}$$

Once again, the verification $\mathfrak{tanh}\left(\frac{1}{2}\mathfrak{log}\left(\frac{1+z}{1-z}\right)\right) = z$ is left as an exercise. □

Trigonometric Functions of Zeons

In this subsection we extend the sine and cosine functions (as defined on $\mathbb{R}$) to general zeon algebras. Using the previously described method for extending functions, we define $\mathfrak{sin}, \mathfrak{cos} : \mathcal{C}\ell_n{}^{\text{nil}} \to \mathcal{C}\ell_n{}^{\text{nil}}$ by

$$\begin{aligned}\mathfrak{sin}\, z &= \sum_{k=0}^{n} \frac{\sin^{(k)}(\Re z)}{k!}(\mathfrak{D}z)^k,\\ \mathfrak{cos}\, z &= \sum_{k=0}^{n} \frac{\cos^{(k)}(\Re z)}{k!}(\mathfrak{D}z)^k.\end{aligned}$$

An equivalent formulation is given by the following lemma.

Lemma 13.2.12. *Let $z \in \mathcal{C}\ell_n{}^{\text{nil}}$. Then,*

$$\begin{aligned}\mathfrak{sin}\, z &= \sin(\Re z)\mathfrak{cos}(\mathfrak{D}z) + \cos(\Re z)\mathfrak{sin}(\mathfrak{D}z),\\ \mathfrak{cos}\, z &= \cos(\Re z)\mathfrak{cos}(\mathfrak{D}z) - \sin(\Re z)\mathfrak{sin}(\mathfrak{D}z),\end{aligned}$$

where

$$\mathfrak{sin}\,(\mathfrak{D}z) = \sum_{k=0}^{\lfloor n/2\rfloor} \frac{(-1)^k(\mathfrak{D}z)^{2k+1}}{(2k+1)!}, \qquad \mathfrak{cos}\,(\mathfrak{D}z) = \sum_{k=0}^{\lfloor n/2\rfloor} \frac{(-1)^k(\mathfrak{D}z)^{2k}}{(2k)!}.$$

Proof. The proof is left as an exercise for the reader. □

We now list some other basic properties of these functions.

Proposition 13.2.13. *Let $z \in \mathcal{C}\ell_n{}^{\text{nil}}$ be given. Then the following hold:*

(i) The sine and cosine functions are 2π-periodic on $\mathcal{C}\ell_n{}^{\text{nil}}$; i.e.,

$$\mathfrak{sin}(z+2\pi) = \mathfrak{sin}\, z, \qquad \mathfrak{cos}(z+2\pi) = \mathfrak{cos}\, z,$$

(ii) $\mathfrak{sin}\, z = 0$ *if and only if* $z = \Re z \equiv 0 \pmod{\pi}$,

(iii) $\mathfrak{cos}\, z = 0$ *if and only if* $z = \Re z \equiv \frac{\pi}{2} \pmod{\pi}$, *and*

(iv) $\mathfrak{sin}\, z = \mathfrak{cos}\, z$ *if and only if* $z = \Re z \equiv \frac{\pi}{4} \pmod{\pi}$.

Proof. Observe that $\Re(z+2\pi) = \Re z + 2\pi$, $\mathfrak{D}(z+2\pi) = \mathfrak{D}z$ and thus

$$\begin{aligned}\mathfrak{sin}(z+2\pi) &= \sin(\Re z + 2\pi)\,\mathfrak{cos}(\mathfrak{D}z) + \cos(\Re z + 2\pi)\,\mathfrak{sin}(\mathfrak{D}z)\\ &= \sin(\Re z)\mathfrak{cos}(\mathfrak{D}z) + \cos(\Re z)\mathfrak{sin}(\mathfrak{D}z) = \mathfrak{sin}\, z.\end{aligned}$$

The proof that $\mathfrak{cos}(z+2\pi) = \mathfrak{cos}\, z$ is similar.

If $z = \Re z \equiv 0 \pmod{\pi}$ then $\sin(\Re z) = \mathfrak{sin}(\mathfrak{D}z) = 0$, and thus $\mathfrak{sin}\, z = 0$. Conversely if $\mathfrak{sin}\, z = 0$, then the grade-0 part of $\mathfrak{sin}\, z$ is zero, and thus $\Re z \equiv 0 \pmod{\pi}$. But then $\cos(\Re z)$ is nonzero, so we must have $\mathfrak{sin}(\mathfrak{D}z) = 0$. Observe that $0 = \langle \mathfrak{sin}(\mathfrak{D}z)\rangle_1 = \langle z\rangle_1$, and by induction it then follows that

$$0 = \langle \mathfrak{sin}(\mathfrak{D}z)\rangle_k = \langle z\rangle_k$$

for $1 \le k \le n$. This proves *(ii)*, and the proofs of *(iii)* and *(iv)* are similar. □

We now aim to show that three fundamental relations of trigonometry hold in this setting. With these results, the majority of elementary trigonometry will follow.

Theorem 13.2.14. *For* $z \in \mathcal{C}\ell_n{}^{\text{nil}}$, $\mathfrak{cos}^2 z + \mathfrak{sin}^2 z = 1$; *i.e., the Pythagorean identity holds.*

Proof. First, observe that for any $z \in \mathcal{C}\ell_n{}^{\text{nil}}$, we have

$$\begin{aligned}\mathfrak{cos}^2 z + \mathfrak{sin}^2 z &= \left(\cos^2(\Re z) + \sin^2(\Re z)\right)\left(\mathfrak{cos}^2\mathfrak{D}z + \mathfrak{sin}^2\mathfrak{D}z\right)\\ &= \mathfrak{cos}^2\mathfrak{D}z + \mathfrak{sin}^2\mathfrak{D}z.\end{aligned}$$

Thus, the problem is reduced to the case of nilpotent zeons. To this end, fix $z \in \mathcal{C}\ell_n{}^{\text{nil}_\circ}$. Then,

$$\begin{aligned}\mathfrak{cos}^2 z &= \left(\sum_{k=0}^{\lfloor n/2\rfloor} \frac{(-1)^k}{(2k)!} z^{2k}\right)\left(\sum_{k=0}^{\lfloor n/2\rfloor} \frac{(-1)^k}{(2k)!} z^{2k}\right)\\ &= \sum_{k=0}^{\lfloor n/2\rfloor} (-1)^k \left(\sum_{\ell=0}^{k} \frac{1}{(2k-2\ell)!(2\ell)!}\right) z^{2k}, \qquad (13.4)\end{aligned}$$

and

$$\begin{aligned}
\mathfrak{sin}^2 z &= \left(\sum_{k=0}^{\lfloor n/2\rfloor} \frac{(-1)^k}{(2k+1)!} z^{2k+1}\right)\left(\sum_{k=0}^{\lfloor n/2\rfloor} \frac{(-1)^k}{(2k+1)!} z^{2k+1}\right)\\
&= \sum_{k=0}^{\lfloor n/2\rfloor} (-1)^k \left(\sum_{\ell=0}^{k} \frac{1}{(2k-2\ell+1)!(2\ell+1)!}\right) z^{2k+2}\\
&= \sum_{k=1}^{\lfloor n/2\rfloor+1} (-1)^{k-1} \left(\sum_{\ell=0}^{k-1} \frac{1}{(2k-2\ell-1)!(2\ell+1)!}\right) z^{2k}\\
&= (-1) \sum_{k=1}^{\lfloor n/2\rfloor} (-1)^k \left(\sum_{\ell=0}^{k-1} \frac{1}{(2k-2\ell-1)!(2\ell+1)!}\right) z^{2k}. \qquad (13.5)
\end{aligned}$$

Combining (13.4) and (13.5),

$$\begin{aligned}
\mathfrak{cos}^2 z + \mathfrak{sin}^2 z &= 1 + \sum_{k=1}^{\lfloor n/2\rfloor} \frac{(-1)^k}{(2k)!}\left(\sum_{l=0}^{k}\binom{2k}{2l} - \sum_{\ell=0}^{k-1}\binom{2k}{2\ell+1}\right) z^{2k}.\\
&= 1 + \sum_{k=1}^{\lfloor n/2\rfloor} \frac{(-1)^k}{(2k)!}\left(\sum_{\ell=0}^{2k}(-1)^\ell\binom{2k}{\ell}\right) z^{2k}\\
&= 1 + \sum_{k=1}^{\lfloor n/2\rfloor} \frac{(-1)^k}{(2k)!}(1-1)^{2k} z^{2k}\\
&= 1.
\end{aligned}$$

□

Theorem 13.2.15. *For $z_1, z_2 \in \mathcal{C}\ell_n{}^{\text{nil}}$, the following identities hold:*

$$\begin{aligned}
\mathfrak{sin}(z_1+z_2) &= \mathfrak{sin}\, z_1 \mathfrak{cos}\, z_2 + \mathfrak{cos}\, z_1 \mathfrak{sin}\, z_2,\\
\mathfrak{cos}(z_1+z_2) &= \mathfrak{cos}\, z_1 \mathfrak{cos}\, z_2 - \mathfrak{sin}\, z_1 \mathfrak{sin}\, z_2.
\end{aligned}$$

Proof. First assume that $\Re z_1 = \Re z_2 = 0$. It follows that

$$\begin{aligned}
\mathfrak{sin}(z_1+z_2) &= \sum_{k=0}^{\lfloor n/2\rfloor} \frac{(-1)^k}{(2k+1)!}(z_1+z_2)^{2k+1}\\
&= \sum_{k=0}^{\lfloor n/2\rfloor} \frac{(-1)^k}{(2k+1)!} \sum_{\ell=0}^{2k+1}\binom{2k+1}{\ell} {z_1}^{2k+1-\ell} {z_2}^{\ell}
\end{aligned}$$

$$= \sum_{k=0}^{\lfloor n/2 \rfloor} (-1)^k \sum_{\ell=0}^{2k+1} \frac{{z_1}^{2k+1-\ell}}{(2k+1-\ell)!} \frac{{z_2}^{\ell}}{\ell!}$$

$$= \sum_{k=0}^{\lfloor n/2 \rfloor} (-1)^k \left[\sum_{\substack{\ell=0 \\ \ell \text{ even}}}^{2k+1} \frac{{z_1}^{2k+1-\ell}}{(2k+1-\ell)!} \frac{{z_2}^{\ell}}{\ell!} + \sum_{\substack{\ell=0 \\ \ell \text{ odd}}}^{2k+1} \frac{{z_1}^{2k+1-\ell}}{(2k+1-\ell)!} \frac{{z_2}^{\ell}}{\ell!} \right]$$

$$= \sum_{k=0}^{\lfloor n/2 \rfloor} (-1)^k \sum_{\ell=0}^{k} \frac{{z_1}^{2k+1-2\ell}}{(2k+1-2\ell)!} \frac{{z_2}^{2\ell}}{(2\ell)!}$$

$$+ \sum_{k=0}^{\lfloor n/2 \rfloor} (-1)^k \sum_{\ell=0}^{k} \frac{{z_1}^{2k-2\ell}}{(2k-2\ell)!} \frac{{z_2}^{2\ell+1}}{(2\ell+1)!}$$

$$= \mathfrak{sin} z_1 \mathfrak{cos} z_2 + \mathfrak{cos} z_1 \mathfrak{sin} z_2.$$

Further,

$$\mathfrak{cos} z_1 \mathfrak{cos} z_2 = \left(\sum_{k=0}^{\lfloor n/2 \rfloor} \frac{(-1)^k {z_1}^{2k}}{(2k)!} \right) \left(\sum_{k=0}^{\lfloor n/2 \rfloor} \frac{(-1)^k {z_2}^{2k}}{(2k)!} \right)$$

$$= \sum_{k=0}^{\lfloor n/2 \rfloor} (-1)^k \sum_{\ell=0}^{k} \frac{{z_1}^{2k-2\ell}}{(2k-2\ell)!} \frac{{z_2}^{2\ell}}{(2\ell)!}$$

$$= \sum_{k=0}^{\lfloor n/2 \rfloor} (-1)^k \sum_{\substack{\ell=0 \\ \ell \text{ even}}}^{2k} \frac{{z_1}^{2k-\ell}}{(2k-\ell)!} \frac{{z_2}^{\ell}}{\ell!}.$$

By similar reasoning,

$$\mathfrak{sin} z_1 \mathfrak{sin} z_2 = \left(\sum_{k=0}^{\lfloor n/2 \rfloor} \frac{(-1)^k {z_1}^{2k+1}}{(2k+1)!} \right) \left(\sum_{k=0}^{\lfloor n/2 \rfloor} \frac{(-1)^k {z_2}^{2k+1}}{(2k+1)!} \right)$$

$$= \sum_{k=0}^{\lfloor n/2 \rfloor - 1} (-1)^k \sum_{\ell=0}^{k} \frac{{z_1}^{2k-2\ell+1}}{(2k-2\ell+1)!} \frac{{z_2}^{2\ell+1}}{(2\ell+1)!}$$

$$= \sum_{k=1}^{\lfloor n/2 \rfloor} (-1)^{k-1} \sum_{\ell=0}^{k-1} \frac{{z_1}^{2k-2\ell-1}}{(2k-2\ell-1)!} \frac{{z_2}^{2\ell+1}}{(2\ell+1)!}$$

$$= (-1) \sum_{k=0}^{\lfloor n/2 \rfloor} (-1)^k \sum_{\substack{\ell=0 \\ \ell \text{ odd}}}^{2k} \frac{{z_1}^{2k-l}}{(2k-\ell)!} \frac{{z_2}^{\ell}}{\ell!}.$$

Taking the difference of these terms,

$$
\begin{aligned}
&\cos z_1 \cos z_2 - \sin z_1 \sin z_2 \\
&= \sum_{k=0}^{\lfloor n/2 \rfloor} (-1)^k \left[\sum_{\substack{\ell=0 \\ \ell \text{ even}}}^{2k} \frac{{z_1}^{2k-\ell}}{(2k-\ell)!} \frac{{z_2}^{\ell}}{\ell!} + \sum_{\substack{\ell=0 \\ \ell \text{ odd}}}^{2k} \frac{{z_1}^{2k-\ell}}{(2k-\ell)!} \frac{{z_2}^{\ell}}{\ell!} \right] \\
&= \sum_{k=0}^{\lfloor n/2 \rfloor} (-1)^k \sum_{\ell=0}^{2k} \frac{{z_1}^{2k-\ell}}{(2k-\ell)!} \frac{{z_2}^{\ell}}{l!} \\
&= \sum_{k=0}^{\lfloor n/2 \rfloor} \frac{(-1)^k}{(2k)!} \sum_{\ell=0}^{2k} \binom{2k}{\ell} {z_1}^{2k-\ell} {z_2}^{\ell} \\
&= \cos(z_1 + z_2).
\end{aligned}
$$

Thus, the result holds for nilpotent elements z_1 and z_2. The result for general elements $z_1, z_2 \in \mathcal{C}\ell_n{}^{\text{nil}}$ then quickly follows.

□

With the results shown thus far, all manner of basic trigonometric identities are readily verifiable for zeons. We list a few here, and their proofs are identical to the real case.

Proposition 13.2.16. *Let $z, z_1, z_2 \in \mathcal{C}\ell_n{}^{\text{nil}}$ be given. The following identities hold:*

(i) $\sin 2z = 2\sin z \cos z$,

(ii) $\cos 2z = \cos^2 z - \sin^2 z = 2\cos^2 z - 1 = 1 - \sin^2 z$,

(iii) $2\cos z_1 \cos z_2 = \cos(z_1 - z_2) + \cos(z_1 + z_2)$,

(iv) $2\sin z_1 \sin z_2 = \cos(z_1 - z_2) - \cos(z_1 + z_2)$,

(v) $2\sin z_1 \cos z_2 = \sin(z_1 + z_2) + \sin(z_1 - z_2)$,

(vi) *if* $\Re z \not\equiv 0 \pmod{\pi}$ *and* $m \in \mathbb{N}$, *then*

$$1 + 2\sum_{k=1}^{m} \cos kz = \frac{\sin\left(\left(m + \frac{1}{2}\right) z\right)}{\sin\left(\frac{1}{2} z\right)}.$$

Exercises

Exercise 13.1: For the following choices of $u, v \in \mathcal{C}\ell_3{}^{\text{nil}}$, apply the zeon division algorithm to find $q, r \in \mathcal{C}\ell_3{}^{\text{nil}}$ such that $u = qv + r$, where $r = 0$ or $\natural r > \natural u$.

(a)

$$u = 2 - 3\zeta_{\{1\}} - \zeta_{\{2\}} + 3\zeta_{\{3\}} - 2\zeta_{\{1,3\}} + \zeta_{\{2,3\}} + 3\zeta_{\{1,2,3\}},$$
$$v = 3 + 2\zeta_{\{1\}} + \zeta_{\{2\}} + 2\zeta_{\{3\}} + 3\zeta_{\{1,3\}} + 2\zeta_{\{2,3\}} - 2\zeta_{\{1,2,3\}}.$$

(b)

$$u = 3 - 3\zeta_{\{1\}} + \zeta_{\{2\}} - 3\zeta_{\{3\}} - 2\zeta_{\{1,2\}} + 3\zeta_{\{1,3\}} - \zeta_{\{2,3\}} + 3\zeta_{\{1,2,3\}},$$
$$v = 1 + 3\zeta_{\{1\}} + \zeta_{\{2\}} - 2\zeta_{\{1,2\}} + 2\zeta_{\{1,3\}} - \zeta_{\{2,3\}} + \zeta_{\{1,2,3\}} + 3\zeta_{\{1\}} + \zeta_{\{2\}}.$$

(c)

$$u = 3 - 2\zeta_{\{1\}} + 2\zeta_{\{2\}} + 2\zeta_{\{3\}} - \zeta_{\{1,2\}} - 3\zeta_{\{1,3\}} - 3\zeta_{\{2,3\}} + \zeta_{\{1,2,3\}},$$
$$v = 3 - 3\zeta_{\{1\}} + 2\zeta_{\{2\}} - 2\zeta_{\{3\}} - 3\zeta_{\{1,2\}} + \zeta_{\{1,3\}} + 2\zeta_{\{2,3\}} - 3\zeta_{\{1,2,3\}}.$$

Exercise 13.2: Compute the elementary factorization of $u = 1 + 10\zeta_{\{1\}} - 2\zeta_{\{2\}} + 51\zeta_{\{1,2\}} + 21\zeta_{\{1,3\}} - 3\zeta_{\{2,3\}} + 20\zeta_{\{1,2,3\}}$ in $\mathcal{C}\ell_3{}^{\text{nil}}$.

Exercise 13.3: Prove Lemma 13.2.12.

Exercise 13.4: Rigorously establish the following identities:

i. $\mathfrak{sin}2z = 2\mathfrak{sin}z\mathfrak{cos}z$,
ii. $\mathfrak{cos}2z = \mathfrak{cos}^2 z - \mathfrak{sin}^2 z = 2\mathfrak{cos}^2 z - 1 = 1 - \mathfrak{sin}^2 z$,
iii. $2\mathfrak{cos}z_1\mathfrak{cos}z_2 = \mathfrak{cos}(z_1 - z_2) + \mathfrak{cos}(z_1 + z_2)$,
iv. $2\mathfrak{sin}z_1\mathfrak{sin}z_2 = \mathfrak{cos}(z_1 - z_2) - \mathfrak{cos}(z_1 + z_2)$,
v. $2\mathfrak{sin}z_1\mathfrak{cos}z_2 = \mathfrak{sin}(z_1 + z_2) + \mathfrak{sin}(z_1 - z_2)$.

Exercise 13.5: Let $u = 1 - 3\zeta_{\{1\}} + 3\zeta_{\{3\}} - 2\zeta_{\{1,2\}}$. Evaluate the following:

i. $\mathfrak{sin}u$,
ii. $\mathfrak{cos}u$,
iii. $\mathfrak{tan}u$.

Exercise 13.6: Let $z_1, z_2 \in (-\pi/2, \pi/2) \oplus \mathcal{C}\ell_n{}^{\text{nil}_\text{o}}$. Establish the following identity:

$$\mathfrak{tan}(z_1 + z_1) = \frac{\mathfrak{tan}z_1 + \mathfrak{tan}z_2}{1 - \mathfrak{tan}z_1\mathfrak{tan}z_2}.$$

Exercise 13.7: Let $z \in \mathcal{C}\ell_n{}^{\text{nil}}$. Use the definitions of zeon sine and cosine to prove that the following identities hold:

$$\begin{aligned}\mathfrak{sin}\, z &= \sin(\Re z)\mathfrak{cos}(\mathfrak{D}z) + \cos(\Re z)\mathfrak{sin}(\mathfrak{D}z),\\ \mathfrak{cos}\, z &= \cos(\Re z)\mathfrak{cos}(\mathfrak{D}z) - \sin(\Re z)\mathfrak{sin}(\mathfrak{D}z),\end{aligned}$$

where

$$\mathfrak{sin}\,(\mathfrak{D}z) = \sum_{k=0}^{\lfloor n/2\rfloor} \frac{(-1)^k(\mathfrak{D}z)^{2k+1}}{(2k+1)!}, \qquad \mathfrak{cos}\,(\mathfrak{D}z) = \sum_{k=0}^{\lfloor n/2\rfloor} \frac{(-1)^k(\mathfrak{D}z)^{2k}}{(2k)!}.$$

Exercise 13.8: Let $z \in \mathcal{C}\ell_n{}^{\text{nil}}$ such that $\Re z \geq 1$. Show that $\mathfrak{cosh}^{-1} z = \left(z + \sqrt{z^2 - 1}\right)$ by verifying the identity $\mathfrak{cosh}\left(z + \sqrt{z^2-1}\right) = z$.

Exercise 13.9: For $z \in \mathcal{C}\ell_n{}^{\text{nil}}$ with $|\Re z| < 1$, show that $\mathfrak{tanh}^{-1}(z) = \frac{1}{2}\mathfrak{log}\left(\frac{1+z}{1-z}\right)$ by verifying the identity $\mathfrak{tanh}\left(\frac{1}{2}\mathfrak{log}\left(\frac{1+z}{1-z}\right)\right) = z$.

Chapter 14

Zeon Differential Calculus

Analogous to real functions, zeon functions are defined as zeon-valued functions of a zeon variable. In this chapter, based on the paper [101], formal criteria for continuity and differentiability of zeon functions are put on a rigorous footing and the "usual" differentiation rules are formally established. As special cases, zeon extensions of real functions and zeon functions of one real variable are considered.

14.1 Zeon Functions

To begin, a *zeon function* (i.e., a zeon-valued function of a zeon variable) is a function $\varphi : \mathfrak{D} \to \mathcal{C}\ell_n{}^{\mathrm{nil}}$, where $\mathfrak{D} \subseteq \mathcal{C}\ell_n{}^{\mathrm{nil}}$. With zeon norms in hand, continuity can be defined in the usual way.

Definition 14.1.1. A zeon function φ is said to be *continuous at* $u \in \mathcal{C}\ell_n{}^{\mathrm{nil}}$ if for every $\varepsilon > 0$, there exists $\delta > 0$ such that $\|u - v\| < \delta$ implies $\|\varphi(u) - \varphi(v)\| < \varepsilon$.

On the other hand, care must be taken in defining differentiability of φ. In order to define the familiar difference quotient $\dfrac{\varphi(u+h) - \varphi(u)}{h}$, it is necessary that h is invertible, i.e., $\Re h \neq 0$. Moreover, as will be seen later, directional independence as $h \to 0$ will require some consideration of the relative rates as $\Re h \to 0$ and $\mathfrak{D} h \to 0$.

For this reason, a *(deleted)* ε *neighborhood of* 0 in $\mathcal{C}\ell_n{}^{\mathrm{nil}\star}$ is defined to be an open connected (in the standard topology of $\mathbb{R}^{2^n} \cong \mathcal{C}\ell_n{}^{\mathrm{nil}}$) subset $\mathcal{N}_\varepsilon$ of invertible elements satisfying the following conditions:

(i.) $0 < |u|_\star < \varepsilon$ for all $u \in \mathcal{N}_\varepsilon$;
(ii.) $0 \leq \|\mathfrak{D} u\| \leq |u|_\star$ for all $u \in \mathcal{N}_\varepsilon$.

Definition 14.1.2. Let φ be a zeon function. The value $w \in {\mathcal{C}\ell_n}^{\rm nil}$ is said to be *the derivative of φ at u* if for every $\varepsilon > 0$, there exists a deleted neighborhood $\mathcal{N}_\varepsilon \subset {\mathcal{C}\ell_n}^{\rm nil\star}$ of 0 such that $u - v \in \mathcal{N}_\varepsilon$ implies $\left\| \dfrac{\varphi(u) - \varphi(v)}{u - v} - w \right\| < \varepsilon$. When such a value w exists, φ is said to be *differentiable at u*, and the derivative w is denoted by $\varphi'(u)$.

In order to reflect this particular consideration, the following notational convention is adopted. When w is the derivative of φ at u, one equivalently writes

$$\begin{aligned} w &= \varphi'(u) \\ &= \lim_{h \twoheadrightarrow 0} \frac{\varphi(u+h) - \varphi(u)}{h} \\ &= \lim_{(u-v) \twoheadrightarrow 0} \frac{\varphi(u) - \varphi(v)}{u - v}. \end{aligned}$$

Writing $h = \Re h + \mathfrak{D}h$, Proposition 9.1.3 allows the multiplicative inverse of h to be expanded as

$$\begin{aligned} h^{-1} &= \frac{1}{\Re h} \sum_{j=0}^{\kappa(\mathfrak{D}h)-1} (-1)^j \left(\frac{\mathfrak{D}h}{\Re h} \right)^j \\ &= \frac{1}{\Re h} + \frac{1}{\Re h} \sum_{j=1}^{\kappa(\mathfrak{D}h)-1} (-1)^j \left(\frac{\mathfrak{D}h}{\Re h} \right)^j. \end{aligned}$$

In the interest of directional independence in the derivative, we note that when $\varphi'(u)$ exists,

$$\begin{aligned} \varphi'(u) &= \lim_{h_\varnothing \to 0} \frac{\varphi(u + h_\varnothing) - \varphi(u)}{h_\varnothing} \\ &= \lim_{h \twoheadrightarrow 0} \frac{\varphi(u + h_\varnothing) - \varphi(u)}{h_\varnothing}. \end{aligned} \tag{14.1}$$

When $\mathfrak{D}h$ is nilpotent of index 2, writing $h = h_\varnothing + \mathfrak{D}h$ gives

$$h^{-1} = \frac{1}{h_\varnothing} - \frac{\mathfrak{D}h}{{h_\varnothing}^2}.$$

In this case, the difference quotient is

$$\begin{aligned} \frac{\varphi(u+h) - \varphi(u)}{h} &= (\varphi(u+h) - \varphi(u)) \left(\frac{1}{h_\varnothing} - \frac{\mathfrak{D}h}{{h_\varnothing}^2} \right) \\ &= \frac{\varphi(u+h) - \varphi(u)}{h_\varnothing} \left(1 - \frac{\mathfrak{D}h}{h_\varnothing} \right). \end{aligned} \tag{14.2}$$

Let $\varepsilon > 0$ be given. Assuming that $\dfrac{\mathfrak{D}h}{h_\varnothing} \to 0$ as $h \to 0$, continuity of φ ensures that for fixed $h_\varnothing$, there exists neighborhood $\mathcal{N}_1$ such that $h \in \mathcal{N}_1$ implies

$$\begin{aligned}\left\|\frac{\varphi(u+h)-\varphi(u+h_\varnothing)}{h_\varnothing}\right\| &= \left\|\frac{\varphi(u+h_\varnothing+\mathfrak{D}h)-\varphi(u+h_\varnothing)}{h_\varnothing}\right\| \\ &< \varepsilon/2.\end{aligned}$$

Similarly, (14.1) implies the existence of deleted neighborhood $\mathcal{N}_2$ such that $h \in \mathcal{N}_2$ implies

$$\left\|\frac{\varphi(u+h_\varnothing)-\varphi(u)}{h_\varnothing} - \varphi'(u)\right\| < \varepsilon/2.$$

Now, $h \in \mathcal{N}_1 \cap \mathcal{N}_2$ implies

$$\begin{aligned}&\left\|\frac{\varphi(u+h)-\varphi(u)}{h_\varnothing} - \varphi'(u)\right\| \\ &= \left\|\frac{\varphi(u+h)-\varphi(u+h_\varnothing)+\varphi(u+h_\varnothing)-\varphi(u)}{h_\varnothing} - \varphi'(u)\right\| \\ &\le \left\|\frac{\varphi(u+h)-\varphi(u+h_\varnothing)}{h_\varnothing}\right\| + \left\|\frac{\varphi(u+h_\varnothing)-\varphi(u)}{h_\varnothing} - \varphi'(u)\right\| \\ &< \varepsilon.\end{aligned}$$

Hence, the assumption $\lim\limits_{h\twoheadrightarrow 0} \dfrac{\mathfrak{D}h}{h_\varnothing} = 0$ together with (14.2) results in

$$\begin{aligned}\lim_{h\twoheadrightarrow 0} \frac{\varphi(u+h)-\varphi(u)}{h} &= \lim_{h\twoheadrightarrow 0} \frac{\varphi(u+h)-\varphi(u)}{h_\varnothing} \\ &= \lim_{h\twoheadrightarrow 0} \frac{\varphi(u+h_\varnothing)-\varphi(u)}{h_\varnothing} \\ &= \varphi'(u).\end{aligned}$$

More generally, when $\kappa(\mathfrak{D}h) > 2$, rewrite $\mathfrak{D}(h^{-1}) = \dfrac{1}{h_\varnothing}\Lambda_h$, where

$$\Lambda_h = \sum_{j=1}^{\kappa(\mathfrak{D}h)-1} (-1)^j \left(\frac{\mathfrak{D}h}{h_\varnothing}\right)^j.$$

The difference quotient now becomes

$$\begin{aligned}\frac{\varphi(u+h)-\varphi(u)}{h} &= (\varphi(u+h)-\varphi(u))\left(\frac{1}{h_\varnothing} + \mathfrak{D}(h^{-1})\right) \\ &= \frac{\varphi(u+h)-\varphi(u)}{h_\varnothing}(1+\Lambda_h).\end{aligned}$$

Assuming convergence of Λ_h, let $\Lambda = \lim_{h \twoheadrightarrow 0} \Lambda_h$. It follows that $\varphi'(u)$ exists only if

$$\varphi'(u) = (1 + \Lambda) \lim_{h \twoheadrightarrow 0} \frac{\varphi(u+h) - \varphi(u)}{h_\varnothing}.$$

When $\Lambda = 0$,

$$\begin{aligned}\varphi'(u) &= \lim_{h \twoheadrightarrow 0} \frac{\varphi(u+h) - \varphi(u)}{h_\varnothing} \\ &= \lim_{h_\varnothing \to 0} \frac{\varphi(u+h_\varnothing) - \varphi(u)}{h_\varnothing}.\end{aligned}$$

Under this definition of differentiability, the ordinary rules of differentiation follow naturally.

Lemma 14.1.3. *For arbitrary $\alpha \in \mathcal{C}\ell_n{}^{\text{nil}}$, the derivative of $\varphi(u) = \alpha u$ is $\varphi'(u) = \alpha$.*

Proof. The proof is straightforward:

$$\begin{aligned}\lim_{h \twoheadrightarrow 0} \frac{(\alpha u + \alpha h) - \alpha u}{h} &= \lim_{h \twoheadrightarrow 0} \frac{\alpha h}{h} \\ &= \alpha.\end{aligned}$$

□

Linearity of the derivative is established as in the real case.

Lemma 14.1.4. *Suppose φ, ψ are differentiable on $\mathfrak{X} \subseteq \mathcal{C}\ell_n{}^{\text{nil}}$. Then, for all $a \in \mathbb{R}$ and $u \in \mathfrak{X}$,*

$$(a\varphi(u) + \psi(u))' = a\varphi'(u) + \psi'(u).$$

Proof. Let $\varepsilon > 0$ be given. By hypothesis, there exist neighborhoods $\mathcal{N}_\varphi$ and $\mathcal{N}_\psi$ such that for $u \in \mathfrak{X}$,

$$\begin{aligned}\left\| \frac{1}{h_1}(\varphi(u+h_1) - \varphi(u)) \right\| &< \varepsilon/2, \\ \left\| \frac{1}{h_2}(\psi(u+h_2) - \psi(u)) \right\| &< \varepsilon/2.\end{aligned}$$

Letting $\mathcal{N}_\varepsilon = \mathcal{N}_\varphi \cap \mathcal{N}_\psi$, it follows that for all $h \in \mathcal{N}_\varepsilon$,

$$\begin{aligned}\left\|\frac{1}{h}\left((a\varphi+\psi)(u+h)-(a\varphi+\psi)(u)\right)\right\| &\le \left\|\frac{a}{h}(\varphi(u+h)-\varphi(u))\right\| \\ &\quad + \left\|\frac{1}{h}(\psi(u+h)-\psi(u))\right\| \\ &\le \left\|\frac{a}{h_1}(\varphi(u+h_1)-\varphi(u))\right\| \\ &\quad + \left\|\frac{1}{h_2}(\psi(u+h_2)-\psi(u))\right\| \\ &< \varepsilon.\end{aligned}$$

□

Lemma 14.1.5. *Let $m \in \mathbb{N}$ be arbitrary and let $\varphi(u) = u^m$. Then, $\varphi'(u) = mu^{m-1}$.*

Proof. Begin by considering the difference quotient:

$$\begin{aligned}\frac{\varphi(u+h)-\varphi(u)}{h} &= \frac{(u+h)^m - u^m}{h} \\ &= h^{-1}\sum_{k=1}^{m}\binom{m}{k}u^{m-k}h^k \\ &= \sum_{k=1}^{m}\binom{m}{k}u^{m-k}h^{k-1} \\ &= m\sum_{k=0}^{m-1}\binom{m-1}{k}u^{m-1-k}h^k \\ &= m(u+h)^{m-1}.\end{aligned}$$

At this point, the development of a rigorous δ-ε proof is complicated by the lack of sub-multiplicativity of the infinity norm:

$$\begin{aligned}\|m(u+h)^{m-1} - mu^{m-1}\| &= |m|\left\|\sum_{\ell=1}^{m-1}\binom{m-1}{\ell}h^\ell u^{m-1-\ell}\right\| \\ &\le |m|2^{n(m-2)}\sum_{\ell=1}^{m-1}\binom{m-1}{\ell}\|h\|^\ell\|u\|^{m-1-\ell}.\end{aligned}$$

Nevertheless, the values $\|u\|$, m, and n are all fixed. For any $\varepsilon > 0$, invertible h can be chosen such that $\|h\|$ is sufficiently small to ensure

$$\|m(u+h)^{m-1} - mu^{m-1}\| < \varepsilon.$$

In other words, letting $h \twoheadrightarrow 0$ completes the proof. □

Two special cases of zeon-valued functions will be considered in detail. First, the most "natural" functions to consider are the zeon extensions of real-valued functions of a real variable. The second special case is the class of zeon-valued functions of one real variable. Having a one-dimensional domain of definition makes differentiation straightforward.

14.2 Zeon Extensions of Real-Valued Functions

As seen in [104], any real-valued function $f : \mathbb{R} \to \mathbb{R}$ n-times differentiable at $\Re u$ extends to ${\mathcal{C}\ell_n}^{\mathrm{nil}}$ via

$$f(u) = \sum_{k=0}^{n} \frac{f^{(k)}(\Re u)}{k!} (\mathfrak{D}u)^k. \tag{14.3}$$

More generally, if f is n-times differentiable on $X \subseteq \mathbb{R}$, the *zeon extension* of f is a function φ defined on $X \oplus {\mathcal{C}\ell_n}^{\mathrm{nil}_\circ} = \left\{u \in {\mathcal{C}\ell_n}^{\mathrm{nil}} : \Re u \in X\right\}$ by

$$\varphi(u) = \sum_{k=0}^{n} \frac{f^{(k)}(\Re u)}{k!} (\mathfrak{D}u)^k.$$

For convenience, write $f \rightsquigarrow \varphi$ when $\varphi : X \oplus {\mathcal{C}\ell_n}^{\mathrm{nil}_\circ} \to {\mathcal{C}\ell_n}^{\mathrm{nil}}$ is a zeon extension of $f : X \to \mathbb{R}$.

Given a subset $\mathfrak{X} \subseteq {\mathcal{C}\ell_n}^{\mathrm{nil}}$, the collection of zeon-valued functions $\mathfrak{X} \to {\mathcal{C}\ell_n}^{\mathrm{nil}}$ obtained as extensions of sufficiently differentiable real-valued functions $\Re\mathfrak{X} \to \mathbb{R}$ will be denoted by $\mathcal{F}(\mathfrak{X})$ for convenience. That is,

$$\mathcal{F}(\mathfrak{X}) = \{\varphi : f \rightsquigarrow \varphi \text{ for some } f : \Re\mathfrak{X} \to \mathbb{R}\}.$$

Given the expansion (14.3), the problem now is to determine the coefficient of ζ_I in $f(u)$. In other words, a closed formula for $\langle f(u), \zeta_I \rangle$ is sought.

Letting $u \in {\mathcal{C}\ell_n}^{\mathrm{nil}}$ be arbitrary, suppose the canonical expansion of u is $u = \sum_{I \in 2^{[n]}} u_I \zeta_I$. It follows that $\Re u = u_\varnothing$ and that $\mathfrak{D}u = \sum_{\substack{I \in 2^{[n]} \\ I \neq \varnothing}} u_I \zeta_I$. For nonnegative integer k and arbitrary multi-index I, one sees (via Lemma 9.1.7) that

$$\begin{aligned}
\langle (\mathfrak{D}u)^k, \zeta_I \rangle &= \sum_{\substack{\pi \in \mathcal{P}(I) \\ |\pi| = k}} k! \prod_{\ell=1}^{k} u_{\pi_\ell} \\
&= \sum_{\substack{\pi \in \mathcal{P}(I) \\ |\pi| = k}} k! u_\pi.
\end{aligned}$$

Hence,

$$\begin{aligned}\langle\varphi(u),\zeta_I\rangle &= \sum_{k=0}^{n}\left\langle\frac{f^{(k)}(\Re u)}{k!}(\mathfrak{D}u)^k,\zeta_I\right\rangle\\ &= \sum_{k=0}^{n}\frac{f^{(k)}(\Re u)}{k!}\left\langle(\mathfrak{D}u)^k,\zeta_I\right\rangle\\ &= \sum_{k=0}^{n} f^{(k)}(\Re u)\sum_{\substack{\pi\in\mathcal{P}(I)\\|\pi|=k}} u_\pi.\end{aligned}$$

For convenience, we recall Lemma 9.1.1, which says that the multiplication of elements of ${\mathcal{C}\ell_n}^{\mathrm{nil}}$ amounts to summing over partitions of subsets of $\{1,\ldots,n\}$.

Lemma 9.1.1. If $z_1,\ldots,z_k\in{\mathcal{C}\ell_n}^{\mathrm{nil}}$ with $z_j=\sum_I a_{j,I}\zeta_I$ for $1\le j\le k$, and $\prod_{j=1}^{k} z_j=\sum_I b_I\zeta_I$, then $b_I=\sum_{(I_1,\ldots,I_k)}\prod_{j=1}^{k} a_{j,I_j}$ where the sum is taken over all k-tuples $(I_1,\ldots,I_k)$ of pairwise disjoint (possibly empty) subsets of I such that $I=\bigcup_j I_j$.

Now we can calculate the coefficient of ζ_I in the expansion of $\varphi(z)$.

Proposition 14.2.1. *Let $f: A\subset\mathbb{R}\to\mathbb{R}$ be n times differentiable on A. If $z=\sum_I a_I\zeta_I\in{\mathcal{C}\ell_n}^{\mathrm{nil}}$ with $a_\varnothing\in A$ and $\varphi(z)=\sum_I b_I\zeta_I$, then*

$$b_I=\sum_{k=0}^{|I|} f^{(k)}(a_\varnothing)\sum_{\substack{\pi\in P(I)\\|\pi|=k}}\prod_{J\in\pi} a_J,$$

where $P(I)$ is the set of partitions of I.

It comes as no surprise that the zeon extension of a real-valued function is unique.

Lemma 14.2.2 (Uniqueness of extensions). *Let f,g be real-valued functions that are n-times differentiable on $\mathcal{D}\subseteq\mathbb{R}$. If $f\rightsquigarrow\varphi$ and $g\rightsquigarrow\psi$ are zeon extensions of f and g, respectively, then $\varphi=\psi$ if and only if $f^{(\ell)}(x)=g^{(\ell)}(x)$ for all $x\in\mathcal{D}$ and for each $\ell=0,1,\ldots,n$.*

Proof. Let $u\in\mathcal{D}\oplus{\mathcal{C}\ell_n}^{\mathrm{nil}_0}$ be arbitrary. If $f^{(\ell)}(\Re u)=g^{(\ell)}(\Re u)$ for all $x\in\mathcal{D}$

and each $\ell = 0, \ldots, n$, it follows immediately that

$$\begin{aligned}\varphi(u) &= \sum_{\ell=0}^{n} \frac{f^{(\ell)}(\Re u)}{\ell!}(\mathfrak{D}u)^{\ell} \\ &= \sum_{\ell=0}^{n} \frac{g^{(\ell)}(\Re u)}{\ell!}(\mathfrak{D}u)^{\ell} \\ &= \psi(u).\end{aligned}$$

On the other hand, suppose that there exists $x \in \mathcal{D}$ and $\ell \in \{0, \ldots, n\}$ such that $f^{(\ell)}(x) \neq g^{(\ell)}(x)$. Choosing $u = x + \sum_{j=1}^{\ell} \zeta_{\{j\}} \in \mathcal{D} \oplus \mathcal{C}\ell_n{}^{\mathrm{nil}_\circ}$, it can be seen that $(\mathfrak{D}u)^{\ell} = \ell! \zeta_{\{1,\ldots,\ell\}}$. It follows that

$$\begin{aligned}\|\varphi(u) - \psi(u)\| &\geq \left\| \frac{f^{(\ell)}(x) - g^{(\ell)}(x)}{\ell!}(\mathfrak{D}u)^{\ell} \right\| \\ &= |f^{(\ell)}(x) - g^{(\ell)}(x)| \\ &> 0.\end{aligned}$$

□

Norms of Zeon Extensions

Norms are easily enough induced on $\mathcal{C}\ell_n{}^{\mathrm{nil}}$ via its natural vector space isomorphism with $\mathbb{R}^n$. For $p \geq 1$, the *p-norm* of $z = \sum_I z_I \zeta_I$ is defined by

$$\|z\|_p = \left(\sum_{I \in 2^{[n]}} |z_I|^p \right)^{1/p},$$

while the *infinity norm* of z is defined by $\|z\|_\infty = \max\{|z_I| : I \in 26[n]\}$. A number of important properties of these norms are established in [65]. Since $\mathcal{C}\ell_n{}^{\mathrm{nil}}$ is finite-dimensional, all norms defined on it are equivalent.

A subset $X \subsetneq \mathcal{C}\ell_n{}^{\mathrm{nil}}$ is said to be *bounded* if there exists a real number $r > 0$ such that $\|x\| \leq r$ for all $x \in X$. The least positive such r is said to be the *radius* of X. Writing $X = \Re X + \mathfrak{D}X \subset \mathbb{R} \oplus \mathcal{C}\ell_n{}^{\mathrm{nil}_\circ}$, it is natural to define $\|\Re X\| := \inf\{|\Re z| : z \in X\}$ and $\|\mathfrak{D}X\| := \inf\{\|\mathfrak{D}z\| : z \in X\}$ when X is bounded.

A zeon function $\varphi : X \to \mathcal{C}\ell_n{}^{\mathrm{nil}}$ is said to be *bounded on* $X \subseteq \mathcal{C}\ell_n{}^{\mathrm{nil}}$ if there exists $r > 0$ such that $\|\varphi(z)\| \leq r$ for all $z \in X$. Given a bounded zeon function $\varphi : X \to \mathcal{C}\ell_n{}^{\mathrm{nil}}$, the *uniform norm* of φ is naturally defined as

$$\|\varphi\| := \sup\{\|\varphi(z)\| : z \in X\}.$$

When φ is induced by a sufficiently differentiable real-valued function f, more can be said. In particular, when $f : A \to \mathbb{R}$ is bounded and n-times continuously differentiable on $A \subseteq \mathbb{R}$, the ℓth derivative $f^{(\ell)}(x)$ is bounded on A for $\ell = 0, \ldots, n$ and the *uniform norm* of $f^{(\ell)}(x)$ is given by

$$\|f^{(\ell)}\| := \sup\{|f^{(\ell)}(x)| : x \in A\}.$$

Taking the supremum of uniform norms over $\ell = 0, \ldots, n$, one sees that

$$\sup_{0\le\ell\le n} \|f^{(\ell)}\| \ge 0, \text{ and } \sup_{0\le\ell\le n} \|f^{(\ell)}\| = 0 \text{ iff } f \equiv 0.$$

Further, linearity of the derivative and the triangle inequality for real numbers guarantees that for sufficiently differentiable functions $f, g : A \to \mathbb{R}$, the following holds for $x \in A$ and each $\ell = 0, \ldots, n$:

$$|(f+g)^{(\ell)}(x)| \le |f^{(\ell)}(x)| + |g^{(\ell)}(x)|.$$

It follows that $\|(f+g)^{(\ell)}\| \le \|f^{(\ell)}\| + \|g^{(\ell)}\|$ and the triangle inequality thus extends to the supremum; i.e.,

$$\begin{aligned}\sup_{0\le\ell\le n} \|(f+g)^{(\ell)}\| &\le \sup_{0\le\ell\le n} \left(\|f^{(\ell)}\| + \|g^{(\ell)}\|\right)\\ &\le \sup_{0\le\ell\le n} \|f^{(\ell)}\| + \sup_{0\le\ell\le n} \|g^{(\ell)}\|.\end{aligned}$$

Hence, the supremum of uniform norms is itself a norm—one that will be particularly useful for establishing properties of zeon functions.

A sufficiently differentiable real-valued function $f : A \to \mathbb{R}$ will be said to be *sufficiently bounded* for $A \oplus {\mathcal{C}\ell_n}^{\mathrm{nil}_\circ}$ if for each $\ell = 0, \ldots, n$, the set $\{|f^{(\ell)}(x)| : x \in A\}$ is bounded.

Definition 14.2.3. Given a real-valued function $f : A \to \mathbb{R}$, sufficiently bounded for $A \oplus {\mathcal{C}\ell_n}^{\mathrm{nil}_\circ}$, the *zeon norm* of f on A is defined by

$$\|f\|_\star := \max_{0\le\ell\le n} \sup_{x\in A}\{|f^{(\ell)}(x)|\}.$$

With this real norm in hand, some zeon function inequalities can be established. First, recall that the nth *Bell number*, B_n, is defined as the number of distinct partitions of the n-set. Similarly, the *Stirling number of the second kind*, $\left\{ {n \atop \ell} \right\}$, enumerates partitions of the n-set into ℓ nonempty subset (or *blocks*). These numbers are clearly related by $B_n = \sum_{\ell=1}^{n} \left\{ {n \atop \ell} \right\}$.

Lemma 14.2.4. *Suppose* $X \subsetneq {\mathcal{C}\ell_n}^{\text{nil}}$ *is bounded. If* $f : \Re X \to \mathbb{R}$ *is sufficiently bounded for* X *and if* $f \rightsquigarrow \varphi$*, then*

$$\|\varphi\| \leq \begin{cases} B_n\|f\|_\star \|\mathfrak{D}X\|^n & \text{if } \|\mathfrak{D}X\| > 1, \\ B_n\|f\|_\star \|\mathfrak{D}X\| & \text{if } \|\mathfrak{D}X\| \leq 1. \end{cases}$$

In particular, $\|\mathfrak{D}X\| \leq 1$ *implies* $\|\varphi\| \leq B_n\|f\|_\star$.

Proof. For clarity and convenience, it is noted that $\left\{ {n \atop 0} \right\} = 0$. Letting $z \in X$ be arbitrary,

$$\begin{aligned} \|\varphi(z)\| &= \left\| \sum_{\ell=0}^{n} \frac{f^{(\ell)}(\Re z)}{\ell!} (\mathfrak{D}z)^\ell \right\| \\ &\leq \sum_{\ell=0}^{n} \left| \frac{f^{(\ell)}(\Re z)}{\ell!} \right| \|(\mathfrak{D}z)^\ell\| \\ &\leq \sum_{\ell=0}^{n} \ell! \left\{ {n \atop \ell} \right\} \left| \frac{f^{(\ell)}(\Re z)}{\ell!} \right| \|\mathfrak{D}z\|^\ell. \end{aligned}$$

When $\|\mathfrak{D}X\| \geq 1$, it is clear that $\|\mathfrak{D}z\|^\ell \leq \|\mathfrak{D}X\|^n$ for $\ell \leq n$ and $z \in X$. Hence,

$$\begin{aligned} \sum_{\ell=0}^{n} \left\{ {n \atop \ell} \right\} \left| f^{(\ell)}(\Re z) \right| \|\mathfrak{D}z\|^\ell &\leq \sum_{\ell=0}^{n} \left\{ {n \atop \ell} \right\} \left| f^{(\ell)}(\Re z) \right| \|\mathfrak{D}X\|^n \\ &\leq B_n\|f\|_\star \|\mathfrak{D}X\|^n. \end{aligned}$$

On the other hand, when $\|\mathfrak{D}X\| < 1$, $\|\mathfrak{D}z\|^\ell \leq \|\mathfrak{D}X\|$ for $0 \leq \ell \leq n$ and $z \in X$. Hence,

$$\begin{aligned} \sum_{\ell=0}^{n} \left\{ {n \atop \ell} \right\} \left| f^{(\ell)}(\Re z) \right| \|\mathfrak{D}z\|^\ell &\leq \sum_{\ell=0}^{n} \left\{ {n \atop \ell} \right\} \left| f^{(\ell)}(\Re z) \right| \|\mathfrak{D}X\| \\ &\leq B_n\|f\|_\star \|\mathfrak{D}X\|. \end{aligned}$$

□

As an immediate consequence, it follows that bounded real functions induce bounded zeon functions.

Corollary 14.2.5. *Suppose* $X \subsetneq {\mathcal{C}\ell_n}^{\text{nil}}$ *is bounded. If* $f : \Re X \to \mathbb{R}$ *is sufficiently bounded for* X *and if* $f \rightsquigarrow \varphi$*, then* $\varphi : X \to {\mathcal{C}\ell_n}^{\text{nil}}$ *is bounded on* X.

Continuity of Zeon Extensions

$$\begin{aligned}|\langle\varphi(u),\zeta_I\rangle| &= \left|\sum_{k=0}^{n} f^{(k)}(\Re u)\sum_{\substack{\pi\in\mathcal{P}(I)\\|\pi|=k}} u_\pi\right|\\ &\le \sum_{k=0}^{n}\left|f^{(k)}(\Re u)\right|\left\{{|I| \atop k}\right\}((2^n-2)\|u\|_\infty)^{k-1}\\ &\le \sum_{k=0}^{n}\left|f^{(k)}(\Re u)\right| B_{|I|}2^{nk}{\|u\|_\infty}^k,\end{aligned}$$

where $B_{|I|}$ denotes the $|I|$th Bell number.

Letting $\omega = \max\{|f^{(k)}(\Re u)| : 0 \le k \le n\}$, observing that $B_{|I|} \le B_n$ for all $I \subseteq [n]$, and assuming $\|u\|_\infty < \dfrac{\varepsilon}{2^n\omega n B_n}$, it becomes evident that

$$\|\varphi(u)\| < \varepsilon.$$

Lemma 14.2.6. *Let $u, v \in {\mathcal{C}\ell_n}^{\rm nil}$. Suppose $\pi \in \mathcal{P}(I)$ and $|\pi| = k$. Then,*

$$\begin{aligned}|u_\pi - v_\pi| &= \left|\prod_{\ell=1}^{k} u_{\pi_\ell} - \prod_{\ell=1}^{k} v_{\pi_\ell}\right|\\ &\le \|u-v\|(\|u\|^k + (1+2\|u\|)^k).\end{aligned}$$

Proof. Letting $u, v \in {\mathcal{C}\ell_n}^{\rm nil}$, observe that

$$\begin{aligned}|u_I u_J - v_I v_J| &= |u_I u_J - u_I v_J + u_I v_J - v_I v_J|\\ &\le |u_I(u_J - v_J)| + |v_J(u_I - v_I)|\\ &\le \|u-v\|(\|u\| + \|v\|)\\ &\le \|u-v\|(1+2\|u\|),\end{aligned}$$

where the last inequality is established by assuming $\|u-v\| \le 1$. Proceeding by induction, assume that for any k-block partition π of any set I, the difference $|u_\pi - v_\pi|$ can be made arbitrarily small. Let $\pi' = \{\pi, K\}$ denote a $(k+1)$-block partition of $I \cup K$ (assuming $I \cap K = \varnothing$). Then,

$$\begin{aligned}|u_{\pi'} - v_{\pi'}| &= |u_\pi u_K - v_\pi v_K|\\ &= |u_\pi u_K - u_\pi v_K + u_\pi v_K - v_\pi v_K|\\ &\le |u_\pi(u_K - v_K)| + |v_K(u_\pi - v_\pi)|\\ &\le \|u-v\||u_\pi| + \|v\||u_\pi - v_\pi|\\ &\le \|u-v\|\|u\|^k + \|v\|\|u-v\|(1+2\|u\|)^{k-1}\\ &= \|u-v\|(\|u\|^k + (1+2\|u\|)^k).\end{aligned}$$

□

It is not difficult to see that for any positive integer k, $\|u\|^k + (1 + 2\|u\|)^k \le (1+3\|u\|)^k$. Indeed, a simple application of the binomial theorem yields

$$\begin{aligned}(1+3\|u\|)^k &= (\|u\| + 1 + 2\|u\|)^k\\ &= \|u\|^k + (1+2\|u\|)^k + \sum_{\ell=1}^{k-1}\binom{k}{\ell}\|u\|^\ell(1+2\|u\|)^{k-\ell}.\end{aligned}$$

It follows immediately that $|u_{\pi'} - v_{\pi'}| \le \|u - v\|(1+3\|u\|)^n$ for any partition π of any nontrivial multi-index in $2^{[n]}$.

Lemma 14.2.7. *Suppose f is n-times continuously differentiable on $D \subseteq \mathbb{R}$. Then, the zeon extension of f is continuous on $X = \{u \in {\mathcal{C}\ell_n}^{\text{nil}} : \Re u \in D\}$.*

Proof. Let $u, v \in X$ and note that for any multi-index I,

$$\langle \varphi(u) - \varphi(v), \zeta_I\rangle = \sum_{k=0}^{n}\left(f^{(k)}(\Re u)\sum_{\substack{\pi\in\mathcal{P}(I)\\|\pi|=k}} u_\pi - f^{(k)}(\Re v)\sum_{\substack{\pi\in\mathcal{P}(I)\\|\pi|=k}} v_\pi\right). \quad (14.4)$$

Adding and subtracting $f^{(k)}(\Re u)\sum_{\substack{\pi\in\mathcal{P}(I)\\|\pi|=k}} v_\pi$ on the right-hand side of (14.4) and applying the triangle inequality, one obtains

$$\begin{aligned}|\langle \varphi(u) - \varphi(v), \zeta_I\rangle| &\le \sum_{k=0}^{n}\left| f^{(k)}(\Re u)\sum_{\substack{\pi\in\mathcal{P}(I)\\|\pi|=k}} u_\pi - f^{(k)}(\Re v)\sum_{\substack{\pi\in\mathcal{P}(I)\\|\pi|=k}} v_\pi\right|\\ &\le \sum_{k=0}^{n}|f^{(k)}(\Re u)|\sum_{\substack{\pi\in\mathcal{P}(I)\\|\pi|=k}}|u_\pi - v_\pi|\\ &\quad + \sum_{k=0}^{n}\left|f^{(k)}(\Re u) - f^{(k)}(\Re v)\right|\sum_{\substack{\pi\in\mathcal{P}(I)\\|\pi|=k}}|v_\pi|\\ &\le \sum_{k=0}^{n}|f^{(k)}(\Re u)|B_n\|u-v\|(1+3\|u\|)^n\\ &\quad + \sum_{k=0}^{n}\left|f^{(k)}(\Re u) - f^{(k)}(\Re v)\right| B_n 2^{n^2}{\|v\|_\infty}^k.\end{aligned}$$

Assuming $\|u-v\|_\infty \le 1$, we have

$$\|v\|_\infty{}^k \le (1+\|u\|)^k \le (1+\|u\|)^n.$$

Let $\varepsilon > 0$ be arbitrary. Letting $\omega = \max\{|f^{(k)}(\Re u)| : 0 \le k \le n\}$, set $\delta_1 = \dfrac{\varepsilon}{2(n+1)\omega B_n(1+3\|u\|)^n}$. Utilizing the n-fold continuous differentiability of f, choose $\delta_2 > 0$ such that $0 < |\Re u - \Re v| < \delta_2$ implies

$$\left|f^{(k)}(\Re u) - f^{(k)}(\Re v)\right| < \frac{\varepsilon}{2(n+1)B_n 2^{n^2}(1+\|u\|)^n}$$

for all $k = 0, \ldots, n$.

Setting $\delta = \min\{1, \delta_1, \delta_2\}$, it follows that for any $v \in \mathcal{C}\ell_n{}^{\text{nil}}$ such that $0 < \|u-v\| < \delta$,

$$\begin{aligned}\|\varphi(u) - \varphi(v)\| &\le \max_{I \in 2^{[n]}} \{|\langle \varphi(u) - \varphi(v), \zeta_I \rangle|\} \\ &\le \sum_{k=0}^{n} \left[\frac{\varepsilon}{2(n+1)} + \frac{\varepsilon}{2(n+1)}\right] \\ &\le \frac{\varepsilon}{2} + \frac{\varepsilon}{2} = \varepsilon.\end{aligned}$$

□

The following result of Faà Di Bruno, recalled without proof, will be useful in the proof of Theorem 14.2.9.

Theorem 14.2.8 (Faà Di Bruno's Formula). *If f and g are functions with a sufficient number of derivatives, then*

$$\frac{d^m}{dt^m} g(f(t)) = \sum \frac{m!}{b_1! \cdots b_m!} g^{(k)}(f(t)) \left(\frac{f'(t)}{1!}\right)^{b_1} \cdots \left(\frac{f^{(m)}(t)}{m!}\right)^{b_m},$$

where $k = b_1 + \cdots + b_m$, and the sum is taken over all nonnegative integers $b_1, \ldots, b_m$ such that $b_1 + \cdots + m b_m = m$.

The next result will simplify much of the work yet to be done.

Theorem 14.2.9. *Suppose f, g are n-times continuously differentiable on $X \subseteq \mathbb{R}$, and suppose $f \rightsquigarrow \varphi$ and $g \rightsquigarrow \psi$. Let $w \in \mathcal{C}\ell_n{}^{\text{nil}}$. Then, the following hold on their respective domains:*

i. $(f + ag) \rightsquigarrow (\varphi + a\psi)$ *for* $a \in \mathbb{R}$;
ii. $fg \rightsquigarrow \varphi\psi$;

iii. $\dfrac{f}{g} \rightsquigarrow \dfrac{\varphi}{\psi}$;
iv. $f \circ g \rightsquigarrow \varphi \circ \psi$.

Proof. Beginning with the definitions of φ and ψ and $u \in X \oplus {\mathcal{C}\ell_n}^{\mathrm{nil_o}}$, we see that

$$\begin{aligned}\varphi(u) + a\psi(u) &= \sum_{\ell=0}^{n} \frac{f^{(\ell)}(\Re u)}{\ell!}(\mathfrak{D}u)^\ell + a\sum_{\ell=0}^{n} \frac{g^{(\ell)}(\Re u)}{\ell!}(\mathfrak{D}u)^\ell \\ &= \sum_{\ell=0}^{n} \frac{f^{(\ell)}(\Re u) + ag^{(\ell)}(\Re u)}{\ell!}(\mathfrak{D}u)^\ell \\ &= \sum_{\ell=0}^{n} \frac{(f+ag)^{(\ell)}(\Re u)}{\ell!}(\mathfrak{D}u)^\ell.\end{aligned}$$

Thus, $(f + ag) \rightsquigarrow (\varphi + a\psi)$ as claimed. Turning now to the series expansion of the product function $\varphi\psi$,

$$\begin{aligned}(\varphi\psi)(u) = \varphi(u)\psi(u) &= \left(\sum_{\ell=0}^{n} \frac{f^{(\ell)}(\Re u)}{\ell!}(\mathfrak{D}u)^\ell\right)\left(\sum_{k=0}^{n} \frac{g^{(k)}(\Re u)}{k!}(\mathfrak{D}u)^k\right) \\ &= \sum_{j=0}^{n} (\mathfrak{D}u)^j \sum_{\ell=0}^{j} \frac{f^{(\ell)}(\Re u)}{\ell!}\frac{g^{(j-\ell)}(\Re u)}{(j-\ell)!} \\ &= \sum_{j=0}^{n} (\mathfrak{D}u)^j \frac{1}{j!}\sum_{\ell=0}^{j} \frac{j!}{\ell!(j-\ell)!} f^{(\ell)}(\Re u) g^{(j-\ell)}(\Re u) \\ &= \sum_{j=0}^{n} (\mathfrak{D}u)^j \frac{(fg)^{(j)}(\Re u)}{j!}.\end{aligned}$$

In other words, if $f \rightsquigarrow \varphi$ and $g \rightsquigarrow \psi$, then $fg \rightsquigarrow \varphi\psi$.

To establish the quotient extension, let $X' = \{x \in X : g(x) \neq 0\}$ and let γ be the zeon extension of $\dfrac{1}{g}$ on $X' \oplus {\mathcal{C}\ell_n}^{\mathrm{nil_o}}$. That is,

$$\gamma(u) = \sum_{\ell=0}^{n} \frac{(1/g)^{(\ell)}(\Re u)}{\ell!}(\mathfrak{D}u)^\ell.$$

Recalling the general Leibniz rule for derivatives:

$$(fg)^{(\ell)}(x) = \sum_{j=0}^{\ell} \binom{\ell}{j} f^{(\ell-j)}(x) g^{(j)}(x),$$

it is easy to verify that

$$\begin{aligned}
\gamma(u)\psi(u) &= \left(\sum_{\ell=0}^{n} \frac{(1/g)^{(\ell)}(\Re u)}{\ell!}(\mathfrak{D}u)^\ell\right)\left(\sum_{j=0}^{n} \frac{g^{(j)}(\Re u)}{j!}(\mathfrak{D}u)^j\right) \\
&= \sum_{\ell=0}^{n} (\mathfrak{D}u)^\ell \sum_{j=0}^{\ell} \frac{g^{(j)}(\Re u)(1/g)^{(\ell-j)}(\Re u)}{j!(\ell-j)!} \\
&= 1 + \sum_{\ell=1}^{n} (\mathfrak{D}u)^\ell \sum_{j=0}^{\ell} \frac{g^{(j)}(\Re u)(1/g)^{(\ell-j)}(\Re u)}{j!(\ell-j)!} \\
&= 1 + \sum_{\ell=1}^{n} \frac{(\mathfrak{D}u)^\ell}{\ell!} \sum_{j=0}^{\ell} \frac{\ell!}{j!(\ell-j)!} g^{(j)}(\Re u)(1/g)^{(\ell-j)}(\Re u) \\
&= 1 + \sum_{\ell=1}^{n} \frac{(\mathfrak{D}u)^\ell}{\ell!} \sum_{j=0}^{\ell} \binom{\ell}{j} g^{(j)}(\Re u)(1/g)^{(\ell-j)}(\Re u) \\
&= 1 + \sum_{\ell=1}^{n} \frac{(\mathfrak{D}u)^\ell}{\ell!} (g \cdot 1/g)^{(\ell)}(\Re u) \\
&= 1.
\end{aligned}$$

It follows that γ is the (unique) reciprocal of ψ. Hence, $g \rightsquigarrow \psi$ implies $1/g \rightsquigarrow 1/\psi$. It then follows from the product extension that $f/g = f \cdot (1/g) \rightsquigarrow \varphi/\psi$.

Moving on to the composite extension, let γ denote the zeon extension of $f \circ g$. Thus,

$$\begin{aligned}
\gamma(u) &= \sum_{\ell=0}^{n} \frac{(f \circ g)^{(\ell)}(\Re u)}{\ell!}(\mathfrak{D}u)^\ell \\
&= \sum_{\ell=0}^{n} \frac{(\mathfrak{D}u)^\ell}{\ell!} \sum_{\substack{0 \le b_1, \dots, b_\ell \\ b_1 + 2b_2 + \cdots + \ell b_\ell = \ell}} \frac{\ell!}{b_1! \cdots b_\ell!} f^{(\ell)}(g(\Re u)) \prod_{k=1}^{\ell} \left(\frac{g^{(k)}(\Re u)}{k!}\right)^{b_k} \\
&= \sum_{\ell=0}^{n} (\mathfrak{D}u)^\ell \sum_{\substack{0 \le b_1, \dots, b_\ell \\ b_1 + 2b_2 + \cdots + \ell b_\ell = \ell}} \frac{f^{(\ell)}(g(\Re u))}{b_1! \cdots b_\ell!} \prod_{k=1}^{\ell} \left(\frac{g^{(k)}(\Re u)}{k!}\right)^{b_k}. \qquad (14.5)
\end{aligned}$$

On the other hand, the zeon composition $\varphi \circ \psi$ is seen to be

$$\begin{aligned}
\varphi(\psi(u)) &= \varphi\left(\sum_{k=0}^{n} \frac{g^{(k)}(\Re u)}{k!}(\mathfrak{D}u)^k\right) \\
&= \sum_{\ell=0}^{n} \frac{f^{(\ell)}(g(\Re u))}{\ell!}\left(\sum_{k=1}^{n} \frac{g^{(k)}(\Re u)}{k!}(\mathfrak{D}u)^k\right)^{\ell} \\
&= \sum_{\ell=0}^{n} \frac{f^{(\ell)}(g(\Re u))}{\ell!} \sum_{\substack{0\le b_1,\ldots,b_n \\ b_1+\cdots+b_n=\ell}} \binom{\ell}{b_1,\ldots,b_n} \prod_{k=1}^{n}\left(\frac{g^{(k)}(\Re u)}{k!}\right)^{b_k} (\mathfrak{D}u)^{kb_k} \\
&= \sum_{\ell=0}^{n} \sum_{\substack{0\le b_1,\ldots,b_n \\ b_1+\cdots+b_n=\ell}} \frac{f^{(\ell)}(g(\Re u))}{b_1!b_2!\cdots b_n!} \prod_{k=1}^{n}\left(\frac{g^{(k)}(\Re u)}{k!}\right)^{b_k} (\mathfrak{D}u)^{kb_k} \\
&= \sum_{\ell=0}^{n} (\mathfrak{D}u)^{\ell} \sum_{\substack{0\le b_1,\ldots,b_n \\ b_1+2b_2+\cdots+nb_n=\ell}} \frac{f^{(\ell)}(g(\Re u))}{b_1!b_2!\cdots b_n!} \prod_{k=1}^{n}\left(\frac{g^{(k)}(\Re u)}{k!}\right)^{b_k}. \qquad (14.6)
\end{aligned}$$

Since $(\mathfrak{D}u)^{\ell} = 0$ for $\ell > n$, one sees that the inner summation of (14.6) can be restricted to a summation over nonnegative integers $b_1, \ldots, b_\ell$. It follows that the equality $\gamma(u) = \varphi(\psi(u))$ is established by the equality of (14.5) and (14.6). □

It is now straightforward to rigorously establish continuity of linear combinations, products, quotients, and compositions of zeon extensions of real-valued functions.

Theorem 14.2.10. *Suppose f, g are n-times continuously differentiable on $X \subseteq \mathbb{R}$ and suppose $f \rightsquigarrow \varphi$ and $g \rightsquigarrow \psi$. Let $w \in {\mathcal{C}\ell_n}^{\mathrm{nil}}$. Then, the following hold for $u \in X \oplus {\mathcal{C}\ell_n}^{\mathrm{nil}_0}$:*

i. $\varphi + w\psi$ *is continuous at* u*;*
ii. $\varphi\psi$ *is continuous at* u*;*
iii. $\dfrac{\varphi}{\psi}$ *is continuous at* u*, provided* $\Re\psi(u) \neq 0$*;*
iv. $\varphi \circ \psi$ *is continuous at* u*, provided* $\Re\psi(u) \in X$*.*

Proof. Let u, w, φ, and ψ be as stated. If $w = 0$, there is nothing to show. Otherwise, let $\varepsilon > 0$ be given and suppose $w \neq 0$. By continuity of φ and ψ, there exists $\delta > 0$ such that $v \in X \oplus {\mathcal{C}\ell_n}^{\mathrm{nil}_0}$ and $\|u - v\| < \delta$ implies

$\|\varphi(u) - \varphi(v)\| < \varepsilon/2$ and $\|\psi(u) - \psi(v)\| < \dfrac{\varepsilon}{2^{n+1}\|w\|}$. Hence,

$$\begin{aligned}\|(\varphi + w\psi)(u) - (\varphi + w\psi)(v)\| &= \|\varphi(u) - \varphi(v) + w(\psi(u) - \psi(v))\| \\ &\leq \|\varphi(u) - \varphi(v)\| + \|w(\psi(u) - \psi(v))\| \\ &\leq \|\varphi(u) - \varphi(v)\| + 2^n\|w\|\|(\psi(u) - \psi(v))\| \\ &< \varepsilon.\end{aligned}$$

Continuity of the product, quotient, and composition follow from Theorem 14.2.9, Lemma 14.2.7, and the usual properties of continuous real-valued functions. □

Derivatives of Zeon Extensions

Lemma 14.2.11. *For any $\varepsilon > 0$, there exists a connected open set $\mathcal{N}_\varepsilon \subset \mathcal{C}\ell_n{}^{\text{nil}\star}$ such that $\|\mathfrak{D}(h^{-1})\| < \varepsilon$ for all $h \in \mathcal{N}_\varepsilon$.*

Proof. Rewriting h^{-1} as in (14.10), the dual part of h^{-1} has the following expansion:

$$\mathfrak{D}(h^{-1}) = \frac{1}{\Re h}\sum_{j=1}^{\kappa(\mathfrak{D}h)-1}(-1)^j\left(\frac{\mathfrak{D}h}{\Re h}\right)^j.$$

In terms of the infinity norm,

$$\begin{aligned}\|\mathfrak{D}(h^{-1})\| &= \frac{1}{|\Re h|}\left\|\sum_{j=1}^{\kappa(\mathfrak{D}h)-1}(-1)^j\left(\frac{\mathfrak{D}h}{\Re h}\right)^j\right\| \\ &\leq \frac{1}{|\Re h|}\sum_{j=1}^{\kappa(\mathfrak{D}h)-1}\left(\frac{\|\mathfrak{D}h\|}{|\Re h|}\right)^j \\ &\leq \frac{1}{|\Re h|}\sum_{j=1}^{\kappa(\mathfrak{D}h)-1}\frac{2^{n(j-1)}\|\mathfrak{D}h\|^j}{|\Re h|^j} \\ &= \frac{1}{2^n|\Re h|}\sum_{j=1}^{\kappa(\mathfrak{D}h)-1}2^{nj}\left\|\frac{\mathfrak{D}h}{\Re h}\right\|^j.\end{aligned}$$

Writing $\eta = 2^n\left\|\frac{\mathfrak{D}h}{\Re h}\right\|$, a little work with geometric series shows that when

$0 < \eta < 1$, the sum satisfies

$$\begin{aligned}\sum_{j=1}^{\kappa(\mathfrak{D}h)-1} 2^{nj} \left\| \frac{\mathfrak{D}h}{\Re h} \right\|^j &= \sum_{j=1}^{\kappa(\mathfrak{D}h)-1} \eta^j \\ &= \frac{\eta - \eta^{\kappa(\mathfrak{D}h)}}{1-\eta} \\ &\leq \frac{\eta}{1-\eta}.\end{aligned}$$

The task at hand now is to show that for any $\varepsilon > 0$, a value of η can be determined so that $\|\mathfrak{D}(h^{-1})\| < \varepsilon$. To that end, first assume that $0 \leq \|\mathfrak{D}h\| < |\Re h|/2^{n+1}$, so that $\eta < 1/2$. With this assumption, the following "nice" inequality is established:

$$\frac{\eta}{1-\eta} < 2\eta.$$

Choosing h such that $\|\mathfrak{D}h\| < \dfrac{(\Re h)^2\varepsilon}{2}$, it now follows that that

$$\begin{aligned}\|\mathfrak{D}(h^{-1})\| &\leq \frac{1}{2^n|\Re h|}\frac{\eta}{1-\eta} \\ &\leq \frac{1}{2^n|\Re h|}2\eta \\ &\leq \frac{1}{2^{n-1}|\Re h|}\frac{2^n\|\mathfrak{D}h\|}{|\Re h|} \\ &\leq \frac{1}{2^{n-1}|\Re h|}\frac{2^n}{|\Re h|}\frac{(\Re h)^2\varepsilon}{2} \\ &< \varepsilon.\end{aligned}$$

Defining the set $\mathcal{N}_\varepsilon = \{h \in \mathcal{C}\ell_n{}^{\text{nil}\star} : 0 \leq \|\mathfrak{D}h\| < \frac{(\Re h)^2\varepsilon}{2}\}$ completes the proof. □

Lemma 14.2.12. *Let $\varphi(u)$ be the zeon extension of sufficiently differentiable real function f to $\mathcal{C}\ell_n{}^{\text{nil}}$. If $\varphi'(u) = \lim_{h\to 0} \dfrac{\varphi(u+h)-\varphi(u)}{h}$ exists as a limit of real values of h, then for any $\varepsilon > 0$, there exists a neighborhood $\mathcal{N}_\varepsilon \subset \mathcal{C}\ell_n{}^{\text{nil}\star}$ of 0 such that $\forall h \in \mathcal{N}_\varepsilon$,*

$$\left\| \frac{\varphi(u+h)-\varphi(u)}{h} - \varphi'(u) \right\| < \varepsilon.$$

Proof. Suppose $h \in \mathcal{C}\ell_n{}^{\mathrm{nil}\star}$ and consider the difference quotient of φ:

$$\begin{aligned}
\frac{\varphi(u+h)-\varphi(u)}{h} &= h^{-1}\sum_{k=0}^{n}\frac{f^{(k)}(\Re u+\Re h)}{k!}(\mathfrak{D}u+\mathfrak{D}h)^k \\
&\quad -h^{-1}\sum_{k=0}^{n}\frac{f^{(k)}(\Re u)}{k!}(\mathfrak{D}u)^k \\
&= h^{-1}\sum_{k=0}^{n}\frac{f^{(k)}(\Re u+\Re h)}{k!}\sum_{\ell=0}^{k}\binom{k}{\ell}(\mathfrak{D}h)^{\ell}(\mathfrak{D}u)^{k-\ell} \\
&\quad -h^{-1}\sum_{k=0}^{n}\frac{f^{(k)}(\Re u)}{k!}(\mathfrak{D}u)^k \\
&= h^{-1}\sum_{k=0}^{n}\frac{f^{(k)}(\Re u+\Re h)-f^{(k)}(\Re u)}{k!}(\mathfrak{D}u)^k \\
&\quad +h^{-1}\sum_{k=0}^{n}\frac{f^{(k)}(\Re u+\Re h)}{k!}\sum_{\ell=1}^{k}\binom{k}{\ell}(\mathfrak{D}h)^{\ell}(\mathfrak{D}u)^{k-\ell}.
\end{aligned}$$

Rewriting $h^{-1} = (\Re h)^{-1} + \mathfrak{D}(h^{-1})$, it follows that

$$\begin{aligned}
\frac{\varphi(u+h)-\varphi(u)}{h} &= (\Re h)^{-1}\sum_{k=0}^{n}\frac{f^{(k)}(\Re u+\Re h)-f^{(k)}(\Re u)}{k!}(\mathfrak{D}u)^k \\
&\quad +\mathfrak{D}(h^{-1})\sum_{k=0}^{n}\frac{f^{(k)}(\Re u+\Re h)-f^{(k)}(\Re u)}{k!}(\mathfrak{D}u)^k \\
&\quad +(\Re h)^{-1}\sum_{k=0}^{n}\frac{f^{(k)}(\Re u+\Re h)}{k!}\sum_{\ell=1}^{k}\binom{k}{\ell}(\mathfrak{D}h)^{\ell}(\mathfrak{D}u)^{k-\ell} \\
&\quad +\mathfrak{D}(h^{-1})\sum_{k=0}^{n}\frac{f^{(k)}(\Re u+\Re h)}{k!}\sum_{\ell=1}^{k}\binom{k}{\ell}(\mathfrak{D}h)^{\ell}(\mathfrak{D}u)^{k-\ell}.
\end{aligned}$$

The difference quotient has been decomposed into a sum of four parts. By hypothesis, the first part converges as $\Re h \to 0$:

$$\begin{aligned}
\lim_{\Re h\to 0}(\Re h)^{-1}\sum_{k=0}^{n}\frac{f^{(k)}(\Re u+\Re h)-f^{(k)}(\Re u)}{k!}(\mathfrak{D}u)^k &= \sum_{k=0}^{n}\frac{f^{(k+1)}(\Re u)}{k!}(\mathfrak{D}u)^k \\
&= \lim_{\substack{h\to 0\\ \mathbb{R}}}\frac{\varphi(u+h)-\varphi(u)}{h}.
\end{aligned}$$

Examining the second part, continuity of f guarantees the existence of $\delta > 0$ such that $|\Re h| < \delta$ implies

$$\left\|\sum_{k=0}^{n}\frac{f^{(k)}(\Re u+\Re h)-f^{(k)}(\Re u)}{k!}(\mathfrak{D}u)^k-\varphi(u)\right\| < 1.$$

By Lemma 14.2.11, for any $\varepsilon > 0$, there exists deleted neighborhood $\mathcal{N}_1$ such that for all $h \in \mathcal{N}_1$,

$$\|\mathfrak{D}(h^{-1})\| < \frac{\varepsilon}{2^n(\|\varphi(u)\|+1)}.$$

It follows that

$$\lim_{\substack{h\to 0\\ \mathcal{N}_1}} \left(\mathfrak{D}(h^{-1}) \sum_{k=0}^{n} \frac{f^{(k)}(\Re u + \Re h) - f^{(k)}(\Re u)}{k!} (\mathfrak{D}u)^k \right) = 0.$$

Rewriting the third part by factoring out $\mathfrak{D}h$, one obtains

$$\frac{\mathfrak{D}h}{\Re h} \sum_{k=0}^{n} \frac{f^{(k)}(\Re u + \Re h)}{k!} \sum_{\ell=1}^{k} \binom{k}{\ell} (\mathfrak{D}h)^{\ell-1} (\mathfrak{D}u)^{k-\ell}.$$

As $h \to 0$ in $\mathcal{C}\ell_n{}^{\text{nil}}$, one notes that

$$\lim_{h\to 0} \sum_{k=0}^{n} \frac{f^{(k)}(\Re u + \Re h)}{k!} \sum_{\ell=1}^{k} \binom{k}{\ell} (\mathfrak{D}h)^{\ell-1} (\mathfrak{D}u)^{k-\ell} = \sum_{k=0}^{n} \frac{f^{(k)}(\Re u)}{(k-1)!} (\mathfrak{D}u)^{k-1}.$$

Further, defining $\mathcal{N}_2 = \{h \in \mathcal{C}\ell_n{}^{\text{nil}\star} : \|\mathfrak{D}h\| < |\Re h|\}$, it follows immediately that

$$\lim_{\substack{h\to 0\\ \mathcal{N}_2}} \left(\frac{\mathfrak{D}h}{\Re h} \sum_{k=0}^{n} \frac{f^{(k)}(\Re u + \Re h)}{k!} \sum_{\ell=1}^{k} \binom{k}{\ell} (\mathfrak{D}h)^{\ell-1} (\mathfrak{D}u)^{k-\ell} \right) = 0.$$

For the fourth part, as $h \to 0$ in $\mathcal{C}\ell_n{}^{\text{nil}}$, one sees that

$$\lim_{h\to 0} \sum_{k=0}^{n} \frac{f^{(k)}(\Re u + \Re h)}{k!} \sum_{\ell=1}^{k} \binom{k}{\ell} (\mathfrak{D}h)^{\ell} (\mathfrak{D}u)^{k-\ell} = 0,$$

and by Lemma 14.2.11, there exists deleted neighborhood $\mathcal{N}_3$ such that

$$\lim_{\substack{h\to 0\\ \mathcal{N}_3}} \left(\mathfrak{D}(h^{-1}) \sum_{k=0}^{n} \frac{f^{(k)}(\Re u + \Re h)}{k!} \sum_{\ell=1}^{k} \binom{k}{\ell} (\mathfrak{D}h)^{\ell} (\mathfrak{D}u)^{k-\ell} \right) = 0.$$

Letting $\mathcal{N} = \mathcal{N}_1 \cap \mathcal{N}_2 \cap \mathcal{N}_3$ completes the proof. □

Theorem 14.2.13 (Derivatives of Zeon Extensions). *Suppose f is a real-valued function with domain $\mathcal{D}$, n-times continuously differentiable at $u_\varnothing$. Let $X = \{u \in \mathcal{C}\ell_n{}^{\text{nil}} : \Re u \in \mathcal{D}\}$, let $\varphi : X \to \mathcal{C}\ell_n{}^{\text{nil}}$ be the zeon extension of f, and let $u \in \{w \in \mathcal{C}\ell_n{}^{\text{nil}} : \Re w = u_\varnothing\}$. Then, φ is differentiable at u and*

$$\varphi'(u) = \sum_{\ell=0}^{n} \frac{f^{(\ell+1)}(\Re u)}{\ell!} (\mathfrak{D}u)^{\ell}.$$

In other words, $f \rightsquigarrow \varphi$ implies $f' \rightsquigarrow \varphi'$.

Proof. For any $\varepsilon > 0$, there exists $\delta > 0$ such that $x \in \mathcal{D}$ with $0 < |x - u_\varnothing| < \delta$ implies $\dfrac{f(x) - f(u_\varnothing)}{x - u_\varnothing} < \varepsilon$. Equivalently, $f'(\Re u) = \lim_{h\to 0} \dfrac{f(\Re u + h) - f(\Re u)}{h}$.

Assume $0 \neq h \in \mathbb{R}$, and consider the difference quotient $\dfrac{\varphi(u+h) - \varphi(u)}{h}$. The numerator has the series expansion

$$\begin{aligned}\varphi(u+h) - \varphi(u) &= f(\Re(u+h)) - f(\Re u) \\ &\quad + \sum_{k=1}^{n} \left[\frac{f^{(k)}(\Re(u+h))}{k!} (\mathfrak{D}(u+h))^k - \frac{f^{(k)}(\Re u)}{k!} (\mathfrak{D}u)^k \right] \\ &= f(\Re(u+h)) - f(\Re u) \\ &\quad + \sum_{k=1}^{n} \left[\frac{f^{(k)}(\Re(u+h))}{k!} - \frac{f^{(k)}(\Re u)}{k!} \right] (\mathfrak{D}u)^k.\end{aligned}$$

Dividing by $h \neq 0$,

$$\begin{aligned}\frac{\varphi(u+h) - \varphi(u)}{h} &= \frac{f(\Re(u+h)) - f(\Re u)}{h} \\ &\quad + \sum_{k=1}^{n} \frac{1}{k!} \left[\frac{f^{(k)}(\Re(u+h)) - f^{(k)}(\Re u)}{h} (\mathfrak{D}u)^k \right].\end{aligned}$$

Letting $h \to 0$, n-fold continuous differentiability of f gives

$$\begin{aligned}\lim_{h\to 0} \frac{\varphi(u+h) - \varphi(u)}{h} &= \lim_{h\to 0} \frac{f(\Re(u+h)) - f(\Re u)}{h} \\ &\quad + \sum_{k=1}^{n} \frac{1}{k!} \left[\lim_{h\to 0} \frac{f^{(k)}(\Re(u+h)) - f^{(k)}(\Re u)}{h} (\mathfrak{D}u)^k \right] \\ &= f'(\Re u) + \sum_{k=1}^{n} \frac{f^{(k+1)}(\Re u)}{k!} (\mathfrak{D}u)^k \\ &= \varphi'(u).\end{aligned}$$

More generally, suppose $h \in {\mathcal{C}\ell_n}^{\mathrm{nil}\star}$ and consider

$$\begin{aligned}\frac{\varphi(u+h) - \varphi(u)}{h} &= h^{-1} \sum_{k=0}^{n} \frac{f^{(k)}(\Re u + \Re h)}{k!} (\mathfrak{D}u + \mathfrak{D}h)^k \\ &\quad - h^{-1} \sum_{k=0}^{n} \frac{f^{(k)}(\Re u)}{k!} (\mathfrak{D}u)^k. \qquad (14.7)\end{aligned}$$

Applying the binomial theorem to $(\mathfrak{D}u + \mathfrak{D}h)^k$ in the first summation on the right-hand side of (14.7) gives

$$\sum_{k=0}^{n} \frac{f^{(k)}(\Re u + \Re h)}{k!}(\mathfrak{D}u + \mathfrak{D}h)^k = \sum_{k=0}^{n} \frac{f^{(k)}(\Re u + \Re h)}{k!}(\mathfrak{D}u)^k + \sum_{k=1}^{n} \frac{f^{(k)}(\Re u + \Re h)}{k!} \sum_{\ell=1}^{k} (\mathfrak{D}u)^{k-\ell}(\mathfrak{D}h)^{\ell}. \tag{14.8}$$

Using (14.8), the difference quotient can now be written as

$$\frac{\varphi(u+h) - \varphi(u)}{h} = h^{-1} \sum_{k=0}^{n} \left[\frac{f^{(k)}(\Re u + \Re h) - f^{(k)}(\Re u)}{k!} \right] (\mathfrak{D}u)^k + h^{-1} \sum_{k=1}^{n} \frac{f^{(k)}(\Re u + \Re h)}{k!} \sum_{\ell=1}^{k} (\mathfrak{D}u)^{k-\ell}(\mathfrak{D}h)^{\ell}. \tag{14.9}$$

Recall that the unique multiplicative inverse of $h = \Re h + \mathfrak{D}h$ is given by

$$\begin{aligned} h^{-1} &= \sum_{j=1}^{n+1} (-1)^{j-1} (\Re h)^{-j} (\mathfrak{D}h)^{j-1} \\ &= \frac{1}{\Re h} + \sum_{j=1}^{n} \frac{(-1)^j}{(\Re h)^{j+1}} (\mathfrak{D}h)^j \\ &= \frac{1}{\Re h} + \frac{1}{\Re h} \sum_{j=1}^{n} \frac{(-1)^j (\mathfrak{D}h)^j}{(\Re h)^j} \\ &= \frac{1}{\Re h} + \mathfrak{D}h \sum_{j=1}^{n} \frac{(-1)^j}{(\Re h)^{j+1}} (\mathfrak{D}h)^{j-1}. \end{aligned} \tag{14.10}$$

Expanding and rewriting the product of h^{-1} with the first summation in (14.9) using (14.10), define

$$\begin{aligned} \xi_u(h) &= h^{-1} \sum_{k=0}^{n} \left[\frac{f^{(k)}(\Re u + \Re h) - f^{(k)}(\Re u)}{k!} \right] (\mathfrak{D}u)^k \\ &= \sum_{k=0}^{n} \frac{1}{k!} \left[\frac{f^{(k)}(\Re u + \Re h) - f^{(k)}(\Re u)}{\Re h} \right] (\mathfrak{D}u)^k \\ &\quad + \sum_{j=1}^{n} \sum_{k=0}^{n} \left(\frac{\mathfrak{D}h}{\Re h} \right)^j \frac{(-1)^j}{k!} \frac{\left(f^{(k)}(\Re u + \Re h) - f^{(k)}(\Re u)\right)}{\Re h} (\mathfrak{D}u)^k. \end{aligned} \tag{14.11}$$

Let the double summation of (14.11) be denoted by $\sigma_u(h)$; that is,

$$\sigma_u(h) = \sum_{\substack{1\le j\le n\\ 0\le k\le n}} \left(\frac{\mathfrak{D}h}{\mathfrak{R}h}\right)^j \frac{(-1)^j}{k!} \frac{\left(f^{(k)}(\mathfrak{R}u+\mathfrak{R}h) - f^{(k)}(\mathfrak{R}u)\right)}{\mathfrak{R}h} (\mathfrak{D}u)^k.$$

As $\mathfrak{R}h \to 0$, one sees immediately that for each $k = 0, \ldots, n$, continuous differentiability of f guarantees that

$$\frac{f^{(k)}(\mathfrak{R}u+\mathfrak{R}h) - f^{(k)}(\mathfrak{R}u)}{\mathfrak{R}h} \to f^{(k+1)}(\mathfrak{R}u).$$

It follows immediately that $\left|\frac{f^{(k)}(\mathfrak{R}u+\mathfrak{R}h) - f^{(k)}(\mathfrak{R}u)}{\mathfrak{R}h}\right|$ is bounded for sufficiently small values of $\mathfrak{R}h$, say $0 < |\mathfrak{R}h| \le \delta_1$. Since u is fixed and finite, so too is $\|(\mathfrak{D}u)^k\|$ for each value of k. Hence, assuming $0 < |\mathfrak{R}h| \le \delta_1$, there exists some real $M > 0$ such that for each $k = 0, \ldots, n$,

$$\left\|\frac{f^{(k)}(\mathfrak{R}u+\mathfrak{R}h) - f^{(k)}(\mathfrak{R}u)}{k!\,\mathfrak{R}h}(\mathfrak{D}u)^k\right\| \le M.$$

With the additional assumption that $\|\mathfrak{D}\| < |\mathfrak{R}h| < \delta_1$, the triangle inequality further guarantees that $\sigma(h)$ satisfies

$$\begin{aligned}
\|\sigma_u(h)\| &\le M \sum_{\substack{1\le j\le n\\ 0\le k\le n}} \left\|\left(\frac{\mathfrak{D}h}{\mathfrak{R}h}\right)^j\right\| \\
&= M(n+1)\sum_{j=1}^{n} \left\|\left(\frac{\mathfrak{D}h}{\mathfrak{R}h}\right)^j\right\| \\
&\le M(n+1)\sum_{j=1}^{n} 2^n \left\|\frac{\mathfrak{D}h}{\mathfrak{R}h}\right\|^j \\
&\le \frac{M(n+1)2^n}{1 - \|\mathfrak{D}h/\mathfrak{R}h\|}.
\end{aligned}$$

In other words, for any $\mu_1 > 0$, there exists $\delta_1 > 0$ such that $0 < \|\mathfrak{D}h\| < |\mathfrak{R}h| < \delta_1$ implies

$$\|\sigma_u(h)\| < \mu_1. \tag{14.12}$$

Next, consider $\xi_u(h)$. More specifically,

$$\begin{aligned}\xi_u(h) &= h^{-1}\sum_{k=1}^{n}\frac{f^{(k)}(\Re u+\Re h)}{k!}\sum_{\ell=1}^{k}(\mathfrak{D}u)^{k-\ell}(\mathfrak{D}h)^{\ell}\\ &= \frac{1}{\Re h}\sum_{k=1}^{n}\sum_{\ell=1}^{k}\frac{f^{(k)}(\Re u+\Re h)}{k!}(\mathfrak{D}u)^{k-\ell}(\mathfrak{D}h)^{\ell}\\ &\quad+\frac{1}{\Re h}\sum_{j=1}^{n}\frac{(-1)^j(\mathfrak{D}h)^j}{(\Re h)^j}\sum_{k=1}^{n}\sum_{\ell=1}^{k}\frac{f^{(k)}(\Re u+\Re h)}{k!}(\mathfrak{D}u)^{k-\ell}(\mathfrak{D}h)^{\ell}\\ &= \frac{\mathfrak{D}h}{\Re h}\sum_{k=1}^{n}\sum_{\ell=1}^{k}\frac{f^{(k)}(\Re u+\Re h)}{k!}(\mathfrak{D}u)^{k-\ell}(\mathfrak{D}h)^{\ell-1}\\ &\quad+\frac{\mathfrak{D}h}{\Re h}\sum_{j=1}^{n}(-1)^j\left(\frac{\mathfrak{D}h}{\Re h}\right)^j\sum_{k=1}^{n}\sum_{\ell=1}^{k}\frac{f^{(k)}(\Re u+\Re h)}{k!}(\mathfrak{D}u)^{k-\ell}(\mathfrak{D}h)^{\ell-1}.\end{aligned}$$

As before, $\|(\mathfrak{D}u)^k\|$ is finite for each k, and sufficiently small values of $|\Re h|$ guarantee boundedness of $|f^{(k)}(\Re u+\Re h)|$. Hence, given $\mu_2>0$, there exists $\delta_2>0$ such that $0<\|\mathfrak{D}h\|<|\Re h|<\delta_2$ implies

$$\|\psi_u(h)\|<\mu_2. \tag{14.13}$$

Now, for $\varepsilon>0$, there exists $\delta>0$ such that $0<\|h\|<\delta$ and $\|\mathfrak{D}h\|<|\Re h|$ imply

$$\begin{aligned}\left\|\frac{\varphi(u+h)-\varphi(u)}{h}-\varphi'(u)\right\| &= \|\sigma_u(h)+\psi_u(h)+\xi_u(h)-\varphi'(h)\|\\ &\le \|\xi_u(h)-\varphi'(u)\|+\|\sigma_u(h)\|+\|\psi_u(h)\|\\ &< \epsilon.\end{aligned}$$

Thus,

$$\begin{aligned}\lim_{\substack{h\to 0\\ \|\mathfrak{D}h\|<|\Re h|}}\sum_{k=0}^{n}\frac{1}{k!}\left[\frac{f^{(k)}(\Re u+\Re h)-f^{(k)}(\Re u)}{\Re h}\right](\mathfrak{D}u)^k &= \sum_{k=0}^{n}\frac{f^{(k+1)}(\Re u)}{k!}(\mathfrak{D}u)^k\\ &= \varphi'(u).\end{aligned}$$

□

Example 14.2.14. Consider $\varphi(u)=u^2$ in $\mathcal{C}\ell_n{}^{\mathrm{nil}}$. Then,

$$\begin{aligned}\frac{\partial}{\partial_\varnothing}\varphi(u) &= \lim_{h\to 0}\frac{(u+h)^2-u^2}{h}\\ &= \lim_{h\to 0}\frac{u^2+2uh-u^2}{h}\\ &= 2u.\end{aligned}$$

For any multi-index I,

$$\begin{aligned}\frac{\partial}{\partial \eta_I}\varphi(u) &= \lim_{h\to 0}\frac{(u+h\eta_I)^2-u^2}{h}\\ &= \lim_{h\to 0}\frac{u^2+2uh\eta_I-u^2}{h}\\ &= 2u\eta_I\\ &= \eta_I\frac{\partial}{\partial_\varnothing}\varphi(u).\end{aligned}$$

Higher-order derivatives of zeon extensions now come naturally. For nonnegative integer k, the *kth derivative of φ at $u \in {\mathcal{C}\ell_n}^{\text{nil}}$* is given by

$$\varphi^{(k)}(u) = \sum_{\ell=0}^{n}\frac{f^{(\ell+k)}(\Re u)}{\ell!}(\mathfrak{D}u)^\ell. \tag{14.14}$$

Corollary 14.2.15. *Let $f \in \mathfrak{C}^{n+\ell}(\mathcal{D})$ for $\mathcal{D} \subseteq \mathbb{R}$. If $\varphi : \mathcal{D} \oplus {\mathcal{C}\ell_n}^{\text{nil}_\circ} \to {\mathcal{C}\ell_n}^{\text{nil}}$ is the zeon extension of f, then for each $j = 0, \ldots, \ell$, $\varphi^{(j)}$ is the zeon extension of $f^{(j)}$.*

The usual rules of differentiation now follow naturally.

Theorem 14.2.16. *Let $\varphi, \psi : {\mathcal{C}\ell_n}^{\text{nil}} \to {\mathcal{C}\ell_n}^{\text{nil}}$ be zeon extensions differentiable at $u \in {\mathcal{C}\ell_n}^{\text{nil}}$ with $\Re(\psi(u)) \neq 0$, and let $\alpha \in \mathbb{R}$. Then, the functions $c\varphi$, $\varphi + \psi$, $\varphi\psi$, and $\dfrac{\varphi}{\psi}$ are differentiable at u and*

$$\begin{aligned}(\alpha\varphi)'(u) &= \alpha\varphi'(u),\\ (\varphi+\psi)'(u) &= \varphi'(u)+\psi'(u),\\ (\varphi\psi)'(u) &= \varphi(u)\psi'(u)+\varphi'(u)\psi(u), \qquad (14.15)\\ \left(\frac{\varphi}{\psi}\right)'(u) &= \frac{\psi(u)\varphi'(u)-\varphi(u)\psi'(u)}{(\psi(u))^2}. \qquad (14.16)\end{aligned}$$

Proof. Let f and g be the real-valued functions satisfying $f \rightsquigarrow \varphi$ and $g \rightsquigarrow \psi$. With Theorem 14.2.13 in hand, the first two rules follow immediately from linearity of the finite power series representations of zeon functions and their derivatives:

$$\begin{aligned}(\varphi+\alpha\psi)'(u) &= \sum_{\ell=0}^{n}\frac{(f+\alpha g)^{(\ell+1)}(\Re u)}{\ell!}(\mathfrak{D}u)^\ell\\ &= \sum_{\ell=0}^{n}\frac{f^{(\ell+1)}(\Re u)}{\ell!}(\mathfrak{D}u)^\ell + \alpha\sum_{\ell=0}^{n}\frac{g^{(\ell+1)}(\Re u)}{\ell!}(\mathfrak{D}u)^\ell\\ &= \varphi'(u)+\alpha\psi'(u).\end{aligned}$$

The product rule (14.15) is established by first recalling that $fg \rightsquigarrow \varphi\psi$, so that $(fg)' \rightsquigarrow (\varphi\psi)'$ by Theorem 14.2.13. Moreover, $f'g+fg' \rightsquigarrow \varphi'\psi+\varphi\psi'$, so that $(\varphi\psi)'$ and $\varphi'\psi+\varphi\psi'$ are zeon extensions of $(fg)' = f'g+fg'$. Uniqueness of zeon extensions then establishes the result.

In accordance with Theorems 14.2.9 and 14.2.13, $(1/\psi)'$ is the extension of $(1/g)' = -g'/g^2$. By the proof of the zeon product rule above, $\dfrac{-g'}{g^2} \rightsquigarrow -\psi'\left(\dfrac{1}{\psi}\right)^2 = \dfrac{-\psi'}{\psi^2}$. Writing $\left(\dfrac{\psi}{\psi}\right)'(u) = \left(\psi\dfrac{1}{\psi}\right)'(u)$, one sees immediately that

$$\begin{aligned}\left(\frac{\varphi}{\psi}\right)'(u) &= \varphi'(u)\frac{1}{\psi(u)} + \varphi(u)\left(\frac{1}{\psi}\right)'(u)\\ &= \frac{\varphi'(u)}{\psi(u)} - \frac{\varphi(u)\psi'(u)}{(\psi(u))^2}\\ &= \frac{\varphi'(u)\psi(u) - \varphi(u)\psi'(u)}{(\psi(u))^2}.\end{aligned}$$

□

Theorem 14.2.17 (Chain Rule). *Let $\varphi, \psi : \mathcal{C}\ell_n{}^{\rm nil} \to \mathcal{C}\ell_n{}^{\rm nil}$, where ψ is differentiable at $u \in \mathcal{C}\ell_n{}^{\rm nil}$ and φ is differentiable at $\psi(u)$. Then, the composite function $\varphi \circ \psi$ is differentiable at u and*

$$(\varphi \circ \psi)'(u) = \varphi'(\psi(u))\psi'(u).$$

Proof. Let f and g be the real-valued functions underlying φ and ψ, respectively. In light of Theorem 14.2.9, $f\circ g \rightsquigarrow \varphi\circ\psi$. By Corollary 14.2.15, it follows that $(f\circ g)' \rightsquigarrow (\varphi\circ\psi)'$. Further, as seen in the proof of the product rule in Theorem 14.2.16, $f'g \rightsquigarrow \varphi'\psi$, so that $(f\circ g)' = (f'\circ g)g' \rightsquigarrow (\varphi'\circ\psi)\psi'$. □

Zeon Power Functions

Given zeons $u, v \in \mathcal{C}\ell_n{}^{\rm nil}$ such that $\Re u > 0$, the general zeon exponential u^v is defined by

$$u^v = \mathfrak{exp}(v\,\mathfrak{log}\,u). \tag{14.17}$$

The condition $\Re u > 0$ is necessary because $\mathfrak{log}\,u$ is undefined when $\Re u \leq 0$.

Continuity and differentiability of u^v now follow naturally from the continuity and differentiability of $\mathfrak{exp}$ and $\mathfrak{log}$. The functions $\varphi(u) = u^v$ and $\psi(v) = u^v$ are considered for fixed v and fixed u, respectively, below.

Proposition 14.2.18. *Let* $u, v \in {\mathcal{C}\ell_n}^{\mathrm{nil}}$*. such that* $\Re u > 0$*. Then,*

$$\frac{d}{du}u^v = vu^{v-1},$$
$$\frac{d}{dv}u^v = u^v \mathfrak{log}\, u.$$

Proof. Fixing v and letting $\varphi(u) = u^v$, note that the zeon exponential and $v\mathfrak{log}\, u$ are differentiable by Corollary 14.2.15 and Theorem 14.2.16, respectively. Thus, the chain rule implies

$$\frac{d}{du}u^v = \varphi'(u) = \mathfrak{exp}(v\mathfrak{log}\, u)\frac{v}{u} = \frac{v}{u}u^v = vu^{v-1}.$$

Similarly, letting $\psi(v) = u^v$ for fixed u gives

$$\frac{d}{dv}u^v = \psi'(u) = \mathfrak{exp}(v\mathfrak{log}\, u)\mathfrak{log}\, u = u^v \mathfrak{log}\, u.$$

□

Let $v \in {\mathcal{C}\ell_n}^{\mathrm{nil}}$ be fixed, and let $f(u) = u^v$ for $u \in \mathcal{D}$. Then, for any $c \in \mathcal{D}$ with $\Re c \neq \Re u$, we define

$$f'(c) = \lim_{\|u-c\| \to 0} \frac{u^v - c^v}{u - c}.$$

Critical Points and Zeros of Induced Zeon Functions

Given a sufficiently differentiable real-valued function $f : A \to \mathbb{R}$ on domain $A \subseteq R$, a zeon-valued function $\varphi : A \oplus {\mathcal{C}\ell_n}^{\mathrm{nil_o}} \to {\mathcal{C}\ell_n}^{\mathrm{nil}}$ is naturally induced via the following:

$$\varphi(u) = \sum_{k=0}^{n} \frac{f^{(k)}(\Re u)}{k!}(\mathfrak{D}u)^k.$$

Given $u \in A \oplus {\mathcal{C}\ell_n}^{\mathrm{nil_o}}$, set $\alpha_k = \dfrac{f^{(k)}(\Re u)}{k!} \in \mathbb{R}$ for each $k = 0, \ldots, n$. Define the polynomial function $\psi : {\mathcal{C}\ell_n}^{\mathrm{nil_o}} \to {\mathcal{C}\ell_n}^{\mathrm{nil}}$ by

$$\psi(x) = \sum_{k=0}^{n} \alpha_k x^k.$$

Recall that when $f \rightsquigarrow \varphi$, the derivative of φ is given by

$$\varphi'(u) = \sum_{k=0}^{n} \frac{f^{(k+1)}(\Re u)}{k!}(\mathfrak{D}u)^k.$$

Proposition 14.2.19 (Critical points). *If f is a sufficiently differentiable real-valued function defined on $A \subseteq \mathbb{R}$ and $f \rightsquigarrow \varphi$, then $\varphi^{(j)}(z) = 0$ if and only if $f^{(\ell)}(\Re z) = 0$ for all nonnegative integers $\ell < \kappa(\mathfrak{D}u) - j$.*

Proof. By Theorem 14.2.13,

$$\varphi^{(j)}(u) = \sum_{\ell=0}^{n} \frac{f^{(\ell+j)}(\Re u)}{\ell!}(\mathfrak{D}u)^{\ell}.$$

Observing that $\natural(\mathfrak{D}u)^{\ell+1} > \natural(\mathfrak{D}u)^{\ell}$ unless one or both are zero, it follows that the sum can only be zero if every term is zero. For $\ell \geq 0$,

$$\frac{f^{(\ell+j)}(\Re u)}{\ell!}(\mathfrak{D}u)^{\ell} = 0$$

implies $f^{(\ell+j)}(\Re u) = 0$ or $(\mathfrak{D}u)^{\ell} = 0$. Since $(\mathfrak{D}u)^{\ell} = 0$ if and only if $\ell \geq \kappa(\mathfrak{D}u)$, the necessary and sufficient condition for all terms of the sum to be zero is therefore $f^{(\ell+j)}(\Re u) = 0$ for all ℓ such that $0 \leq \ell < \kappa(\mathfrak{D}u) - j$. □

As a corollary, it is possible to characterize the real-valued functions inducing zeon functions having a zero of prescribed multiplicity at a particular point $u \in {\mathcal{C}\ell_n}^{\text{nil}}$.

Corollary 14.2.20. *Let $A \subseteq \mathbb{R}$, let $u \in {\mathcal{C}\ell_n}^{\text{nil}}$ such that $\Re u \in A$, and let j be a nonnegative integer. The collection X of functions $f : A \to \mathbb{R}$ inducing function φ such that u is a zero of multiplicity j for φ, i.e., $\varphi^{(\ell)}(u) = 0$ for $\ell = 0, 1, \ldots, j-1$ and $\varphi^{(j)}(u) \neq 0$, is given by*

$$X = \{(x - \Re u)^{\kappa(\mathfrak{D}u)+j} g(x) : g \in \mathfrak{C}^{\kappa(\mathfrak{D}u)+j} A, g(\Re u) \neq 0\}.$$

Proof. Suppose $f(x) = (x - \Re u)^{\kappa(\mathfrak{D}u)+j} g(x)$, where $g(x)$ is at least $\kappa(\mathfrak{D}u) + j$-times continuously differentiable on A. Then f has a zero of multiplicity $\kappa(\mathfrak{D}u) + j$ at $\Re u$, so that $f^{(\ell)}(u) = 0$ for $\ell = 0, 1, \ldots, \kappa(\mathfrak{D}u) + j - 1$. Thus, $f \rightsquigarrow \varphi$ implies $\varphi^{(j)}(u) = 0$ by Proposition 14.2.19. □

Example 14.2.21. Let $u = 1 + \zeta_{\{1\}} + 5\zeta_{\{2,3\}} - 7\zeta_{\{4\}} \in {\mathcal{C}\ell_4}^{\text{nil}}$, and note that $\kappa(\mathfrak{D}u) = 4$. The zeon functions with zero of multiplicity 2 at u, i.e., $\{\varphi : \varphi(u) = \varphi'(u) = 0; \varphi^{(2)}(u) \neq 0\}$, are induced by the following real-valued functions:

$$X = \{(x - 1)^5 g(x) : g \in C^5(\mathbb{R})\}.$$

Given $f(x) = (x-1)^5 g(x)$, where g is at least 5-times continuously differentiable on $\mathbb{R}$, we see that

$$\begin{aligned} f'(x) &= 5(x-1)^4 g(x) + (x-1)^5 g'(x), \\ f''(x) &= 20(x-1)^3 g(x) + 10(x-1)^4 g'(x) + (x-1)^5 g''(x), \\ f'''(x) &= 60(x-1)^2 g(x) + 60(x-1)^3 g'(x) + (x-1)^5 g'''(x), \\ f^{(4)}(x) &= 120(x-1) g(x) + 240(x-1)^2 g'(x) + (x-1)^5 g^{(4)}(x), \\ f^{(5)}(x) &= 120 g(x) + 600(x-1) g'(x) + (x-1)^5 g^{(5)}(x). \end{aligned}$$

Thus, $f^{(\ell)}(1) = 0$ for $\ell = 0, 1, \ldots, 4$ and $f^{(5)}(1) \neq 0$. In particular, if $f \in X$ and $f \rightsquigarrow \varphi$, then

$$\begin{aligned} \varphi(u) = f(1) &+ f'(1)(\zeta_{\{1\}} + 5\zeta_{\{2,3\}} - 7\zeta_{\{4\}}) \\ &+ f''(1)(5\zeta_{\{1,2,3\}} - 7\zeta_{\{1,4\}} - 35\zeta_{\{2,3,4\}}) \\ &+ f'''(1)(-35/2)\zeta_{\{1,2,3,4\}}. \end{aligned}$$

Similarly,

$$\begin{aligned} \varphi'(u) = f'(1) &+ f''(1)(\zeta_{\{1\}} + 5\zeta_{\{2,3\}} - 7\zeta_{\{4\}}) \\ &+ f'''(1)(5\zeta_{\{1,2,3\}} - 7\zeta_{\{1,4\}} - 35\zeta_{\{2,3,4\}}) \\ &+ f^{(4)}(1)(-35/2)\zeta_{\{1,2,3,4\}}, \end{aligned}$$

and

$$\begin{aligned} \varphi''(u) = f''(1) &+ f'''(1)(\zeta_{\{1\}} + 5\zeta_{\{2,3\}} - 7\zeta_{\{4\}}) \\ &+ f^{(4)}(1)(5\zeta_{\{1,2,3\}} - 7\zeta_{\{1,4\}} - 35\zeta_{\{2,3,4\}}) \\ &+ f^{(5)}(1)(-35/2)\zeta_{\{1,2,3,4\}}. \end{aligned}$$

Partial Derivatives

Viewing the zeon extension φ as a zeon-valued function of 2^n real variables $\{u_I : I \in 2^{[n]}\}$, one can consider properties of partial derivatives. With this in mind, the partial derivative of φ with respect to u_I is defined by

$$\frac{\partial}{\partial u_I}\varphi(u) = \lim_{h \to 0} \frac{\varphi(u + h\zeta_I) - \varphi(u)}{h}.$$

Lemma 14.2.22. *Let $\varphi(u)$ be a zeon elementary function on ${\mathcal{C}\ell_n}^{\mathrm{nil}}$ and let $I \in 2^{[n]}$ be arbitrary. The partial derivative $\frac{\partial}{\partial_{u_I}}\varphi$ is given by*

$$\frac{\partial}{\partial_{u_I}}\varphi(u) = \zeta_I \varphi'(u).$$

Proof. When $I = \varnothing$, the result follows naturally from previous results. Assuming $I \neq \varnothing$ and examining the difference quotient, we see that the $k = 0$ terms disappear and that no power of $h\zeta_I$ greater than one survives from the binomial theorem. Thus,

$$\begin{aligned}\frac{\varphi(u+h\zeta_I)-\varphi(u)}{h} &= \frac{1}{h}\sum_{k=0}^{n}\frac{f^{(k)}(\Re u)}{k!}(\mathfrak{D}u+h\zeta_I)^k - \frac{1}{h}\sum_{k=0}^{n}\frac{f^{(k)}(\Re u)}{k!}(\mathfrak{D}u)^k\\ &= \frac{1}{h}\sum_{k=1}^{n}\frac{f^{(k)}(\Re u)}{k!}\sum_{\ell=1}^{k}\binom{k}{\ell}(h\zeta_I)^\ell(\mathfrak{D}u)^{k-\ell}\\ &= \frac{1}{h}\sum_{k=1}^{n}\frac{f^{(k)}(\Re u)}{(k-1)!}(h\zeta_I)(\mathfrak{D}u)^{k-1}\\ &= \zeta_I\sum_{k=0}^{n}\frac{f^{(k+1)}(\Re u)}{k!}(\mathfrak{D}u)^k.\end{aligned}$$

The difference quotient is independent of h, so the limit is trivial; hence the result. □

Corollary 14.2.23. *Let $\varphi(u)$ be a zeon elementary function on ${\mathcal{C}\ell_n}^{\text{nil}}$ and let $I, J \in 2^{[n]}$ be arbitrary. Then,*

$$\frac{\partial^2}{\partial_J\partial_I}\varphi(u) = \begin{cases}\zeta_{I\cup J}\varphi''(u) & \text{if } I\cap J=\varnothing,\\ 0 & \text{otherwise.}\end{cases}$$

Proof. When either I or J is empty, the result follows naturally from Lemma 14.2.22. Assuming $I, J \neq \varnothing$, the difference quotient of interest is determined from

$$\frac{\partial}{\partial_J}(\zeta_I\varphi'(u)) = \lim_{h\to 0}\frac{\zeta_I\varphi'(u+h\zeta_J)-\zeta_I\varphi'(u)}{h}.$$

Again we see that the $k = 0$ terms disappear and that no power of $h\zeta_J$ greater than one survives from the binomial theorem. Thus,

$$\begin{aligned}\frac{\zeta_I(\varphi(u+h\zeta_J)-\varphi(u))}{h} &= \frac{\zeta_I}{h}\sum_{k=0}^{n}\frac{f^{(k+1)}(\Re u)}{k!}(\mathfrak{D}u+h\zeta_J)^k\\ &\quad -\frac{\zeta_I}{h}\sum_{k=0}^{n}\frac{f^{(k+1)}(\Re u)}{k!}(\mathfrak{D}u)^k\end{aligned}$$

$$= \frac{\zeta_I}{h} \sum_{k=1}^{n} \frac{f^{(k+1)}(\Re u)}{k!} \sum_{\ell=1}^{k} \binom{k}{\ell} (h\zeta_J)^\ell (\mathfrak{D}u)^{k-\ell}$$
$$= \frac{\zeta_I}{h} \sum_{k=1}^{n} \frac{f^{(k+1)}(\Re u)}{(k-1)!} (h\zeta_J)(\mathfrak{D}u)^{k-1}$$
$$= \zeta_J \zeta_I \sum_{k=0}^{n} \frac{f^{(k+2)}(\Re u)}{k!} (\mathfrak{D}u)^k.$$

The difference quotient is independent of h, so the limit is trivial; hence the result. □

Corollary 14.2.24. *Let $\varphi(u)$ be a zeon elementary function on ${\mathcal{C}\ell_n}^{\text{nil}}$ and let $I \in 2^{[n]}$ be arbitrary. Then,*

$$\frac{\partial^2}{{\partial_I}^2} \varphi(u) = \begin{cases} 0 & \text{if } I \neq \varnothing, \\ \varphi''(u) & \text{if } I = \varnothing. \end{cases}$$

In light of these results, it becomes evident that the partial differential operators $\{\partial/\partial_I : I \in 2^{[n]}\}$ satisfy the Zeon Commutation Relations (ZCR):

$$[\partial/\partial_I, \partial/\partial_J] = 0 \text{ if } I \neq J$$
$$(\partial/\partial_I)^2 = 0 \text{ if } I \neq \varnothing.$$

In particular,

$$\frac{\partial}{\partial_I}\frac{\partial}{\partial_J} = \begin{cases} \frac{\partial}{\partial_{I \cup J}} & \text{if } I \cap J = \varnothing, \\ 0 & \text{otherwise.} \end{cases}$$

Hence, the real algebra generated by the partial differential operators on $\mathcal{F}({\mathcal{C}\ell_n}^{\text{nil}})$ is isomorphic to ${\mathcal{C}\ell_n}^{\text{nil}}$ itself.

Antiderivatives of Zeon Extensions

As one might expect, a general theory of zeon antiderivatives follows naturally.

Definition 14.2.25. Let $X \subset {\mathcal{C}\ell_n}^{\text{nil}}$. A zeon function $\Phi : X \to {\mathcal{C}\ell_n}^{\text{nil}}$ is said to be a *zeon antiderivative of φ on X* if $\Phi'(u) = \varphi(u)$ for all $u \in X$.

Since $f \rightsquigarrow \varphi$ implies $f' \rightsquigarrow \varphi'$, it is not difficult to see that $f \rightsquigarrow \varphi$ also implies $F \rightsquigarrow \Phi$ when F is an antiderivative of f.

Theorem 14.2.26. *Suppose $F \in \mathfrak{C}^{n+1}(A)$ is an antiderivative of f on $A \subseteq \mathbb{R}$. If $F \rightsquigarrow \Phi$ and $f \rightsquigarrow \varphi$, then Φ is an antiderivative of φ on $X = A \oplus \mathcal{C}\ell_n{}^{\text{nil}}$.*

Proof. Starting with $\Phi(u)$ and writing $\Phi(u) = \sum_{\ell=0}^{n} \frac{F^{(\ell)}(\Re u)}{\ell!}(\mathfrak{D}u)^\ell$, Theorem 14.2.13 implies

$$\begin{aligned}\Phi'(u) &= \sum_{\ell=0}^{n} \frac{F^{(\ell+1)}(\Re u)}{\ell!}(\mathfrak{D}u)^\ell \\ &= \sum_{\ell=0}^{n} \frac{f^{(\ell)}(\Re u)}{\ell!}(\mathfrak{D}u)^\ell \\ &= \varphi(u).\end{aligned}$$

□

14.3 Zeon-Valued Functions of a Real Variable

In this section, we consider zeon-valued functions of a single real variable, i.e., functions of the form $\varphi : X \to \mathcal{C}\ell_n{}^{\text{nil}}$ for $X \subseteq \mathbb{R}$.

Definition 14.3.1. Let $X \subseteq \mathbb{R}$ and $n \in \mathbb{N}$ be fixed. For each $I \in 2^{[n]}$, let $f_I : X \to \mathbb{R}$ be a real-valued function defined on X. The function $\varphi : X \to \mathcal{C}\ell_n{}^{\text{nil}}$ defined by

$$\varphi(t) = \sum_{I \in 2^{[n]}} f_I(t)\zeta_I$$

is said to be a *zeon-valued function of one real variable* determined by the collection $\{f_I : I \in 2^{[n]}\}$ having domain X.

It is not difficult to see that φ is continuous at $c \in X$ if and only if each coordinate function f_I is continuous at $c \in X$. Moreover, if each coordinate function is differentiable at $c \in X$, the derivative of φ at c is defined in the natural way.

Lemma 14.3.2. *Let $X \subseteq \mathbb{R}$, let $c \in X$, and suppose $\varphi : X \to \mathcal{C}\ell_n{}^{\text{nil}}$ is defined on X as in Definition 14.3.1. If $f_I : X \to \mathbb{R}$ is differentiable at c for each $I \in 2^{[n]}$, then φ is differentiable at c and the derivative is given by*

$$\varphi'(c) = \sum_{I \in 2^{[n]}} f_I{}'(c)\zeta_I.$$

Proof. The proof is straightforward. Assuming each f_I is differentiable at c, one sees that

$$\begin{aligned}\lim_{h\to 0}\frac{\varphi(c+h)-\varphi(c)}{h} &= \lim_{h\to 0}\sum_{I\in 2^{[n]}}\frac{(f_I(c+h)-f_I(c))}{h}\zeta_I\\ &= \sum_{I\in 2^{[n]}}\lim_{h\to 0}\frac{(f_I(c+h)-f_I(c))}{h}\zeta_I\\ &= \sum_{I\in 2^{[n]}} f_I{}'(c).\end{aligned}$$

□

The sum and difference rules follow naturally from linearity.

Lemma 14.3.3 (Linearity of differentiation). *Let $X\subseteq\mathbb{R}$, let $t\in X$, let $c\in\mathbb{R}$, and suppose the collection $\{f_I, g_I : I\in 2^{[n]}\}$ are real-valued functions defined on X, differentiable at t. Writing $\varphi(t)=\sum_{I\in 2^{[n]}} f_I(t)\zeta_I$ and $\gamma(t)=\sum_{I\in 2^{[n]}} g_I(t)\zeta_I$, it follows that φ, γ, $c\varphi$, and $\varphi+\gamma$ are zeon-valued functions $X\to \mathcal{C}\ell_n{}^{\mathrm{nil}}$ that are differentiable at t. Further,*

$$(c\varphi(t)+\gamma(t))' = c\varphi'(t)+\gamma'(t).$$

Let $\varphi(t)=\sum_{I\in 2^{[n]}} f_I(t)\zeta_I$ and $\gamma(t)=\sum_{I\in 2^{[n]}} g_I(t)\zeta_I$ be zeon-valued functions on $X\subseteq\mathbb{R}$. The product function $\psi=\varphi\gamma$ has the following canonical expansion:

$$\psi(t)=\sum_{I\in 2^{[n]}}\sum_{J\subseteq I} f_J(t)g_{I\setminus J}(t)\zeta_I. \tag{14.18}$$

Proposition 14.3.4 (Product Rule). *Let φ and γ be zeon-valued functions on $X\subseteq\mathbb{R}$, differentiable at $t\in X$. Then, the product function $\psi=\varphi\gamma$ is differentiable at t and*

$$\psi'(t)=\varphi(t)\gamma'(t)+\varphi'(t)\gamma(t).$$

Proof. In light of (14.18), the result follows naturally from the usual product rule; i.e.,

$$\begin{aligned}\psi'(t) &= \sum_{I\in 2^{[n]}}\sum_{J\subseteq I}(f_J(t)g_{I\setminus J}{}'(t)+f_J{}'(t)g_{I\setminus J}(t))\zeta_I\\ &= \sum_{I\in 2^{[n]}}\sum_{J\subseteq I} f_J(t)g_{I\setminus J}{}'(t)\zeta_I+\sum_{I\in 2^{[n]}}\sum_{J\subseteq I} f_J{}'(t)g_{I\setminus J}(t)\zeta_I\\ &= \varphi(t)\gamma'(t)+\varphi'(t)\gamma(t).\end{aligned}$$

□

Before moving on to a quotient rule, it will be informative to develop the series expansion formalism for reciprocal functions. In order to develop a zeon quotient rule, it is necessary to first be able to differentiate the product function $g_\pi(t)$ for a partition π of multi-index I.

Lemma 14.3.5 (The derivative of g_π). *Let $\{g_I : I \in 2^{[n]}\}$ denote a collection of real-valued functions defined on $X \subseteq \mathbb{R}$, differentiable at $t \in X$. Given partition $\pi = (b_1|\cdots|b_{|\pi|})$ of $I \in 2^{[n]}$, let $g_\pi(t) = \prod_{b\in\pi} g_b(t)$. Then,*

$$g_\pi{}'(t) = \sum_{b\in\pi} g_{\pi\setminus b}(t) g_b{}'(t).$$

Proof. The result is an immediate consequence of the generalized product rule. □

Proposition 9.1.3 is useful for expressing reciprocal functions as power series. The following result is an immediate corollary.

Lemma 14.3.6 (Reciprocal derivatives). *Let $\{g_I : I \in 2^{[n]}\}$ denote a collection of real-valued functions differentiable on $X \subseteq \mathbb{R}$, with $g_\varnothing$ nonzero on X. Let $\gamma : X \to \mathcal{C}\ell_n{}^{\text{nil}}$ be defined by $\gamma(t) = \sum_{I\in 2^{[n]}} g_I(t)\zeta_I$ and $c \in X$ such that $\Re(\gamma(c)) \neq 0$. Then,*

$$\frac{d}{dt}\left(\frac{1}{\gamma(t)}\right) = \frac{-\gamma'(t)}{(\gamma(t))^2}.$$

Proof. The hypothesis that $g_\varnothing(t) \neq 0$ for all $t \in X$ guarantees that $\frac{1}{\gamma(t)}$ is defined on X. To show differentiability of $\frac{1}{\gamma(t)}$, first consider the canonical expansion

$$\begin{aligned}
\frac{1}{\gamma(t)} &= \sum_{j=0}^{n}(-1)^j\left(\frac{1}{g_\varnothing(t)}\right)^{j+1}\left(\sum_{I\neq\varnothing} g_I(t)\zeta_I\right)^j \\
&= \frac{1}{g_\varnothing(t)}\sum_{j=0}^{n}(-1)^j\left(\frac{1}{g_\varnothing(t)}\right)^{j}\left(\sum_{I\neq\varnothing} g_I(t)\zeta_I\right)^j \\
&= \frac{1}{g_\varnothing(t)} + \frac{1}{g_\varnothing(t)}\sum_{I\neq\varnothing}\sum_{j=1}^{n} j!\left(\frac{-1}{g_\varnothing(t)}\right)^{j}\left(\sum_{\substack{\pi\in\mathcal{P}(I)\\|\pi|=j}} g_\pi(t)\right)\zeta_I,
\end{aligned}$$

where g_π denotes a product of functions indexed by blocks of the partition π (i.e., $g_\pi(t) = \prod_{b\in\pi} g_b(t)$ for $t \in X$). Existence of the reciprocal derivative follows from differentiability of $\gamma(t)$ by applying the product rule and real chain rule to the canonical expansion of the reciprocal $\frac{1}{\gamma(t)}$ as follows

$$\begin{aligned}\left(\frac{1}{\gamma}\right)'(t) &= -\frac{g_\varnothing'(t)}{(g_\varnothing(t))^2}\\ &\quad+\sum_{I\neq\varnothing}\sum_{j=1}^{n}(j+1)!\left(\frac{-1}{g_\varnothing(t)}\right)^j\frac{-g_\varnothing'(t)}{(g_\varnothing(t))^2}\left(\sum_{\substack{\pi\in\mathcal{P}(I)\\|\pi|=j}} g_\pi(t)\right)\zeta_I\\ &\quad-\sum_{I\neq\varnothing}\sum_{j=1}^{n}\left(\frac{-1}{g_\varnothing(t)}\right)^{j+1} j!\left(\sum_{\substack{\pi\in\mathcal{P}(I)\\|\pi|=j}}\sum_{b\in\pi} g_{\pi\setminus b}(t)g_b'(t)\right)\zeta_I.\end{aligned}$$

Having established existence, the desired form is obtained by applying the product rule to both sides of $1 = \gamma(t)\frac{1}{\gamma(t)}$. In particular, one obtains

$$0 = \gamma'(t)\frac{1}{\gamma(t)} + \gamma(t)\left(\frac{1}{\gamma}\right)'(t),$$

which implies the result. □

Proposition 14.3.7 (Quotient Rule). *Let φ and γ be zeon-valued functions on $X \subseteq \mathbb{R}$, differentiable at $c \in X$, and assume that $\Re(\gamma(t)) \neq 0$ for all $t \in X$. Then, the quotient function $\psi = \frac{\varphi}{\gamma}$ is differentiable on X and*

$$\psi'(t) = \frac{\gamma(t)\varphi'(t) - \gamma'(t)\varphi(t)}{(\gamma(t))^2} \quad (t \in X).$$

Proof. The proof is left as an exercise. □

The chain rule for zeon-valued functions of a real-variable is straightforward.

Proposition 14.3.8 (Chain Rule). *For interval $X \subseteq \mathbb{R}$, let $g : X \to \mathbb{R}$ be differentiable at $c \in X$ and suppose $\varphi : \mathcal{C}\ell_n{}^{\text{nil}}$ is differentiable at $g(c)$. Then, the composite function $\varphi \circ g : X \to \mathbb{R}$ is differentiable at c and*

$$(\varphi\circ g)'(c) = \varphi'(g(c))g'(c).$$

Proof. The result follows from definitions and the hypotheses using real-valued functions. Details are left as an exercise. □

Exercises

Exercise 14.1: Find an upper bound on $|\langle\varphi(u), \zeta_I\rangle|$ in terms of $\|u\|_\infty$.

Exercise 14.2: Prove that

$$\lim_{u \twoheadrightarrow 1+2\zeta_{\{1\}}} u^2 - 3u + 1 = -1 - 2\zeta_{\{1\}}.$$

Exercise 14.3: Find all critical points of $\varphi(u) = u^3 - 1$.

Exercise 14.4: Prove Proposition 14.3.7.

Exercise 14.5: Prove Proposition 14.3.8.

Chapter 15

Graph Enumeration Problems

Examples of graph enumeration problems include counting structures like paths, trails, cycles, circuits, spanning trees, matchings, cliques, and independent sets in a given graph. The null-square property of zeon generators makes them especially convenient for symbolic computations associated with enumeration problems on finite graphs.

15.1 Nilpotent Adjacency Matrices

As seen in Proposition 2.2.4, powers of the adjacency matrix of a graph provide a simple and convenient tool for counting walks in finite graphs. What the adjacency matrix fails to provide is a method of counting paths and cycles in graphs. For that, a special type of adjacency matrix is needed.

Definition 15.1.1. Given a graph $G = (V, E)$ on $n = |V|$ vertices, let ${\mathcal{C}\ell_V}^{\text{nil}}$ be the zeon algebra of dimension 2^n whose generators are in one-to-one correspondence with the vertices of G. The *nilpotent adjacency matrix* associated with G is defined by

$$\Phi_{ij} = \begin{cases} \zeta_{v_j}, & \text{if } \{v_i, v_j\} \in E(G) \\ 0, & \text{otherwise.} \end{cases}$$

Example 15.1.2. Figure 15.1 depicts a graph on 10 vertices and its associated nilpotent adjacency matrix. Here, $\text{tr}(\mathfrak{A}^{10}) = 80\zeta_{\{1,2,3,4,5,6,7,8,9,10\}}$. Hence, $H_c = 4$ (ignoring orientation).

The next theorem will be used to derive results throughout the rest of the book. To make referencing easy, it will be called the "nil-structure theorem."

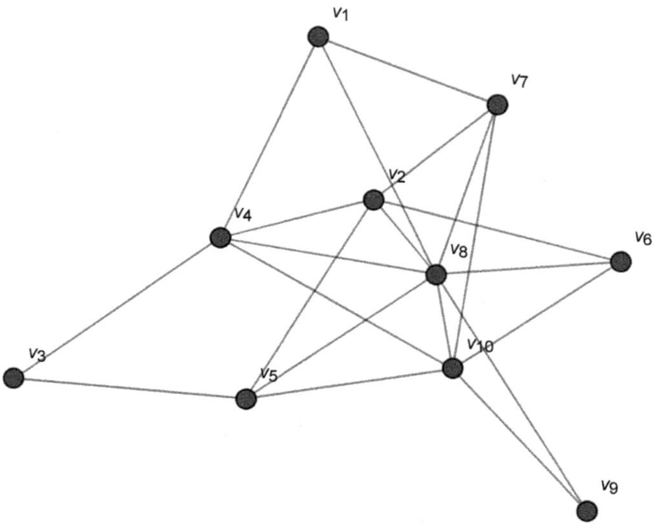

$$\begin{pmatrix}
0 & 0 & 0 & \zeta_{\{4\}} & 0 & 0 & \zeta_{\{7\}} & \zeta_{\{8\}} & 0 & 0 \\
0 & 0 & 0 & \zeta_{\{4\}} & \zeta_{\{5\}} & \zeta_{\{6\}} & \zeta_{\{7\}} & \zeta_{\{8\}} & 0 & 0 \\
0 & 0 & 0 & \zeta_{\{4\}} & \zeta_{\{5\}} & 0 & 0 & 0 & 0 & 0 \\
\zeta_{\{1\}} & \zeta_{\{2\}} & \zeta_{\{3\}} & 0 & 0 & 0 & 0 & \zeta_{\{8\}} & 0 & \zeta_{\{10\}} \\
0 & \zeta_{\{2\}} & \zeta_{\{3\}} & 0 & 0 & 0 & 0 & \zeta_{\{8\}} & 0 & \zeta_{\{10\}} \\
0 & \zeta_{\{2\}} & 0 & 0 & 0 & 0 & 0 & \zeta_{\{8\}} & 0 & \zeta_{\{10\}} \\
\zeta_{\{1\}} & \zeta_{\{2\}} & 0 & 0 & 0 & 0 & 0 & \zeta_{\{8\}} & 0 & \zeta_{\{10\}} \\
\zeta_{\{1\}} & \zeta_{\{2\}} & 0 & \zeta_{\{4\}} & \zeta_{\{5\}} & \zeta_{\{6\}} & \zeta_{\{7\}} & 0 & \zeta_{\{9\}} & \zeta_{\{10\}} \\
0 & 0 & 0 & 0 & 0 & 0 & 0 & \zeta_{\{8\}} & 0 & \zeta_{\{10\}} \\
0 & 0 & 0 & \zeta_{\{4\}} & \zeta_{\{5\}} & \zeta_{\{6\}} & \zeta_{\{7\}} & \zeta_{\{8\}} & \zeta_{\{9\}} & 0
\end{pmatrix}$$

Figure 15.1 A graph on 10 vertices and its nilpotent adjacency matrix.

Theorem 15.1.3 (Nil-Structure Theorem). *Let Φ be the nilpotent adjacency matrix of an n-vertex graph G. For any $k > 1$ and $1 \leq i, j \leq n$,*

$$\langle v_i|\Phi^k|v_j\rangle = \sum_{\substack{(w_1,\ldots,w_k)\in V^k \\ (w_k=v_j)\wedge(m\neq\ell\Rightarrow w_m\neq w_\ell)}} \zeta_{\{w_1,\ldots,w_k\}} = \sum_{\substack{I\subseteq V \\ |I|=k}} \omega_I \zeta_I,$$

where ω_I denotes the number of k-step walks from v_i to v_j visiting each vertex in I exactly once when initial vertex $v_i \notin I$ and revisiting v_i exactly once when $v_i \in I$. In particular, for any $k \geq 3$ and $1 \leq i \leq n$,

$$\langle v_i|\Phi^k|v_i\rangle = \sum_{\substack{I\subseteq V \\ |I|=k}} \omega_I \zeta_I,$$

where ω_I denotes the number of k-cycles on vertex set I based at $v_i \in I$.

Proof. Because the generators of ${\mathcal{C}\ell_n}^{\text{nil}}$ square to zero, a straightforward inductive argument shows that the nonzero terms of $\langle v_i|\Phi^k|v_j\rangle$ are multivectors corresponding to two types of k-walks from v_i to v_j: self-avoiding walks (i.e., walks with no repeated vertices) and walks in which v_i is repeated exactly once at some step but are otherwise self-avoiding. Walks of the second type are zeroed in the kth step when the walk is closed. Hence, terms of $\langle v_i|\Phi^k|v_i\rangle$ represent the collection of k-cycles based at v_i. □

In light of this theorem, the name "nilpotent adjacency matrix" is justified by the following corollary.

Corollary 15.1.4. *Let Φ be the nilpotent adjacency matrix of a simple graph on n vertices. For any positive integer $k \leq n$, the entries of Φ^k are homogeneous elements of grade k in ${\mathcal{C}\ell_n}^{\text{nil}}$. Moreover, $\Phi^k = \mathbf{0}$ for all $k > n$.*

Fixing a basis $\{\mathbf{u}_I : I \in 2^{[n]}\}$ of blades for a combinatorial algebra $\mathcal{A}$, the *scalar sum* map $\mathcal{A} \to \mathbb{R}$ is defined by the following:

$$\left\langle\left\langle \sum_{I\in 2^{[n]}} \alpha_I \mathbf{u}_I \right\rangle\right\rangle = \sum_{I\in 2^{[n]}} \alpha_I. \tag{15.1}$$

Corollary 15.1.5. *Let Φ be the nilpotent adjacency matrix of a graph G. Then,*

$$\langle\langle \operatorname{tr}\left(\Phi^k\right)\rangle\rangle = k\ |\{k\text{-cycles in } G\}|. \tag{15.2}$$

Proof. The result follows immediately from Theorem 15.1.3 since each k-cycle appears at k choices of base point along the main diagonal of Φ^k. □

Remark 15.1.6. In an undirected graph, a distinction must be made about cycle orientation. Each cycle recovered along the main diagonal of Φ^k appears with two orientations, doubling the associated scalar coefficients. From this point forward, cycles differing by orientation are considered distinct unless specified otherwise.

It should be clear that since Φ^k is the $n \times n$ zero matrix for all $k > n$, $(I - t\Phi)^{-1} = \displaystyle\sum_{k=0}^{n} t^k\, \Phi^k$ exists as a finite sum, and one can recover

$$\operatorname{tr}\left(\Phi^k\right) = \operatorname{tr}\left(I - t\Phi\right)^{-1}\Big|_{t^k}.$$

In other words, the trace of Φ^k is the ${\mathcal{C}\ell_n}^{\mathrm{nil}}$-valued coefficient of t^k in the power series expansion of $(I - t\Phi)^{-1}$.

Moreover, $\exp(t\Phi)$ is a finite sum, and

$$\operatorname{tr}\left(\Phi^k\right) = \frac{d^k}{dt^k}\operatorname{tr}\left[\exp\left(t\Phi\right)\right]\Big|_{t=0}.$$

Corollary 15.1.7 (Counting Hamiltonian Cycles). *Let Φ be the nilpotent adjacency matrix of an n-vertex graph G. Let H_n denote the number of Hamiltonian cycles appearing in the graph G. Then*

$$\langle\langle \operatorname{tr}\left(\Phi^n\right)\rangle\rangle = nH_n.$$

Proof. Proof is by setting $k = n$ in (15.2). □

Example 15.1.8. Returning to the graph of Example 15.1.2, $\operatorname{tr}(\mathfrak{A}^{10}) = 80\zeta_{\{1,2,3,4,5,6,7,8,9,10\}}$. Hence, the number of Hamiltonian cycles contained in the graph is $H_c = 4$ (ignoring orientation).

Corollary 15.1.9 (Tuples of Cycles). *Let Φ be the nilpotent adjacency matrix of an n-vertex graph G. Let $X_{m,\ell}$ denote the number of ℓ-tuples of pairwise disjoint m-cycles appearing in the graph G, where $m > 1$ and $\ell \geq 1$. Then*

$$\left\langle\left\langle \left(\operatorname{tr}\left(\Phi^m\right)\right)^\ell \right\rangle\right\rangle = m^\ell \ell! X_{m,\ell}.$$

Proof. Note that $\dfrac{\operatorname{tr}\Phi^m}{m}$ is a sum of nilpotent multivectors associated with m-cycles in the graph. By nilpotency, the nonzero terms of $\left(\dfrac{\operatorname{tr}\Phi^m}{m}\right)^\ell$ represent pairwise disjoint m-cycles, and each term occurs $\ell!$ times. □

Example 15.1.10. Consider the nilpotent adjacency matrix of the three-dimensional hypercube of Fig. 15.2.

$$\Phi = \begin{pmatrix} 0 & \zeta_{\{2\}} & \zeta_{\{3\}} & 0 & \zeta_{\{5\}} & 0 & 0 & 0 \\ \zeta_{\{1\}} & 0 & 0 & \zeta_{\{4\}} & 0 & \zeta_{\{6\}} & 0 & 0 \\ \zeta_{\{1\}} & 0 & 0 & \zeta_{\{4\}} & 0 & 0 & \zeta_{\{7\}} & 0 \\ 0 & \zeta_{\{2\}} & \zeta_{\{3\}} & 0 & 0 & 0 & 0 & \zeta_{\{8\}} \\ \zeta_{\{1\}} & 0 & 0 & 0 & 0 & \zeta_{\{6\}} & \zeta_{\{7\}} & 0 \\ 0 & \zeta_{\{2\}} & 0 & 0 & \zeta_{\{5\}} & 0 & 0 & \zeta_{\{8\}} \\ 0 & 0 & \zeta_{\{3\}} & 0 & \zeta_{\{5\}} & 0 & 0 & \zeta_{\{8\}} \\ 0 & 0 & 0 & \zeta_{\{4\}} & 0 & \zeta_{\{6\}} & \zeta_{\{7\}} & 0 \end{pmatrix}.$$

The 96 Hamiltonian cycles in $\mathcal{Q}_3$ are recovered from

$$\frac{1}{8!}\frac{\partial^8}{\partial t^8}\mathrm{tr}(I - t\Phi)^{-1}\bigg|_{t=0} = 96\zeta_{\{1,2,3,4,5,6,7,8\}}t^8.$$

It is worth noting that these 96 circuits include all possible orientations and base point selections.

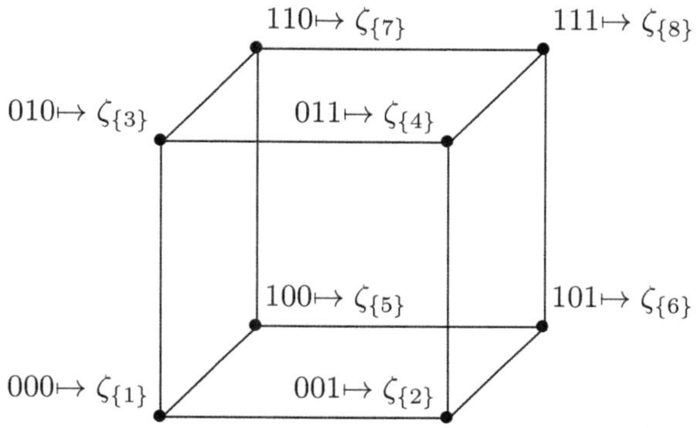

Figure 15.2 Three-dimensional hypercube.

Euler Circuits

In contrast to the preceding section, graphs are now allowed to have multiple edges between pairs of vertices.

Definition 15.1.11. Let G be any finite graph having n vertices and $|E|$ edges. Labeling edges of G with integers $1, 2, \ldots, |E|$, and utilizing ${\mathcal{C}\ell_{|E|}}^{\mathrm{nil}}$, define the $n \times n$ *edge-labeled nilpotent adjacency matrix* of G by

$$i, j \in V(G) \Rightarrow \mathcal{E}_{ij} = \sum_{\text{edges } k: i \to j} \zeta_k.$$

Proposition 15.1.12. *Let $\mathcal{E}$ be the edge-labeled nilpotent adjacency matrix of a finite graph G having $|E|$ edges and n vertices. Summing the coefficients of* $\mathrm{tr}\left(\mathcal{E}^{|E|}\right)$ *yields n times the number of Euler circuits occurring in G.*

Proof. As in the proof of Theorem 15.1.3, entries of $\mathcal{E}^k$ are algebraic polynomials corresponding to k-walks in the graph G. Given such a walk $i \to j$ and its corresponding polynomial $\left(\mathcal{E}^k\right)_{ij}$, the only nonzero terms correspond to self-avoiding walks $i \to j$. In the context of edge-labeled nilpotent adjacency matrices, this means no *edge* appears more than once in the walk. Thus $\left(\mathcal{E}^{|E|}\right)_{ii}$ represents the collection of all self-avoiding $|E|$-circuits based at vertex i (i.e., the collection of all Euler circuits $i \to i$). Since every vertex appears in each such circuit, a representation of each circuit appears at every diagonal entry in $\mathcal{E}^{|E|}$, and the nonzero terms in the diagonal elements of $\mathcal{E}^{|E|}$ are identical.

$$\left\langle\left\langle \operatorname{tr}\left(\mathcal{E}^{|E|}\right)\right\rangle\right\rangle = \sum_{i=1}^{n}\left\langle\left\langle\left(\mathcal{E}^{|E|}\right)_{ii}\right\rangle\right\rangle = n\cdot\left\langle\left\langle\left(\mathcal{E}^{|E|}\right)_{kk}\right\rangle\right\rangle$$
$$= n\cdot|\{\text{Euler circuits in } G\}|$$

where $|\cdot|$ denotes cardinality and $1 \le k \le n$ is arbitrary and fixed. □

15.2 Matchings, Cliques, and Independent Sets

Given a graph $G = (V, E)$ on n vertices, a *cycle cover* of G is a collection of cycles $\{C_1, \ldots, C_k\}$ contained as subgraphs of G such that each vertex is contained in exactly one of the cycles. Note that a cycle cover of a graph of size n contains n edges.

Given a graph $G = (V, E)$, a *matching* of G is a subset $E_1 \subset E$ of the edges of G having the property that no pair of edges in E_1 shares a common vertex. A *k-matching* is a matching containing k edges.

Note that in a graph on n vertices, the maximum size of any matching is $\lfloor n/2\rfloor$. An n-matching on a graph with $2n$ vertices is called a *perfect matching*, while an n-matching on a graph with $2n+1$ vertices is said to be a *critical perfect matching*.

An *independent set* of a graph is a set of vertices that are pairwise nonadjacent. A *clique* of a graph is a set of vertices such that every vertex is adjacent to every other in the set. Thus, a clique is a complete subgraph. Note that independent sets of a graph G correspond to cliques in $\overline{G}$.

Given a simple graph $G = (V, E)$ on n vertices, the adjacency structure of G is represented uniquely within ${\mathcal{C}\ell_n}^{\text{nil}}$ by

$$\Gamma = \sum_{\{v_i,v_j\}\in E(G)} \zeta_{\{v_i,v_j\}}. \tag{15.3}$$

In particular, Γ is a sum of bivectors representing edges of G by using each edge's incident vertices as indices.

Proposition 15.2.1. *Let G be a graph on n vertices with Γ defined as in*

(15.3), *and let* $\mathfrak{z} = \sum_{i=1}^{n} \zeta_{\{i\}}$. *Then, for* $k \in \mathbb{N}$,

$$\Gamma^k = k! \sum_{I \in V_k} \alpha_I \, \zeta_I,$$

where α_I *denotes the number of* k*-matchings on the vertex set* $I \subset V$.

Proof. From construction of Γ, it follows that Γ^k is a sum of blades representing k-subsets of edges of G. Moreover, by the null-square property of the vertex labels ζ_j, such k-subsets of edges must represent k-matchings of G since two edges incident with a common vertex would result in a blade with a squared generator.

Let V_k denote the collection of subsets of V representing the vertices incident with edges in a k-matching of G, and note that $|V_k| = 2k$. For $I \in V_k$, letting α_I denote the number of k-matchings on the vertex set I, it follows that $\Gamma^k = k! \sum_{I \in V_k} \alpha_I \, \zeta_I$. □

In light of Proposition 15.2.1, it is clear that $\Gamma^\ell = 0$ for all $\ell > \frac{\lfloor n \rfloor}{2}$.

Corollary 15.2.2. *Let* G *be a graph on* n *vertices with* Γ *defined as in* (15.3), *and let* $\mathfrak{z} = \sum_{i=1}^{n} \zeta_{\{i\}}$. *Let* μ *denote the number of perfect matchings of* G. *Then,*

$$\mu = \begin{cases} \dfrac{1}{(n/2)!} \displaystyle\int \Gamma^{n/2} \, d\zeta_{[n]} & \text{when } n \equiv 0 \pmod 2, \\ \dfrac{1}{((n-1)/2)!} \displaystyle\int \mathfrak{z}\Gamma^{(n-1)/2} \, d\zeta_{[n]} & \text{when } n \equiv 1 \pmod 2. \end{cases}$$

Proof. When n is even, all vertices are accounted for in any perfect matching. Hence, the Berezin integral reveals the number of perfect matchings, with each matching overcounted by permutations of edges taken in the product. When n is odd, each perfect matching is represented by an element multi-indexed by set of size $n-1$; that is, one generator of ${\mathcal{C}\ell_n}^{\mathrm{nil}}$ is missing. Multiplying $\Gamma^{(n-1)/2}$ by $\mathfrak{z}$ transforms all grade-$(n-1)$ basis blades into $\zeta_{[n]}$ so that the Berezin integral reveals the required information. □

Denote by ${\mathcal{C}\ell_n}^{\mathrm{nil}} \otimes \mathbb{R}[t]$ the algebra of polynomials in the unknown t with ${\mathcal{C}\ell_n}^{\mathrm{nil}}$ coefficients. In particular,

$${\mathcal{C}\ell_n}^{\mathrm{nil}} \otimes \mathbb{R}[t] = \{\psi_n t^n + \psi_{n-1} t^{n-1} + \cdots + \psi_1 t + \psi_0 : \psi_i \in {\mathcal{C}\ell_n}^{\mathrm{nil}} (i \le n \in \mathbb{N})\}.$$

Definition 15.2.3. Given a simple graph G, let Γ be defined as in (15.3). Define the *nilpotent graph polynomial* of G by

$$e^{t\Gamma} = \sum_{k=0}^{\lfloor n/2 \rfloor} \frac{t^k \Gamma^k}{k!}. \tag{15.4}$$

Formally, $e^{t\Gamma}$ represents the exponential function with argument $t\Gamma$. However, nilpotency of $\Gamma \in \mathcal{C}\ell_n{}^{\text{nil}}$ reduces the infinite power series to the finite sum seen in (15.4).

Corollary 15.2.4. *Let G be a graph on n vertices with nilpotent graph polynomial $e^{t\Gamma}$ as defined in* (15.3)*. Let M denote the number of edges in a maximal matching of G. Then, $e^{t\Gamma}$ is a polynomial in $\mathcal{C}\ell_n{}^{\text{nil}} \otimes \mathbb{R}[t]$, and*

$$M = \deg_t e^{t\Gamma}.$$

Proof. The proof follows from Proposition 15.2.1 by observing that

$$e^{t\Gamma} = \sum_{k=0}^{\infty} \frac{(t\Gamma)^k}{k!} = \sum_{k=0}^{n/2} \frac{(t\Gamma)^k}{k!} = \sum_{k=0}^{n/2} t^k \sum_{I \in V_k} \alpha_I \,\zeta_I.$$

It follows immediately that $\deg_t e^{t\Gamma} = k$ if and only if k is the greatest integer for which a k-matching of G exists. □

Cliques and Independent Sets

Cliques in a graph G correspond to independent sets in the complement G'. More specifically, the vertices appearing in a clique of G are independent in G'.

Let edges of $G' = (V, E')$ be labeled with zeon generators by a bijection $\phi : E' \to \{\zeta_1, \ldots, \zeta_{|E'|}\}$, and let $\lambda : V \to \{x_i : 1 \le i \le n\}$ be a bijective labeling of the vertices of G' by commuting variables. For each $v \in V$, define

$$\begin{aligned}\psi(v) &= \lambda(v) \prod_{\substack{u \in E' \\ u \text{ is incident to } v}} \phi(u) \\ &= \lambda(v)\zeta_{\mathcal{N}(v)},\end{aligned}$$

where $\zeta_{\mathcal{N}(v)}$ is a zeon basis blade indexed by the edges incident[1] to v in G'. Define $\Phi_{G'} \in \mathcal{C}\ell_{|E'|}{}^{\text{nil}}$ by

$$\Phi_{G'} = \sum_{v \in V} \psi(v) = \sum_{v \in V} \lambda(v)\zeta_{\mathcal{N}(v)}.$$

[1] In other words, $\mathcal{N}(v)$ represents the "edge-neighborhood" of v.

Definition 15.2.5. The element $\Phi_{G'} \in \{x_i : 1 \leq i \leq n\} \otimes \mathcal{C}\ell_{|E'|}{}^{\rm nil}$ is called the *zeon clique formulation* of the graph G.

Theorem 15.2.6. *Let $G = (V, E)$ be a finite simple graph on n vertices, and let $\Phi_{G'} \in \{x_i : 1 \leq i \leq n\} \otimes \mathcal{C}\ell_{|E'|}{}^{\rm nil}$ be the zeon clique formulation of G. Then,*

$$\Phi_{G'}{}^m = m! \sum_{\substack{\{I \subseteq V, |I|=m\} \\ I \text{ independent in } G'}} x_I \zeta_{\mathcal{N}(I)},$$

where, for each independent set I in G', $x_I = \prod_{i \in I} x_i$ is a monomial representing an independent subset I of vertices in G', and equivalently the vertex set of a clique in G.

Proof. Let $G = (V, E)$ be as stated. To show

$$\Phi_{G'}{}^k = k! \sum_{\substack{\{I \subseteq V, |I|=m\} \\ I \text{ independent in } G'}} x_I \zeta_{\mathcal{N}(I)},$$

consider the multinomial expansion

$$\begin{aligned} \Phi_{G'}{}^m &= \left(\sum_{v \in V} \psi(v) = \sum_{v \in V} \lambda(v) \zeta_{\mathcal{N}(v)} \right)^m \\ &= \sum_{t_1 + \cdots + t_n = m} \binom{m}{t_1, t_2, \ldots, t_n} \prod_{i=1}^{n} (\lambda(v_i))^{t_i} {\zeta_{\mathcal{N}(v_i)}}^{t_i}. \end{aligned}$$

By the null-square property of zeon generators, the only surviving terms of the sum correspond to n-tuples $(t_1, \ldots, t_n) \in \{0,1\}^n$. Further, any product

$$\zeta_{\mathcal{N}(v_i)} \zeta_{\mathcal{N}(v_j)} = \begin{cases} \zeta_{\mathcal{N}(\{v_i, v_j\})} & \{v_i, v_j\} \notin E, \\ 0 & \{v_i, v_j\} \in E. \end{cases}$$

Hence, the only nonzero terms of the product $\prod_{i=1}^{n} (\lambda(v_i))^{t_i} {\zeta_{\mathcal{N}(v_i)}}^{t_i}$ are of the form

$$\prod_{i=1}^{n} (\lambda(v_i))^{t_i} {\zeta_{\mathcal{N}(v_i)}}^{t_i} = \prod_{i \in I} (\lambda(v_i)) \zeta_{\mathcal{N}(I)}$$

where I is an independent set of m vertices in G. Setting $x_I = \prod_{i \in I} \lambda(v_i)$ and recalling the idempotent properties of the labeling λ completes the proof. □

15.3 Minimal Path Algorithms

Nilpotent adjacency matrices provide symbolic tools for sieving out paths, cycles, circuits, and other self-avoiding structures in finite graphs. However, computing powers of a nilpotent matrix can be unwieldy and often unnecessary for enumerating paths in real-world applications.

In this section, we consider algorithms for enumerating minimal paths from a fixed initial vertex to a fixed target without the computational overhead of computing powers of a nilpotent adjacency matrix. The savings in computational complexity are substantial.

Definition 15.3.1. A *hop-minimal* path from initial vertex v_0 to terminal vertex v_∞ is a sequence $(v_0, v_1, \ldots, v_k = v_\infty)$ of $k+1$ vertices such that $(v_i, v_{i+1}) \in E(G)$ for each $i = 0, \ldots, k-1$ and there exists no path $v_0 \to v_\infty$ containing fewer vertices. The *length* of the path $(v_0, v_1, \ldots, v_k = v_\infty)$ is k.

Given a graph $G = (V, E)$ on n vertices[2], let Ψ be the graph's nilpotent adjacency matrix.

Recalling Dirac notation, the ith row of Ψ is conveniently denoted by $\langle v_i|\,\Psi$ while the jth column is denoted by $\Psi\,|v_j\rangle$. In this way, Ψ is completely determined by

$$\langle v_i|\Psi|v_j\rangle = \begin{cases} \zeta_j & \text{if there exists a directed edge } v_i \to v_j \text{ in } G, \\ 0 & \text{otherwise,} \end{cases} \tag{15.5}$$

for all vertex pairs $(v_i, v_j) \in E$.

Theorem 15.3.2. *Let Ψ be the nilpotent adjacency matrix of n-vertex graph G. For any $k > 1$ and $1 \le i \ne j \le n$,*

$$\begin{aligned} \zeta_i \langle v_i|\Psi^k|v_j\rangle &= \sum_{k\text{-paths } \mathbf{w}: v_i \to v_j} \zeta_{\mathbf{w}} \\ &= \sum_{|I|=k+1} \alpha_I \zeta_I, \end{aligned}$$

where α_I denotes the number of k-paths from v_i to v_j on vertices indexed by I. Hence, all paths of length k with initial vertex v_i and terminal vertex v_j are enumerated by the multi-indices of the terms in the summation.

Proof. The result follows from straightforward mathematical induction on k using properties of the multiplication in $\mathcal{C}\ell_n{}^{\text{nil}}$ with the observation that the initial vertex of the walk (i.e., v_i) is unaccounted for in $\langle v_i|\Psi^k|v_j\rangle$, as

[2]The graph may be either simple or directed with no multiple edges.

seen in (15.6) of the matrix definition. Hence, each term of $\langle v_i|\Psi^k|v_j\rangle$ is indexed by the vertex sequence of a k-walk from v_i to v_j with no repeated vertices, except possibly v_i at some intermediate step. Left multiplication by ζ_i thus sieves out the k-paths.

Considering entries along the main diagonal of Ψ^k, note that the final step of a k-cycle based at v_i returns to v_i so that left multiplication by ω_i is not required for cycle enumeration. □

While Theorem 15.3.2 is useful for counting paths between a pair of vertices, the next theorem will allow us to list the paths themselves. First, for fixed $n \in \mathbb{N}$, let $\{x_1, \ldots, x_n\}$ be a family of non-commuting variables and let $\mathbb{R}[x_1, \ldots, x_n]$ denote the *algebra of polynomials in* $\{x_1, \ldots, x_n\}$. Adopting *multi-exponent notation*, a typical element of $\mathbb{R}[x_1, \ldots, x_n]$ is any finite sum of the following form:

$$\begin{aligned} f(x) &= \sum_{\vec{j}=(j_1,\ldots,j_n)\in J} \alpha_{\vec{j}}\, x^{\vec{j}} \\ &= \sum_{\vec{j}=(j_1,\ldots,j_n)\in J} \alpha_{\vec{j}}\, {x_1}^{j_1} \cdots {x_n}^{j_n} \end{aligned}$$

where J is a finite subset of ${\mathbb{N}_0}^n$ and $\alpha_{\vec{j}} \in \mathbb{R}$ for each $\vec{j}$.

It is important to note that the lack of commutativity in the variables implies

$$\begin{aligned} (\alpha x^{\vec{j}})(\beta x^{\vec{\ell}}) &= \alpha\beta x^{\vec{j}} x^{\vec{\ell}} \\ &= \alpha\beta {x_1}^{j_1} \cdots {x_n}^{j_n} {x_1}^{\ell_1} \cdots {x_n}^{\ell_n}. \end{aligned}$$

The strategy is to construct a new nilpotent adjacency matrix that preserves path information while also sieving out paths and cycles. This can be accomplished by defining a matrix whose entries are elements of the algebra ${\mathcal{C}\ell_n}^{\text{nil}} \otimes \mathbb{R}[x_1, \ldots, x_n]$. Elements of this tensor algebra are polynomials in the non-commuting variables $x_1, \ldots, x_n$ whose coefficients are elements of the zeon algebra. Multiplication in the tensor algebra is defined by linear extension of the following for $uf(x), wg(x) \in {\mathcal{C}\ell_n}^{\text{nil}} \otimes \mathbb{R}[x_1, \ldots, x_n]$:

$$uf(x)wg(x) = uwf(x)g(x).$$

Definition 15.3.3. Given a directed graph $G = (V, E)$ on n vertices with no multiple edges, the *path-identifying nilpotent adjacency matrix* of G is the $n \times n$ matrix $\Psi = (\psi_{ij})$ with entries in the non-commutative tensor algebra ${\mathcal{C}\ell_n}^{\text{nil}} \otimes \mathbb{R}[x_1, \ldots, x_n]$ defined by

$$\psi_{ij} = \begin{cases} \zeta_j x_j & \text{if } (v_i, v_j) \in E, \\ 0 & \text{otherwise.} \end{cases} \tag{15.6}$$

For convenience, we set $\omega_j = \zeta_j x_j \in \mathcal{C}\ell_n{}^{\text{nil}} \otimes \mathbb{R}[x_1, \ldots, x_n]$ for each $j = 1, \ldots, n$. It is then evident that multiplication of the ωs satisfies the following: for two paths (i.e., finite sequences of indices) $\mathbf{u}$ and $\mathbf{w}$,

$$\omega_{\mathbf{u}}\omega_{\mathbf{w}} = \begin{cases} \omega_{\mathbf{u}\centerdot\mathbf{w}} & \mathbf{u} \cap \mathbf{w} = \varnothing, \\ 0 & \text{otherwise.} \end{cases}$$

Here, $\mathbf{u} \centerdot \mathbf{w}$ denotes the *concatenation* of paths $\mathbf{u}$ and $\mathbf{w}$; i.e., if $\mathbf{u} = u_1 u_2 \cdots u_k$ and $\mathbf{w} = w_1 w_2 \cdots w_\ell$, then $\mathbf{u} \centerdot \mathbf{w} = u_1 \cdots u_k w_1 \cdots w_\ell$.

With all tools now in hand, the desired theorem can now be stated.

Theorem 15.3.4. *Let Ψ be the path-identifying nilpotent adjacency matrix of an n-vertex graph G. For any $k > 1$ and $1 \leq i \neq j \leq n$,*

$$\omega_i \left\langle v_i | \Psi^k | v_j \right\rangle = \sum_{k\text{-paths } \mathbf{w}: v_i \to v_j} \omega_{\mathbf{w}}. \tag{15.7}$$

Moreover,

$$\left\langle v_i | \Psi^k | v_i \right\rangle = \sum_{k\text{-cycles } \mathbf{w} \text{ based at } v_i} \omega_{\mathbf{w}}. \tag{15.8}$$

Thus, all paths of length k with initial vertex v_i and terminal vertex v_j are enumerated by the multi-indices of the terms in the summation.

Proof. The result follows from straightforward mathematical induction on k using properties of the multiplication in $\mathcal{C}\ell_n{}^{\text{nil}} \otimes \mathbb{R}[x_1, \ldots, x_n]$. Each term of $\left\langle v_i | \Psi^k | v_j \right\rangle$ is indexed by the vertex sequence of a k-walk from v_i to v_j with no repeated vertices, except possibly v_i at some intermediate step. Left multiplication by ω_i thus sieves out the k-paths. Again, the final step of a k-cycle based at v_i returns to v_i so that left multiplication by ω_i is not required for cycle enumeration. □

The nilpotent adjacency matrix construction allows one to list all paths and cycles in a finite graph by considering powers of the matrix. The associated tree structure underlying the cycle/path enumeration problem is automatically "pruned" by the inherent properties of the algebra. An immediate consequence of Theorem 15.3.4 is that Algorithm 4 enumerates all hop-minimal paths from vertex v_0 to vertex v_∞ in a fixed finite graph G.

input : Nilpotent adjacency matrix Ψ, initial vertex v_0, terminal vertex v_∞
output: All hop-minimal paths from v_0 to v_∞ in graph associated with Ψ

Prepend starting vertex (to prune cycles) and recover all one-step paths emanating from v_0.
$\langle \xi | = \omega_{\{v_0\}} \langle v_0 | \Psi$
Continue as long as no path to terminal vertex found and paths have not all self-intersected.
while $(\langle \xi | v_\infty \rangle = 0 \wedge \langle \xi | \neq \langle 0 |)$ **do**
  Extend all partial paths by one step.
  $\langle \xi | = \langle \xi | \Psi$
end
Return $[\langle \xi | v_\infty \rangle]$

Algorithm 4: HopMin

Example 15.3.5. In Figure 15.3, all hop-minimal paths from v_3 to v_{50} are enumerated in a randomly-generated 100 vertex graph.

Graph Processes

Given a set V of vertices, a *graph process* on V is a sequence of graphs (G_j) having common vertex set V. In particular, there exists a sequence of edge sets (E_j) such that for each $j \in \mathbb{N}$, $G_j = (V, E_j)$.

The problem of interest is the enumeration of paths in a graph process. To accomplish this, one begins by constructing the associated sequence (Ψ_j) of nilpotent adjacency matrices for the graph sequence (G_j). Further, let $\{s_j : j \in \mathbb{N}\}$ be a collection of commuting variables.

Given a k-path $\mathbf{p}$ occurring in a finite segment $(G_\ell)_{\ell=1}^{\mathfrak{f}}$ of graph process $(\mathcal{G}_t)$, a *frame partition* of $\mathbf{p}$ associated with the segment is an ordered $\mathfrak{f}$-tuple of nonnegative integers, $(d_1|\cdots|d_{\mathfrak{f}})$, where $k = \sum_\ell d_\ell$, such that d_ℓ steps of $\mathbf{p}$ occur in the ℓth graph (or "frame") of the process.

Theorem 15.3.6. *Let* $(G_j) = ((V, E_j))$ *be a graph process on vertex set* V. *For each* $j \in \mathbb{N}$, *let* Ψ_j *be the path-identifying nilpotent adjacency matrix of* G_j. *The collection of paths of length* $k \geq 1$ *from* $v_0 \to v_\infty$ *occurring within*

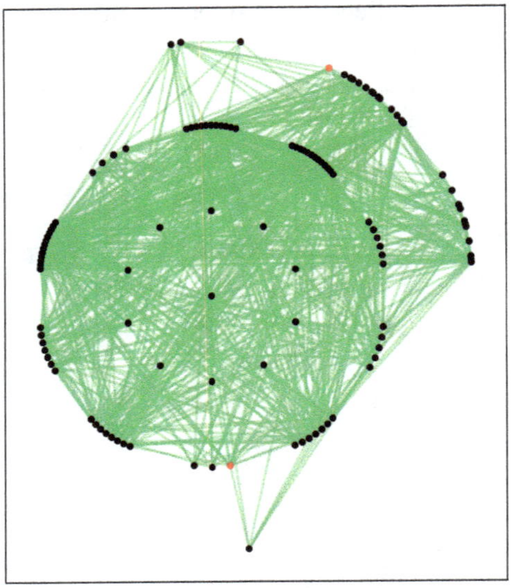

```
Hop-minimal paths from v3 to v50 :
```

$\omega_{\{3,40,8,50\}} + \omega_{\{3,40,58,50\}} + \omega_{\{3,40,86,50\}} + \omega_{\{3,59,73,50\}} + \omega_{\{3,98,86,50\}}$

```
Computation time: 0.049076 seconds.
```

Figure 15.3 Hop-minimal paths in a 100 vertex graph.

the first $\mathfrak{f}$ *frames of the process is given by the following:*

$$\omega_{v_0} \sum_{\substack{0 \le \ell_1, \ell_2, \ldots, \ell_{\mathfrak{f}} \\ \ell_1 + \cdots + \ell_{\mathfrak{f}} = k}} \langle v_0 | (s_1 \Psi_1)^{\ell_1} \cdots (s_{\mathfrak{f}} \Psi_{\mathfrak{f}})^{\ell_{\mathfrak{f}}} | v_\infty \rangle$$

$$= \sum_{k\text{-paths } \mathbf{p}: v_0 \to v_\infty} f(\mathbf{p}) \omega_{\mathbf{p}}, \tag{15.9}$$

where $f(\mathbf{p}) \in \mathbb{R}[s_1, \ldots, s_k]$ *is a polynomial of the form*

$$f(\mathbf{p}) = \sum_{\mathbf{d} \in \mathbb{N}_0{}^k} \alpha_{\mathbf{d}} s_1{}^{d_1} \cdots s_k{}^{d_k}$$

such that $\alpha_{\mathbf{d}} = 1$ *if* $\mathbf{p}$ *is a path whose frame partition is* $\mathbf{p} = (d_1 | \cdots | d_k)$ *and* $\alpha_{\mathbf{d}} = 0$ *otherwise.*

Proof. Note that $k \geq 1$ ensures that at least one of the integers ℓ_i is nonzero.

For fixed nonnegative integers ℓ_1, ℓ_2 with $\ell_1 + \ell_2 = k$,

$$\begin{aligned}
\omega_{v_0}\langle v_0|{\Psi_1}^{\ell_1}{\Psi_2}^{\ell_2}|v_\infty\rangle &= \omega_{v_0}\sum_{v_j\neq v_0}\langle v_0|{\Psi_1}^{\ell_1}|v_j\rangle\langle v_j|{\Psi_2}^{\ell_2}|v_\infty\rangle \\
&= \sum_{v_j\neq v_0}\sum_{\substack{\ell_1\text{-paths } \mathbf{p}:v_0\to v_j\\ \text{In frame 1}}}\alpha_{\mathbf{p}}\omega_{\mathbf{p}}\sum_{\substack{\ell_2\text{-paths } \mathbf{q}=:v_j\to v_\infty\\ \text{In frame 2}}}\alpha_{\mathbf{q}}\omega_{\mathbf{q}} \\
&= \sum_{\substack{k\text{-paths } \mathbf{p}:v_0\to v_\infty\\ \in\mathcal{P}^{\ell_1,\ell_2}}}\alpha_{\mathbf{p}}\omega_{\mathbf{p}},
\end{aligned}$$

where $\mathcal{P}^{\ell_1,\ell_2}$ denotes the collection of paths from v_0 to v_∞ in which ℓ_1 steps occur in frame 1 and ℓ_2 steps occur in frame 2. Proceeding by induction, the result is established for fixed m-tuple of nonnegative integers $(\ell_1, \ldots, \ell_m)$ with $\ell_1 + \cdots + \ell_m = k$. Assume that for positive integer m_0 and fixed nonnegative integers $\ell_1, \ldots, \ell_{m_0}$ with $\ell_1 + \cdots + \ell_{m_0} = k' \leq k$, the following holds:

$$\omega_{v_0}\langle v_0|{\Psi_1}^{\ell_1}\cdots{\Psi_{m_0}}^{\ell_{m_0}}|v_\infty\rangle = \sum_{\substack{k'\text{-paths } \mathbf{p}:v_0\to v_\infty\\ \in\mathcal{P}^{\ell_1,\ldots,\ell_{m_0}}}}\alpha_{\mathbf{p}}\omega_{\mathbf{p}}.$$

It follows that for $\ell_{m_0+1} = k - k'$, one has

$$\begin{aligned}
&\omega_{v_0}\langle v_0|{\Psi_1}^{\ell_1}\cdots{\Psi_{m_0}}^{\ell_{m_0}}{\Psi_{m_0+1}}^{\ell_{m_0+1}}|v_\infty\rangle \\
&= \omega_{v_0}\sum_{v_j\neq v_0}\langle v_0|{\Psi_1}^{\ell_1}\cdots{\Psi_{m_0}}^{\ell_{m_0}}|v_j\rangle\langle v_j|{\Psi_{m_0+1}}^{\ell_{m_0+1}}|v_\infty\rangle \\
&= \sum_{v_j\neq v_0}\sum_{\substack{k'\text{-paths } \mathbf{p}:v_0\to v_j\\ \text{in } \mathcal{P}^{\ell_1,\ldots,\ell_{m_0}}}}\alpha_{\mathbf{p}}\omega_{\mathbf{p}}\sum_{\substack{(k-k')\text{-paths } \mathbf{q}:v_j\to v_\infty\\ \text{in } \mathcal{P}^{0,\ldots,\ell_{m_0+1}}}}\alpha_{\mathbf{q}}\omega_{\mathbf{q}} \\
&= \sum_{\substack{k\text{-paths } \mathbf{p}:v_0\to v_\infty\\ \in\mathcal{P}^{\ell_1,\ldots,\ell_{m_0+1}}}}\alpha_{\mathbf{p}}\omega_{\mathbf{p}}.
\end{aligned}$$

Hence, the result is established for positive integer m:

$$\omega_{v_0}\langle v_0|{\Psi_1}^{\ell_1}\cdots{\Psi_m}^{\ell_m}|v_\infty\rangle = \sum_{\substack{k\text{-paths } \mathbf{p}=(v_0,\ldots,v_\infty)\\ \in\mathcal{P}^{\ell_1,\ldots,\ell_m}}}\alpha_{\mathbf{p}}\omega_{\mathbf{p}}.$$

The proof is thus completed by summing over all such m-tuples. □

It is worth noting that while products of powers of matrices appear in the formulation of Theorem 15.3.6, only products of row vectors with matrices need be computed in the implementation. Beginning with row vector $\omega_{v_1}\langle v_1|$, all multiplications are sequential right-multiplication of a row vector by an appropriate nilpotent adjacency matrix.

Performing symbolic computations in *Mathematica* allows one to work over algebras of arbitrarily high dimension. The complexity of the computations, however, depends on the complexity of the graph itself. For example, to extend by one step all partial paths leaving a fixed initial vertex, 400 multiplications are performed (multiplying a 1×20 row with a 20×20 matrix). The complexity of each multiplication depends on the complexity of the entries being multiplied. In turn, the number of nonzero products obtained and the complexity of the entries in the row vector depend only on the number of partial paths leaving the fixed vertex. In useful real-world applications, this complexity can be quite manageable.

Exercises

Exercise 15.1: Defining $\omega_j = \zeta_j x_j \in \mathcal{C}\ell_n{}^{\text{nil}} \otimes \mathbb{R}[x_1, \ldots, x_n]$ for each $j = 1, \ldots, n$, verify that the tensor algebra multiplication satisfies

$$\omega_{\mathbf{u}}\omega_{\mathbf{w}} = \begin{cases} \omega_{\mathbf{u}\triangle\mathbf{w}} & \mathbf{u} \cap \mathbf{w} = \varnothing, \\ 0 & \text{otherwise.} \end{cases}$$

Exercise 15.2: Again defining $\omega_j = \zeta_j x_j \in \mathcal{C}\ell_n{}^{\text{nil}} \otimes \mathbb{R}[x_1, \ldots, x_n]$, let Ω_n be the associative $\mathbb{R}$-algebra generated by the collection $\{\omega_j : 1 \le j \le n\}$ along with unit scalar $\omega_\varnothing = 1$. The algebra $\mathcal{C}\ell_n{}^{\text{nil}} \otimes \mathbb{R}[x_1, \ldots, x_n]$ is called the *n-particle path-identifying zeon algebra.* For $n \in \mathbb{N}$, determine the dimension of Ω_n.

Exercise 15.3: Let $G = (V, E)$ be a graph on vertices $V = \{1, \ldots, 7\}$ with edges $E = \{(1,2), (1,3), (1,4), (2,3), (2,4), (2,6), (3,4), (3,6), (4,6), (4,7), (5,6), (6,7)\}$. Construct the nilpotent adjacency matrix of G, and use it to count the 3-paths $1 \to 6$.

Exercise 15.4: Let $(\Psi_1, \Psi_2, \ldots)$ be a sequence of path-identifying nilpotent adjacency matrices associated with a graph process on a fixed set of vertices. Give an interpretation of

$$\xi = \omega_1 \langle v_1 | {\Psi_1}^3 \Psi_2 {\Psi_4}^4 | v_5 \rangle.$$

Chapter 16

Graph Colorings and Chromatic Structures

The paper [103] was the first extension of nilpotent adjacency matrix methods to graph colorings. The extension of nilpotent matrix methods to graph colorings allows one to count heterochromatic and monochromatic self-avoiding walks in colored graphs. Further, the zeon-algebraic formalism allows one to quickly verify whether a given graph coloring is proper, and it provides a convenient framework for implementing greedy coloring algorithms.

Monochromatic paths and cycles have been objects of interest for decades. Notable works include the papers of Erdös and Tuza [36, 37]; Albert, Frieze and Reed [2, 49]; and Broersma, *et al.* [15].

In 1973, Raynaud proved a conjecture by Lehel that a 2-colored complete symmetric directed graph with at least two vertices contains a simple directed (monochromatic) Hamiltonian cycle. In a 1983 paper, Gyáfás surveyed results covering the vertices of 2-colored complete graphs by two paths or two cycles of different color [53]. In [64], Li, Zhang, and Broersma established some sufficient conditions for the existence of monochromatic and heterochromatic paths and cycles.

In [21], Chen and Li assume that the *color degree*[1] of a graph's vertices is bounded below by some integer k, and they show that if $3 \leq k \leq 7$, then G has a heterochromatic path of length at least $k-1$. They also show that if $k > 8$, then G has a heterochromatic path of length at least $\lceil \frac{3k}{5} \rceil + 1$. More recently, Babu, Chandran, and Rajendraprasad [7] established lower bounds for the length of a maximum heterochromatic path in an edge-colored graph without small cycles.

[1] The color degree of a vertex is the cardinality of the distinct colors among the vertex's neighbors.

16.1 Graph Colorings

Definition 16.1.1. A *(vertex) coloring* of a graph $G = (V, E)$ is a mapping $\phi : V \to \{1, \ldots, \varkappa\}$. The set $\{1, \ldots, \varkappa\}$ is referred to as the *palette* of the coloring, and its elements are referred to as *colors*. A coloring ϕ is said to be *proper* if $(v_i, v_j) \in E$ implies $\phi(v_i) \neq \phi(v_j)$.

In light of Definition 16.1.1, a *colored graph* is a pair (G, ϕ), where ϕ is a coloring of the graph G. For our purposes, a *surjective* mapping $\phi : V \to \{1, \ldots, \varkappa\}$ will be referred to as a vertex $\varkappa$-coloring.

Definition 16.1.1 extends naturally to *edge colorings*. A *proper edge coloring* is a mapping $\theta : E \to \{1, \ldots, \varkappa\}$ such that $(v_i, v_\ell), (v_\ell, v_j) \in E$ implies $\theta((v_i, v_\ell)) \neq \theta((v_\ell, v_j))$. In other words, no pair of coincident edges can be associated with the same color in a proper edge coloring.

A $\varkappa$-coloring that is proper will be referred to specifically as a *proper $\varkappa$-coloring*. A graph G will be said to be *properly $\varkappa$-colorable* if there exists a proper coloring of G having a palette of cardinality $\varkappa$. The minimal $\varkappa$ for which a proper $\varkappa$-coloring exists is called the *chromatic number* of G.

16.2 Proper Colorings and Heterochromatic Walks

The task at hand is to define nilpotent matrices associated with vertex- and edge-colorings of a finite graph. Properties of the matrices can then be used to quickly determine whether or not a given graph coloring is proper and to count the heterochromatic self-avoiding walks (i.e., paths, cycles, trails, & circuits) in a finite graph.

Definition 16.2.1. Let $G = (V, E)$ be a graph on n vertices with vertex $\varkappa$-coloring ϕ. The *zeon vertex-coloring matrix* Ψ *associated with* (G, ϕ) is the $n \times n$ matrix having entries in ${\mathcal{C}\ell_\varkappa}^{\text{nil}}$ defined by

$$\langle v_i | \Psi | v_j \rangle = \begin{cases} \zeta_{\phi(v_j)} & \text{if } (v_i, v_j) \in E, \\ 0 & \text{otherwise.} \end{cases}$$

Theorem 16.2.2. *Let Ψ be the zeon vertex-coloring matrix of a colored graph (G, ϕ) on n vertices. Then, for $1 \leq i, j \leq n$ and $k \in \mathbb{N}$,*

$$\langle v_i | \Psi^k | v_j \rangle = \sum_{|I|=k} \alpha_I \zeta_I$$

where α_I is the number of walks from $v_i \to v_j$ in the graph such that the vertex colors are indexed by I and no color is repeated with the possible *exception of $\phi(v_i)$; in particular, no vertex can be repeated, except the initial vertex can repeated at most once in an intermediate step.*

Proof. Proof is by using induction on $k \geq 1$. When $k = 1$, the result holds by definition of Ψ. Let a *good walk* $v_i \to v_\ell$ on I be a walk from v_i to v_ℓ such that no vertex color is repeated except possibly $\phi(v_i)$ one time. Now, assume the result holds for some $k \geq 1$.

$$\begin{aligned}\langle v_i|\Psi^{k+1}|v_j\rangle &= \langle v_i|\Psi^k\Psi|v_j\rangle \\ &= \sum_{l=1}^{n}\langle v_i|\Psi^k|v_\ell\rangle\langle v_\ell|\Psi|v_j\rangle \\ &= \sum_{l=1}^{n}\sum_{|I|=k}\sharp\{k \text{ good walks } v_i \to v_\ell \text{ on } I\}\zeta_I\langle v_\ell|\Psi|v_j\rangle.\end{aligned}$$

For fixed ℓ, $\langle v_i|\Psi^k|v_\ell\rangle$ is a linear combination of ζ_I's representing good k-walks from v_i to v_ℓ. The product of an arbitrary term from this linear combination with $\langle v_\ell|\Psi|v_j\rangle$ will be zero if there is no edge from v_ℓ to v_j or if v_j is the same color as another vertex in the good k-walk $v_i \to v_\ell$. Hence, the product $\langle v_i|\Psi^k|v_\ell\rangle\langle v_\ell|\Psi|v_j\rangle$ represents a sum of good $(k+1)$-walks $v_i \to v_j$ of the following form:

$$\begin{aligned}&\sum_{|I|=k}\sharp\{\text{good } k\text{-walks } v_i \to v_\ell \text{ on I}\}\sharp\{1\text{-walks } v_\ell \to v_j\}\zeta_I\zeta_{\phi(v_j)} \\ &= \sum_{|I|=k+1}\sharp\{\text{good } (k+1)\text{-walks } v_i \to v_j \text{ on I visiting } v_\ell \text{ in step } k\}\zeta_{I\cup\phi(v_j)}.\end{aligned}$$

Thus, summing over all ℓ represents the colors along a path as long as no colors are repeated and v_i can be visited only once after starting the path. □

Counting heterochromatic *cycles* is accomplished by the following corollary.

Corollary 16.2.3. *Let Ψ be the zeon vertex-coloring matrix of a colored graph (G, ϕ) on n vertices. Let $k \in \mathbb{N}$ be arbitrary and let $\mathfrak{h}_k$ denote the number of heterochromatic k-cycles in (G, ϕ). Then, $\langle\langle \mathrm{tr}(\Psi^k)\rangle\rangle = k\mathfrak{h}_k$.*

Proof. The result follows from Theorem 16.2.2 by making two observations. First, the last vertex visited in a cycle is the initial vertex; therefore, any walk revisiting the initial vertex in an intermediate step will be annihilated by closing the walk. Secondly, each k-cycle appears along the main diagonal with multiplicity k due to various choices of basepoint for the cycle. □

Theorem 16.2.2 also reveals an algebraic method for determining whether a coloring is proper.

Corollary 16.2.4. *A zeon vertex-coloring matrix* Ψ *represents a proper coloring of a graph* $G = (V, E)$ *if and only if*

$$\langle\langle \mathrm{tr}(\Psi^2)\rangle\rangle = 2|E|.$$

Proof. By Theorem 16.2.2, $\langle v_i|\Psi^2|v_i\rangle$ is a linear combination of products of zeon pairs representing heterochromatic 2-cycles based at v_i. Each 2-cycle appears twice in the trace by choice of basepoint. It follows immediately that the scalar sum $\langle\langle \mathrm{tr}\,\Psi^2\rangle\rangle$ is twice the number of heterochromatic pairs of adjacent vertices in the graph. By definition, the graph is properly colored if and only if every adjacent pair of vertices is heterochromatic. □

Corollary 16.2.4 can be restated to provide the following nilpotent adjacency matrix formulation of k-colorability of a graph.

Theorem 16.2.5 (Proper $\varkappa$-colorability). *A graph* $G = (V, E)$ *is* properly $\varkappa$-colorable *if and only if there exists a nilpotent coloring matrix* Ψ *having entries in* ${\mathcal{C}\ell_\varkappa}^{\mathrm{nil}}$ *such that*

$$\langle\langle \mathrm{tr}(\Psi^2)\rangle\rangle = 2|E|.$$

Example 16.2.6. Vertices of the graph seen in Figure 16.1 were colored with 8 randomly assigned colors. The trace of ψ^2 as computed by *Mathematica* is

$$\begin{aligned}&6\zeta_{\{1,2\}} + 8\zeta_{\{1,3\}} + 2\zeta_{\{1,5\}} + 2\zeta_{\{1,6\}} + 2\zeta_{\{1,7\}} + 4\zeta_{\{1,8\}} + 2\zeta_{\{2,3\}}\\ &\quad + 4\zeta_{\{2,4\}} + 2\zeta_{\{2,5\}} + 4\zeta_{\{2,6\}} + 4\zeta_{\{2,7\}} + 2\zeta_{\{2,8\}} + 6\zeta_{\{3,4\}} + 4\zeta_{\{3,5\}}\\ &\quad + 2\zeta_{\{3,7\}} + 6\zeta_{\{3,8\}} + 2\zeta_{\{4,5\}} + 2\zeta_{\{4,6\}} + 2\zeta_{\{4,7\}} + 4\zeta_{\{4,8\}} + 2\zeta_{\{5,6\}}\\ &\quad + 4\zeta_{\{5,7\}} + 2\zeta_{\{5,8\}} + 4\zeta_{\{6,8\}} + 8\zeta_{\{7,8\}}.\end{aligned}$$

The scalar sum of the trace is 90, while the graph contains 54 edges. Hence, the coloring represented by ψ is not a proper coloring.

The previous theorems and corollaries can now be extended from vertex colorings to edge colorings.

Definition 16.2.7. Let $G = (V, E)$ be a graph (either simple or directed with no multiple edges) on n vertices with edge $\varkappa$-coloring θ. Define the *zeon edge-coloring matrix* Λ associated with (G, θ) to be the $n \times n$ matrix having entries in ${\mathcal{C}\ell_\varkappa}^{\mathrm{nil}}$ given by

$$\langle v_i|\Lambda|v_j\rangle = \begin{cases}\zeta_{\theta(v_i,v_j)} & (v_i, v_j) \in E,\\ 0 & \text{otherwise.}\end{cases}$$

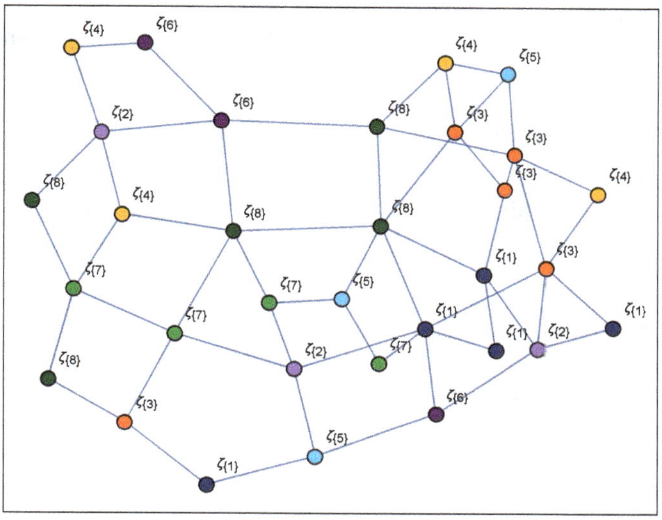

Figure 16.1 A vertex-colored graph on 32 vertices.

Theorem 16.2.8. *Let Λ be the zeon edge-coloring matrix of a simple graph G on n vertices. Then, for $1 \leq i, j \leq n$ and $m \in \mathbb{N}$,*

$$\langle v_i|\Lambda^m|v_j\rangle = \sum_{|I|=m} \alpha_I \zeta_I$$

where α_I is the number of trails from $v_i \to v_j$ in the graph such that the edge colors are indexed by I and no color is repeated. In particular, α_I is the number of heterochromatic trails from $v_i \to v_j$ on colors indexed by I.

Proof. Proof is by using induction on $m \geq 1$. When $m = 1$, the result holds by definition of Λ. Assume the result holds for some $m \geq 1$.

$$\begin{aligned}
\langle v_i|\Lambda^{m+1}|v_j\rangle &= \langle v_i|\Lambda^m\Lambda|v_j\rangle \\
&= \sum_{l=1}^{n} \langle v_i|\Lambda^m|v_\ell\rangle\langle v_\ell|\Lambda|v_j\rangle \\
&= \sum_{l=1}^{n} \sum_{|I|=m} \sharp\{\text{good } m\text{-trails } v_i \to v_\ell \text{ on I}\}\zeta_I \langle v_\ell|\Lambda|v_j\rangle.
\end{aligned}$$

For fixed ℓ, $\langle v_i|\Lambda^m|v_\ell\rangle$ is a linear combination of ζ_I's representing good (i.e., heterochromatic) m-trails from v_i to v_ℓ. The product of any term from this linear combination with $\langle v_\ell|\Lambda|v_j\rangle$ will be zero if there is no edge from v_ℓ to v_j or if the edge (v_ℓ, v_j) is the same color as another edge in the m-trail $v_i \to v_\ell$. Hence, the product $\langle v_i|\Lambda^k|v_\ell\rangle\langle v_\ell|\Lambda|v_j\rangle$ represents a sum of $(m+1)$-trails $v_i \to v_j$ of the following form:

$$\sum_{|I|=m} \sharp\{m\text{-trails } v_i \to v_\ell \text{ on colors indexed by } I\}\zeta_I\zeta_{\theta(v_\ell,v_j)} =$$

$$\sum_{|I|=m+1} \sharp\{(m+1)\text{-trails } v_i \to v_j \text{ on colors } I\text{; last step } (v_\ell, v_j)\ \}\zeta_{I\cup\theta(v_\ell,v_j)}.$$

Summing over all ℓ gives a representation of all heterochromatic $(m+1)$-trails $v_i \to v_j$ in G. □

Corollary 16.2.9. *Let Λ be a zeon edge-coloring matrix of a graph G. For any $m \in \mathbb{N}$,*

$$\langle\langle \operatorname{tr}(\Lambda^m)\rangle\rangle = m\rho,$$

where ρ is the number of heterochromatic m-circuits in G.

Proof. From Theorem 16.2.8, element $\langle v_i|\Lambda^m|v_i\rangle$ is a linear combination of heterochromatic m-circuits based at v_i. Each m-circuit appears with multiplicity m along the main diagonal due to the possible choices of basepoint. Hence, the result. □

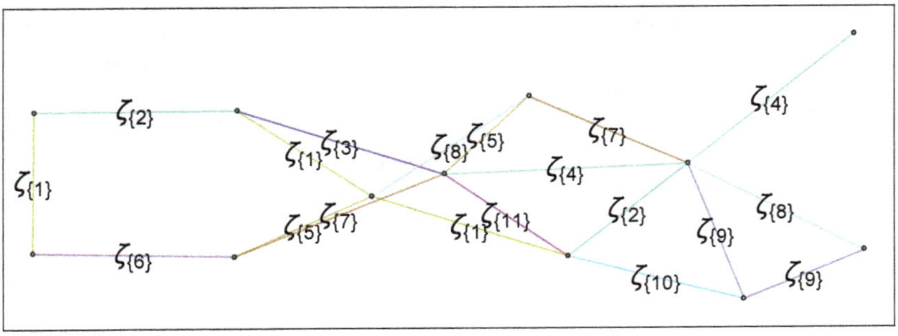

Figure 16.2 A zeon 11-edge-colored graph on 12 vertices.

In order to formulate a way to determine proper colorings, there needs to be a checking of all coincident pairs of edges to be sure no two coincident edges are the same color.

Lemma 16.2.10. *The total number of pairs of coincident edges in a graph G is given by*

$$\sum_{v\in V}\binom{\deg(v)}{2}.$$

Proof. First, note that if $\deg(v) \leq 1$ for some $v \in V$, there is no pair of edges coincident with v. By definition, $\binom{\deg(v)}{2} = 0$ in this case. Otherwise, $\binom{\deg(v)}{2}$ represents the number of pairs of edges coincident with v. Summing over all vertices thus gives the result. □

Recalling that entries of Λ^2 represent heterochromatic 2-trails and 2-circuits in G, off-diagonal elements correspond to heterochromatic pairs of coincident edges. For convenience, set $\beta = (1, 1, \ldots, 1) \in \mathbb{R}^n$ for appropriate dimension n, determined henceforth by context.

Theorem 16.2.11. *The zeon edge-coloring matrix Λ represents a proper edge coloring of a graph $G = (V, E)$ if and only if*

$$\langle\langle \beta\Lambda^2\beta^\dagger\rangle\rangle = 2\sum_{v\in V}\binom{\deg(v)}{2}.$$

Proof. By Theorem 16.2.8, $\langle v_i|\Lambda^2|v_j\rangle$ is a linear combination of ζ_I's representing the sum of all heterochromatic 2-trails $v_i \to v_j$. It is clear that heterochromatic 2-circuits cannot exist in any edge-colored graph, so the diagonal entries of Λ^2 are all zero. Summing coefficients of all off-diagonal entries is accomplished by computing $\langle\langle \beta\Lambda^2\beta^\dagger\rangle\rangle$, which counts the number of heterochromatic pairs of coincident vertices in G. By definition, G is properly edge-colored if and only if every pair of coincident vertices is heterochromatic. Hence, the result. □

Greedy Coloring

The matrix-based algorithm developed here works from right to left across columns of the adjacency matrix, so the vertex ordering is inferred from the construction of the adjacency matrix. To represent a proper vertex coloring, the zeon generators appearing in columns associated with adjacent vertices must be distinct. To this end, Algorithm 5 operates as follows.

As input, the algorithm accepts the ordinary adjacency matrix $A = (a_1|\cdots|a_n)$ and constructs a nilpotent coloring matrix $\Psi = (\psi_1|\cdots|\psi_n)$.

After assigning $\Psi \leftarrow A$ as an initialization[2], the algorithm proceeds from left to right.

Considering the jth column ψ_j, let M denote the indices of all zeon generators appearing in ψ_j. Observing that the graph is properly n-colorable, it follows immediately that setting $\chi = \min\{[n] \setminus M\}$ makes ζ_χ the least-indexed generator available to color vertex v_j. This coloring is accomplished by setting $\psi_j = \zeta_\chi a_j$. To make this color unavailable to the remaining uncolored vertices, the algorithm sets $\psi_{ji} = \psi_{ij}$ for $j < i \leq n$. This is repeated as j runs from 1 to n. At the end, the matrix Ψ represents a proper $\varkappa$-coloring of G, where $\varkappa \leq n$ is the maximum index appearing among zeon generators in Ψ.

16.3 Orthozeons and Monochromatic Walks

While zeons lend themselves nicely to counting heterochromatic self-avoiding walks, some new algebraic tools are required for the monochromatic case.

Given a $\varkappa$-dimensional vector space V equipped with inner product $\langle \cdot, \cdot \rangle : V \to \mathbb{R}$, it is not difficult to see that for any unit column vector $\mathbf{u} \in V$, the outer product $\mathbf{u}\mathbf{u}^\dagger$ is an order-$\varkappa$ matrix that acts on V as orthogonal projection onto span$(\{\mathbf{u}\})$ via matrix multiplication.

Denoting such a rank-one projection by $\tau_{\mathbf{u}}$, it is also not difficult to see that the product $\tau_{\mathbf{u}}\tau_{\mathbf{v}}$ is the zero matrix when $\langle \mathbf{u}, \mathbf{v} \rangle = 0$. Being a projection, $\tau_{\mathbf{u}}$ is obviously idempotent. Hence, the rank-one projections associated with any orthonormal basis $\{\mathbf{u}_i : 1 \leq i \leq \varkappa\}$ for V generate a commutative $\varkappa$-dimensional algebra satisfying

$$\tau_{\mathbf{u}_i}\tau_{\mathbf{u}_j} = \begin{cases} 0 & \text{when } i \neq j, \\ \tau_{\mathbf{u}_i} & \text{when } i = j. \end{cases}$$

For notational convenience, the generators will be denoted by $\{\tau_i : 1 \leq i \leq \varkappa\}$. The algebra generated by these projections will be denoted $\mathcal{P}_\varkappa$, and it is isomorphic to the algebra of diagonal matrices with real coefficients.

The goal of the current section is to develop methods for counting monochromatic self-avoiding walks in finite graphs. To that end, nilpotent coloring matrices will be defined having entries in the tensor algebra $\mathcal{P}_\varkappa \otimes \mathcal{C}\ell_n{}^{\mathrm{nil}}$. Generators of this algebra will be referred to as *orthozeons*. For notational convenience, define $\overset{j}{\zeta}_X = \tau_j \otimes \zeta_X$ for $1 \leq j \leq \varkappa$ and $X \subseteq 2^{[n]}$.

[2] Ψ can be initialized as any $n \times n$ matrix. The assignment $\Psi \leftarrow A$ is expedient.

input : Adjacency matrix $A = (a_1|\cdots|a_n)$ of a simple graph G on n vertices.
output: Proper zeon vertex-coloring matrix $\Psi = (\psi_1|\cdots|\psi_n)$ associated with graph G.

Initialize matrix Ψ.;
$\Psi \leftarrow A$;

Begin with first vertex (i.e., first column of Ψ*).*;
$j \leftarrow 1$;

while $j \leq n$ **do**
  Get indices of any ζ*'s appearing in current column of* Ψ.;
  $M \leftarrow$ {Indices of generators in ψ_j};
  Choose minimum available color index.;
  $\chi \leftarrow \min([n] \setminus M)$;
  Set j*th column of* Ψ *to represent color.*;
  $\psi_j \leftarrow \zeta_\chi a_j$;
  Make this color unavailable to neighbors yet to be evaluated.;
  for $i \leftarrow j+1$ **to** n **do**
    $\psi_{ji} \leftarrow \psi_{ij}$;
  end
  $j \leftarrow j+1$;
end
return Ψ;

Algorithm 5: Proper Zeon Vertex Coloring Matrix of a Graph

Orthozeons thereby satisfy the following multiplication rules:

$$\overset{i}{\zeta}_X \overset{j}{\zeta}_Y = \overset{j}{\zeta}_Y \overset{i}{\zeta}_X = \begin{cases} \overset{i}{\zeta}_{X\cup Y} & (i=j)\wedge(X\cap Y=\varnothing), \\ 0 & (i\neq j)\vee(X\cap Y\neq\varnothing). \end{cases} \tag{16.1}$$

Multiplication in the algebra is defined by associative linear extension of the action (16.1) defined on generators. The dimension of the algebra is readily seen to be $\varkappa 2^n$.

Constructing an adjacency matrix with orthozeon generators now allows one to count monochromatic self-avoiding walks in colored graphs.

Definition 16.3.1. Let $G = (V, E)$ be a simple graph on n vertices with vertex coloring $\phi : V \to \{1, \ldots, \varkappa\}$. The *orthozeon vertex-coloring matrix*

Φ associated with G is the $n \times n$ matrix whose entries are generators of $\mathcal{P}_\varkappa \otimes {\mathcal{C}\ell_n}^{\text{nil}}$ defined for $1 \le i, j \le n$ by

$$\langle v_i|\Phi|v_j\rangle = \begin{cases} \overset{\phi(j)}{\zeta}_{\{j\}} & \text{if } (v_i, v_j) \in E, \\ 0 & \text{otherwise.} \end{cases}$$

Theorem 16.3.2. *Let Φ be the orthozeon vertex-coloring matrix of a graph of G on n vertices. Then, for $1 \le i, j \le n$ and $m \in \mathbb{N}$,*

$$\langle v_i|\Phi^m|v_j\rangle = \sum_{\ell=1}^{\varkappa} \sum_{|I|=m} \alpha_{\ell,I} \overset{\ell}{\zeta}_I$$

where $\alpha_{\ell,I}$ is the number of m-walks from $v_i \to v_j$ in the graph on vertices indexed by I, each of color ℓ, such that no vertex is repeated with the possible exception of v_i exactly once. In particular, the coefficient of $\overset{\ell}{\zeta}_{I\cup\{i\}}$ in $\overset{\ell}{\zeta}_{\{i\}}\langle v_i|\Phi^m|v_j\rangle$ is the number of monochromatic m-paths of color ℓ from v_i to v_j on vertices indexed by I.

Proof. Proof is by induction on m, using the inherent properties of the algebra $\mathcal{P}_\varkappa \otimes {\mathcal{C}\ell_n}^{\text{nil}}$. The structure is similar to the proof of Theorem 16.2.2. □

Example 16.3.3. Figure 16.3 depicts an orthozeon 5-coloring of a graph on 30 vertices. Letting Φ denote an orthozeon coloring matrix of the graph, one finds

$$\begin{aligned} \operatorname{tr}(\Phi^5) &= 10\overset{3}{\zeta}_{\{2,3,19,26,27\}} + 10\overset{3}{\zeta}_{\{2,4,19,26,27\}} + 10\overset{5}{\zeta}_{\{1,5,15,16,22\}} \\ &\quad + 10\overset{5}{\zeta}_{\{1,5,15,16,28\}} + 10\overset{5}{\zeta}_{\{1,5,15,22,28\}} + 10\overset{5}{\zeta}_{\{1,5,16,22,28\}}. \end{aligned}$$

Observing that $\langle\langle \operatorname{tr}(\Phi^5)\rangle\rangle = 60$, one concludes that G contains 12 monochromatic 5-cycles. Four of the 5-cycles are on vertices of color 3, and eight are on vertices of color 5. Their respective vertex sets are seen in the subscripts.

Definition 16.3.4. Let G be a simple (possibly directed) graph with edge $\varkappa$-coloring θ. Define the *orthozeon edge-coloring matrix* Υ of (G, θ) having entries in $\mathcal{P}_\varkappa \otimes {\mathcal{C}\ell_n}^{\text{nil}}$ by

$$\langle v_i|\Upsilon|v_j\rangle = \begin{cases} \overset{\theta(i,j)}{\zeta}_{\{v_i,v_j\}} & \text{if } (v_i, v_j) \in E, \\ 0 & \text{otherwise.} \end{cases}$$

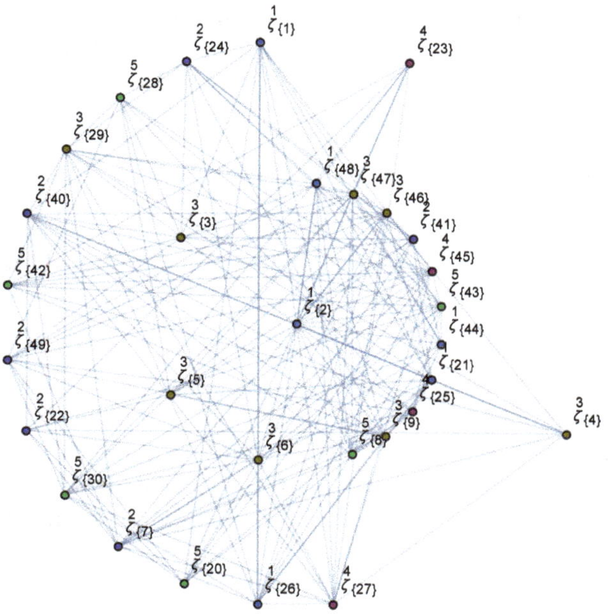

Figure 16.3 An orthozeon 5-colored 30-vertex graph.

Theorem 16.3.5. *Let* Υ *be the orthozeon edge-coloring matrix of* (G,θ)*, where* θ *is an edge* $\varkappa$*-coloring of* G*. Then, for* $1 \leq i, j \leq n$ *and* $m \in \mathbb{N}$*,*

$$\langle v_i|\Upsilon^m|v_j\rangle = \sum_{\ell=1}^{\varkappa} \sum_{|I|=m} \alpha_{\ell,I} \overset{\ell}{\zeta}_I,$$

where $\alpha_{\ell,I}$ *denotes the number of monochromatic* m*-trails* $v_i \to v_j$ *on edges of color* ℓ *indexed by* I*.*

Proof. Proof is by induction on m, using the inherent properties of the algebra $\mathcal{P}_\varkappa \otimes \mathcal{C}\ell_{[E]}{}^{\text{nil}}$. The structure is similar to the proof of Theorem 16.2.8. □

16.4 Zeon and Orthozeon Coloring Polynomials

By introducing polynomials having zeon or orthozeon coefficients, details of a graph's hetero- or monochromatic cliques, independent sets, matchings, girth, or circumference can be revealed.

Definition 16.4.1. The *heterochromatic circumference* of a colored graph G is the size of a maximal heterochromatic cycle in G. The *monochromatic circumference* of a colored graph G is the size of a maximal monochromatic cycle in G.

Dual to the notion of circumference is the notion of girth. Hence, the following definition.

Definition 16.4.2. The *heterochromatic girth* of a colored graph G is the size of a minimal heterochromatic cycle in G. The *monochromatic girth* of a colored graph G is the size of a minimal monochromatic cycle in G.

Two other (dual) structures of interest are independent sets and matchings in graphs. In a graph $G = (V, E)$, a vertex subset $U \subseteq V$ is said to be an *independent set* if no pair of vertices in U is adjacent in the graph; i.e., $u, v \in U$ implies $(u, v) \notin E$ and $(v, u) \notin E$. These definitions extend in the obvious way to define *heterochromatic or monochromatic independent sets.*

Definition 16.4.3. A *heterochromatic independent set* of a graph $G = (V, E)$ with vertex coloring ϕ is a subset of distinctly colored vertices of G that are pairwise non-adjacent. More specifically, $U \subseteq V$ is a heterochromatic independent set of G if $u, v \in U$ implies $\phi(u) \neq \phi(v)$, $(u, v) \notin E$, and $(v, u) \notin E$. Similarly, a *monochromatic independent set* of a graph $G = (V, E)$ with vertex coloring ϕ is a single-colored subset of vertices of G that are pairwise non-adjacent. In other words, $U \subseteq V$ is a heterochromatic independent set of G if $u, v \in U$ implies $\phi(u) = \phi(v)$, $(u, v) \notin E$, and $(v, u) \notin E$.

Dual to the notion of an independent set, a *matching* of G is a subset $F \subseteq E$ such that no pair of edges in F is coincident in G. This definition extends naturally to *heterochromatic or monochromatic matchings.*

Definition 16.4.4. A *heterochromatic matching* of a graph $G = (V, E)$ with edge coloring θ is a subset of distinctly colored edges of G that are pairwise non-coincident. Equivalently, $F \subseteq E$ is a heterochromatic matching of G if $(a, b), (c, d) \in F$ implies $\theta((a, b)) \neq \theta((c, d))$, and that the sets $\{a, b\}$ and $\{c, d\}$ are disjoint. A *monochromatic matching* of a graph $G = (V, E)$ with edge coloring θ is a single-colored subset of edges of G that are pairwise non-coincident. That is, $F \subseteq E$ is a heterochromatic matching of G if $(a, b), (c, d) \in F$ implies $\theta((a, b)) = \theta((c, d))$, and that the sets $\{a, b\}$ and $\{c, d\}$ are disjoint.

With these concepts established, it is now possible to define polynomials that reveal more information about structures contained in colored graphs. The notation $\deg_t(u)$ will be used to indicate the *degree of u regarded as a polynomial in t*.

Proposition 16.4.5. *Let Ψ be a zeon vertex-coloring matrix of G, and define the* zeon coloring polynomial of G, $\mathfrak{z}(t)$, *by*

$$\mathfrak{z}(t) = \mathrm{tr}\left(e^{t\Psi}\right).$$

Then, the coefficient of t^k in $\mathfrak{z}(t)$ is of the form

$$\langle \mathfrak{z}(t), t^k \rangle = \sum_{|I|=k} k\alpha_I \zeta_I,$$

where α_I is the number of heterochromatic k-cycles in G on colors indexed by I. In particular, the graph is acyclic if $\mathfrak{z}(t) = 0$; otherwise, $\deg_t(\mathfrak{z}(t))$ is the heterochromatic circumference of G.

Proof. Given the nilpotent structure of Ψ as a matrix having generators of ${\mathcal{C}\ell_{\varkappa}}^{\mathrm{nil}}$ as entries, it is clear that the matrix exponential can be written as a finite sum. Further, by linearity of trace, $\mathfrak{z}(t) = \sum_{\ell=0}^{\varkappa} \frac{t^\ell}{\ell!}\mathrm{tr}(\Psi^\ell)$. The result now follows immediately from Corollary 16.2.3. □

The following corollary is immediate.

Corollary 16.4.6. *Let Ψ be a zeon vertex-coloring matrix of G on n vertices, and let $\mathfrak{z}(t)$ be the zeon coloring polynomial of G as defined in Proposition 16.4.5. If $\mathfrak{z}(t) \neq 0$, then the* heterochromatic girth *of G is given by $n - \deg_t(t^n \mathfrak{z}(1/t))$.*

Defining polynomials with orthozeon coefficients allow consideration of a graph's monochromatic subgraphs.

Proposition 16.4.7. *Let Φ be an orthozeon vertex-coloring matrix of G, and define the* orthozeon coloring polynomial of G, $\mu(t)$, *by*

$$\mu(t) = \mathrm{tr}\left(e^{t\Phi}\right).$$

Then $\deg_t(\mu(t))$ is the monochromatic circumference of G.

Proof. As a polynomial in t, the degree of $\operatorname{tr}\left(e^{t\Phi}\right)$ is the maximum exponent k for which Φ^k is nonzero. By Theorem 16.3.2, k is the length of a maximal monochromatic cycle of (G, ϕ). □

These results extend naturally to edge-colored graphs in order to reveal details of heterochromatic and monochromatic matchings.

Proposition 16.4.8. *Let $G = (V, E)$ be a simple graph on n vertices with edge-coloring $\phi : E \to \{n+1, \ldots, n+\varkappa\}$. Setting $\Gamma = \sum_{(v_i,v_j)\in E\subset V\times V} \zeta_{\{v_i,v_j,\phi(v_i,v_j)\}}$, the exponential $e^{t\Gamma}$ is a polynomial in t with coefficients in ${\mathcal{C}\ell_{n+\varkappa}}^{\text{nil}}$. Furthermore $\deg_t(e^{t\Gamma})$ is the size of a maximal heterochromatic matching in G.*

Proof. The key here is that the product $\zeta_{\{v_i,v_j,\phi(v_i,v_j)\}}\zeta_{\{v_\ell,v_m,\phi(v_\ell,v_m)\}}$ is only nonzero if $\{v_i, v_j, v_\ell, v_m, \phi(v_i, v_j), \phi(v_\ell, v_m)\}$ is a pairwise-disjoint set. Given that Γ is clearly nilpotent, the exponential is a finite sum (i.e., a polynomial in t), and the degree of t is the maximal number of factors $\zeta_{\{v_i,v_j,\phi(v_i,v_j)\}}$ appearing in any nonzero product taken over all edges in G. For the product to be nonzero, endpoints of the edges are disjoint and colors of the edges are disjoint. Hence, the result. □

Proposition 16.4.9. *Letting $G = (V, E)$ be a simple graph on n vertices with edge $\varkappa$-coloring $\phi : E \to \{1, \ldots, \varkappa\}$. Define $\omega : E \to \mathcal{P}_\varkappa \otimes {\mathcal{C}\ell_n}^{\text{nil}}$ by*

$$\omega(v_i, v_j) = {}^{\phi(v_i,v_j)}\zeta_{\{v_i,v_j\}}.$$

Setting $\Gamma = \sum_{\varepsilon\in E} \omega(\varepsilon)$, the exponential $e^{t\Gamma}$ is seen to be a polynomial in t with orthozeon coefficients such that $\deg_t(e^{t\Gamma})$ is the size of a maximal monochromatic matching in G.

Proof. Nilpotent properties of $\mathcal{P}_\varkappa \otimes {\mathcal{C}\ell_n}^{\text{nil}}$ guarantee that the exponential $e^{t\Gamma}$ is a finite sum of the form $e^{t\Gamma} = \sum_{m=0}^{n} \frac{t^m}{m!}\Gamma^m$. Further, for a given m, straightforward application of the multinomial theorem yields the following:

$$\begin{aligned}
\Gamma^m &= \left(\sum_{(v_i,v_j)\in E} \zeta^{\phi(v_i,v_j)}{}_{\{v_i,v_j\}} \right)^m \\
&= \left(\sum_{\ell=1}^{\varkappa} \sum_{\substack{(v_i,v_j)\in E \\ \phi(v_i,v_j)=\ell}} \zeta^{\ell}_{\{v_i,v_j\}} \right)^m \\
&= \left(\sum_{\ell=1}^{\varkappa} \sum_{\substack{(v_i,v_j)\in E \\ \phi(v_i,v_j)=\ell}} \tau_\ell \otimes \zeta_{\{v_i,v_j\}} \right)^m \\
&= \sum_{\ell_1+\cdots+\ell_\varkappa=m} \binom{m}{\ell_1,\ldots,\ell_\varkappa} \prod_{q=1}^{\varkappa} \left({\tau_q}^{\ell_q} \otimes \left(\sum_{\substack{(v_i,v_j)\in E \\ \phi(v_i,v_j)=q}} \zeta_{\{v_i,v_j\}} \right)^{\ell_q} \right),
\end{aligned}$$

where orthogonality of the τ_q's guarantees that the product taken over q from 1 to $\varkappa$ is nonzero only if $(\varkappa-1)$ of the ℓ_q's are zero. Hence, for some $q' \in \{1,\ldots,\varkappa\}$,

$$\begin{aligned}
\prod_{q=1}^{\varkappa} \left({\tau_q}^{\ell_q} \otimes \left(\sum_{\substack{(v_i,v_j)\in E \\ \phi(v_i,v_j)=q}} \zeta_{\{v_i,v_j\}} \right)^{\ell_q} \right) &= {\tau_{q'}}^{\ell_{q'}} \otimes \left(\sum_{\substack{(v_i,v_j)\in E \\ \phi(v_i,v_j)=q'}} \zeta_{\{v_i,v_j\}} \right)^{\ell_{q'}} \\
&= \tau_{q'} \otimes \left(\sum_{\substack{(v_i,v_j)\in E \\ \phi(v_i,v_j)=q'}} \zeta_{\{v_i,v_j\}} \right)^{\ell_{q'}}.
\end{aligned}$$

The multinomial theorem now further implies that $\left(\sum_{\substack{(v_i,v_j)\in E \\ \phi(v_i,v_j)=q'}} \zeta_{\{v_i,v_j\}} \right)^{\ell_{q'}}$ is nonzero if and only if there exists a matching of size ℓ_q (i.e., a collection of ℓ_q edges whose endpoints form a pairwise disjoint collection) in the graph. Further, the edges of this matching are monochromatic of color q'. The largest exponent $\ell_{q'}$ for which the expression is nonzero is thereby the size of a maximal monochromatic matching in the graph. One sees immediately that this maximal exponent is the degree of $e^{t\Gamma}$ as a polynomial in t. □

Exercise

Exercise 16.1: Let Φ be an orthozeon vertex-coloring matrix of G on n vertices, and let $\mu(t)$ be the orthozeon coloring polynomial of G, as defined in Proposition 16.4.7. If $\mu(t) \neq 0$, show that the *monochromatic girth* of G is given by $n - \deg_t(t^n \mu(1/t))$.

Chapter 17

Boolean Satisfiability

The Boolean satisfiability problem, or SAT, is the problem of determining whether the variables of a given Boolean formula can be consistently replaced by true or false in such a way that the formula evaluates to be true. In fact, SAT was the first known NP-complete problem, as proved by Stephen Cook in 1971 [26]. Boolean satisfiability has many real-world applications, including model checking, classical planning, design of experiments, scheduling, optimal control, e-commerce, etc. [13]. The continuing advances of SAT solvers are the driving force of model checking tools used to check the correctness of hardware designs [108].

In [17], Budinich utilized properties of primitive idempotents in Clifford algebras to give a formulation for Boolean satisfiability as a decision problem, in that he provides an algebraic test for determining whether a given formula is satisfiable.

The basis for this chapter was the joint work with Amanda Davis [28]. The idem-Clifford approach detailed in Section 17.2 provides an algebraic test for satisfiability comparable to that developed by Budinich. The test can be implemented directly in *Mathematica* using the CliffMath package [102]. The zeon and "ortho-idem" approaches detailed in Sections 17.3 and 17.4 not only determine satisfiability but generate explicit solutions to the satisfiability problem. In this way, the zeon and ortho-idem approaches form the basis of a Clifford-algebraic SAT solver.

A Boolean formula is built from *literals* (Boolean variables), operators: AND (conjunction); OR (disjunction); NOT (negation); and parentheses. A formula is said to be satisfiable if it can be made true (T) by assigning appropriate logical values (i.e. T, F) to its variables. The *Boolean satisfiability problem* (SAT) is to check whether a given formula is satisfiable. The decision problem is central to numerous areas of computer science, including complexity theory, cryptography, and artificial intelligence.

Given a family of Boolean variables $\{x_1, \ldots, x_k\}$, let $\overline{x_i}$ denote the logical negation[1] (NOT) of x_i. That is, $x_i = \mathrm{T} \Leftrightarrow \overline{x_i} = \mathrm{F}$. Further, for variables x and y, the logical conjunction (AND) $x \wedge y$ and disjunction (OR) $x \vee y$ are determined as in Table 17.1.

Table 17.1 Conjunction and Disjunction.

x	y	$x \vee y$	$x \wedge y$	$\overline{x \wedge y}$	$\overline{x} \vee \overline{y}$
T	T	T	T	F	F
T	F	T	F	T	T
F	T	T	F	T	T
F	F	F	F	T	T

The last two columns of Table 17.1 illustrate one of DeMorgan's Laws showing the logical equivalence $\overline{x \wedge y} \equiv \overline{x} \vee \overline{y}$; i.e., the negation of a conjunction is the disjunction of negations. In similar fashion, one can illustrate the DeMorgan's Law $\overline{x \vee y} \equiv \overline{x} \wedge \overline{y}$.

A literal x_i is said to be *positive* if it is not negated. Negated literals are said to be *negative*.

Definition 17.0.1. Let $\mathbb{B}[X]$ denote the *Boolean algebra* generated by literals $X = \{x_i : 1 \leq i \leq n\}$ with the operators of conjunction, disjunction, and negation. In particular, for $a, b, c \in X$, the algebra is defined by the following properties:

$$a \wedge a \equiv a \qquad a \vee a \equiv a \tag{17.1}$$
$$a \wedge \overline{a} \equiv F \qquad a \vee \overline{a} \equiv T \tag{17.2}$$
$$\overline{\overline{a}} \equiv a \tag{17.3}$$
$$a \wedge b \equiv b \wedge a \qquad a \vee b \equiv b \vee a \tag{17.4}$$
$$(a \wedge b) \wedge c \equiv a \wedge (b \wedge c) \tag{17.5}$$
$$(a \vee b) \vee c \equiv a \vee (b \vee c) \tag{17.6}$$
$$a \wedge (a \vee b) \equiv a \qquad a \vee (a \wedge b) \equiv a \tag{17.7}$$
$$a \vee (b \wedge c) \equiv (a \vee v) \wedge (a \vee c) \tag{17.8}$$

[1] Often, the negation of x_i is denoted by $\neg x_i$.

$$a \wedge (b \vee c) \equiv (a \wedge b) \vee (a \wedge c) \tag{17.9}$$

$$a \wedge F \equiv F \qquad a \vee F \equiv a \tag{17.10}$$

$$a \wedge T \equiv a \qquad a \vee T \equiv T \tag{17.11}$$

$$\overline{a \wedge b} \equiv \bar{a} \vee \bar{b} \tag{17.12}$$

$$\overline{a \vee b} \equiv \bar{a} \wedge \bar{b}. \tag{17.13}$$

The idempotent law is seen in (17.1), while the complementary properties of negation are seen in (17.2). Commutativity of conjunction and disjunction is seen in (17.4), while the equivalences (17.5) and (17.6) represent associativity of conjunction and disjunction, respectively. The logical equivalence in (17.7) is known as the "absorption law." Equivalences (17.8) and (17.9) describe the distributive properties of conjunction and disjunction. Properties of the universal elements "F" and "T" are seen in (17.10) and (17.11). Finally, the equivalences (17.12) and (17.13) represent DeMorgan's Laws.

17.1 Conjunctive Normal Form

A satisfiability problem is said to be in *conjunctive normal form* (CNF) if it is a conjunction of one or more statements (or clauses), each of which being a disjunction of one or more literals. It is important to point out that any satisfiability problem can be converted to conjunctive normal form.

Example 17.1.1. One can directly apply logical equivalences to obtain the following transformation:

$$\begin{aligned}(\overline{a \wedge b}) \rightarrow c &\equiv \overline{(\overline{a \wedge b})} \vee c \\ &\equiv (a \wedge b) \vee c \\ &\equiv (a \vee c) \wedge (b \vee c).\end{aligned}$$

The following satisfiability problem is known as k-**SAT** (or k-**CNFSAT**): Given a collection $C = \{\mathfrak{c}_1, \ldots, \mathfrak{c}_m\}$ of clauses on a finite set U of variables such that $|\mathfrak{c}_i| \leq k$ for $1 \leq i \leq m$, is there a truth assignment for U that satisfies all the clauses in C? Here, $|\mathfrak{c}_i|$ denotes the number of variables appearing in the ith clause.

For example, given clauses $\mathfrak{c}_1 = (x_1 \vee x_2)$ and $\mathfrak{c}_2 = (\overline{x_1} \vee x_3)$, does there exist an assignment of the variables x_1, x_2, and x_3 such that the evaluation of the conjunction $\mathfrak{c}_1 \wedge \mathfrak{c}_2$ is TRUE? In this very simple example, it is not difficult to see that two such assignments exist: $x_2 = x_3 =$ T, $x_1 \in \{T, F\}$. However, the problem grows quickly as the number and size of clauses increases. In fact, while **2-SAT** is known to be of complexity class P (solvable in polynomial time), k-**SAT** is NP-complete for $k > 2$ [59].

Knowing that any arbitrary SAT problem can be converted to CNF, we can express a new notation for the SAT problem. A SAT problem written in CNF such that each clause has at most (or no more than) k literals is said to be a k-CNF problem. Such a problem can also be denoted as k-SAT.

17.2 The "Idem-Clifford" Algebra $\mathcal{C}\ell_n{}^{\text{idem}}$

Let $\mathcal{C}\ell_n{}^{\text{idem}}$ denote the real abelian algebra generated by the collection $\{\varepsilon_{\{i\}} : 1 \le i \le n\}$ along with the scalar $1 = \varepsilon_\varnothing$ subject to the following multiplication rules:

$$\varepsilon_{\{i\}}\,\varepsilon_{\{j\}} = \varepsilon_{\{j\}}\,\varepsilon_{\{i\}} \text{ for } i \neq j, \text{ and}$$
$$\varepsilon_{\{i\}}{}^2 = \varepsilon_{\{i\}} \text{ for } 1 \le i \le n.$$

It is evident that a general element $\alpha \in \mathcal{C}\ell_n{}^{\text{idem}}$ has canonical expansion of the form $\alpha = \sum_{I \in 2^{[n]}} \alpha_I\,\varepsilon_I$. Here, $I \in 2^{[n]}$ is a subset of $[n] = \{1, 2, \ldots, n\}$, used as a multi-index, $\alpha_I \in \mathbb{R}$, and $\varepsilon_I = \prod_{i \in I} \varepsilon_{\{i\}}$. The algebra $\mathcal{C}\ell_n{}^{\text{idem}}$ is called the (n-generator) *idem-Clifford algebra*.

Remark 17.2.1. The generators of $\mathcal{C}\ell_n{}^{\text{idem}}$ are denoted $\varepsilon_{\{i\}}$ with multi-index notation in mind. This also syntactically matches the requirements of the CliffMath package for *Mathematica* [102].

As a vector space, this 2^n-dimensional algebra has a canonical basis of *basis blades* of the form $\{\varepsilon_I : I \subseteq [n]\}$. The idempotent property of the generators $\{\varepsilon_i : 1 \le i \le n\}$ guarantees that the product of two basis blades satisfies the following:

$$\varepsilon_I \varepsilon_J = \varepsilon_{I \cup J}.$$

The algebra $\mathcal{C}\ell_n{}^{\text{idem}}$ is isomorphic to a commutative subalgebra of $\mathcal{C}\ell_{n,n}$ by associating its generators with commuting idempotents; i.e.,

$$\varepsilon_{\{i\}} \mapsto \frac{1}{2}\left(1 - \mathbf{e}_{\{i,n+i\}}\right).$$

Here, $\mathcal{C}\ell_{n,n}$ is the Clifford algebra generated by the collection $\{\mathbf{e}_i : 1 \le i \le 2n\}$ satisfying

$$\mathbf{e}_i\mathbf{e}_j + \mathbf{e}_j\mathbf{e}_i = \begin{cases} 0 & i \neq j \\ 1 & 1 \le i = j \le n, \\ -1 & n+1 \le i = j \le 2n, \end{cases}$$

and $\mathbf{e}_{\{i,n+1\}} = \mathbf{e}_i\mathbf{e}_{n+i}$.

The idempotent generators of $C\ell_n{}^{\text{idem}}$ can be used to represent literals of a Boolean satisfiability problem. Using the algebra $C\ell_n{}^{\text{idem}}$, one can represent positive literals with generators $\varepsilon_{\{i\}}$ and negative literals by $\widetilde{\varepsilon_{\{i\}}} = (1 - \varepsilon_{\{i\}})$.

The extension of multi-index notation to the elements $\widetilde{\varepsilon_{\{i\}}} = 1 - \varepsilon_{\{i\}}$ is natural; however, the interpretation of expressions relative to symbolic logic requires careful attention. In particular, multi-index notation is defined as follows:

$$\begin{aligned}\widetilde{\varepsilon_I} &:= \prod_{i\in I} \widetilde{\varepsilon_{\{i\}}} \\ &= \prod_{i\in I} \left(1 - \varepsilon_{\{i\}}\right).\end{aligned}$$

Proposition 17.2.2. *Suppose $\mathcal{F} = \mathfrak{c}_1 \wedge \mathfrak{c}_2 \wedge \cdots \wedge \mathfrak{c}_m$ is a Boolean k-SAT formula with n literals $X = \{x_i : 1 \leq i \leq n\}$. Define $\Psi : \mathbb{B}[X] \to C\ell_n{}^{\text{idem}}$ by extension of the following for $x_i \in X$ and $a, b \in \{x_i, \overline{x_i} : 1 \leq i \leq n\}$:*

$$\begin{aligned}\Psi(F) = 0 \quad \Psi(T) &= 1 \\ \Psi(x_i) &= \varepsilon_{\{i\}}, \\ \Psi(\overline{x_i}) &= \widetilde{\varepsilon_{\{i\}}} = 1 - \varepsilon_{\{i\}}, \\ \Psi(a \vee b) &= \Psi(a) + \Psi(b) - \Psi(a)\Psi(b), \\ \Psi(a \wedge b) &= \Psi(a)\Psi(b).\end{aligned}$$

Then, $\mathcal{F}$ is satisfiable if and only if $\Psi(\mathcal{F}) \neq 0$.

Proof. To prove the result, we show that

$$\Psi(\mathcal{F}) = \prod_{j=1}^{m} \Psi(\mathfrak{c}_j) = 0$$

if and only if $\mathcal{F}$ is a contradiction; i.e., the formula reduces to F regardless of the truth values of its variables.

First, we verify that the mapping Ψ preserves the properties of the Boolean algebra generated by the literals. Defining $\Psi(T) = 1$ and $\Psi(F) = 0$ will guarantee that tautologies map to 1 and contradictions map to 0.

A straightforward calculation reveals that for any literal x_i,

$$\begin{aligned}\Psi(x_i \wedge x_i) &= \Psi(x_i)^2 = \Psi(x_i), \\ \Psi(x_i \wedge \overline{x_i}) &= \Psi(x_i)(1 - \Psi(x_i)) = 0, \\ \Psi(x_i \vee \overline{x_i}) &= \Psi(x_i) + (1 - \Psi(x_i)) - \Psi(x_i)(1 - \Psi(x_i)) = 1, \\ \Psi(\overline{\overline{x_i}}) &= 1 - (1 - \Psi(x_i)) = \varepsilon_{\{i\}}.\end{aligned}$$

Commutativity and associativity of the conjunction is trivially preserved by multiplication in ${\mathcal{C}\ell_n}^{\rm idem}$. To see that the associative property of disjunction is preserved under Ψ, consider

$$\begin{aligned}\Psi((a \vee b) \vee c) &= \Psi(a \vee b) + \Psi(c) - \Psi(a \vee b)\Psi(c)\\ &= \Psi(a) + \Psi(b) - \Psi(a)\Psi(b) + \Psi(c) - \Psi(a)\Psi(c) - \Psi(b)\Psi(c)\\ &\quad +\Psi(a)\Psi(b)\Psi(c)\\ &= \Psi(a) + \Psi(b \vee c) - \Psi(a)\Psi(b \vee c)\\ &= \Psi(a \vee (b \vee c)).\end{aligned}$$

The absorption laws are easily verified by noting that

$$\Psi(a)(\Psi(a) + \Psi(b) - \Psi(a)\Psi(b)) = \Psi(a),$$
$$\Psi(a) + \Psi(a)\Psi(b) - \Psi(a)\Psi(a)\Psi(b) = \Psi(a).$$

Preservation of the distributive properties is verified directly by

$$\begin{aligned}\Psi((a \vee b) \wedge \Psi(a \vee c)) &= \Psi(a \vee b)\Psi(a \vee c)\\ &= (\Psi(a) + \Psi(b) - \Psi(a)\Psi(b))(\Psi(a) + \Psi(c)\\ &\quad - \Psi(a)\Psi(c))\\ &= \Psi(a) + \Psi(b)\Psi(c) - \Psi(a)\Psi(b)\Psi(c)\\ &= \Psi(a) + \Psi(b \wedge c) - \Psi(a)\Psi(b \wedge c)\\ &= \Psi(a \vee (b \wedge c)) \qquad (17.14)\end{aligned}$$

and

$$\begin{aligned}\Psi((a \wedge b) \vee \Psi(a \wedge c)) &= \Psi(a)\Psi(b) + \Psi(a)\Psi(c) - \Psi(a)\Psi(b)\Psi(c)\\ &= \Psi(a)(\Psi(b) + \Psi(c) - \Psi(b)\Psi(c))\\ &= \Psi(a \wedge (b \vee c)). \qquad (17.15)\end{aligned}$$

We extend the distributive properties (17.14) and (17.15) to conjunctions of arbitrary finite clauses by induction. Assume that for a clause $\mathfrak{c}_1$ satisfying $1 \leq |\mathfrak{c}_1| < k$ and arbitrary finite clause $\mathfrak{c}_2$, the result holds; i.e., assume $\Psi(\mathfrak{c}_1 \wedge \mathfrak{c}_2) = \Psi(\mathfrak{c}_1)\Psi(\mathfrak{c}_2)$. Then, for literal $x_i \notin \mathfrak{c}_1$, the clause $x_i \vee \mathfrak{c}_1$ is a clause on k literals and, writing $\mathcal{F} = (x_i \vee \mathfrak{c}_1) \wedge \mathfrak{c}_2$, we have

$$\begin{aligned}\Psi(\mathcal{F}) &= \Psi((x_i \vee \mathfrak{c}_1) \wedge \mathfrak{c}_2)\\ &= \Psi((x_i \wedge \mathfrak{c}_2) \vee (\mathfrak{c}_1 \wedge \mathfrak{c}_2))\\ &= \Psi(x_i \wedge \mathfrak{c}_2) + \Psi(\mathfrak{c}_1 \wedge \mathfrak{c}_2) - \Psi(x_i \wedge \mathfrak{c}_2)\Psi(\mathfrak{c}_1 \wedge \mathfrak{c}_2)\\ &= \Psi(x_i)\Psi(\mathfrak{c}_2) + \Psi(\mathfrak{c}_1)\Psi(\mathfrak{c}_2) - \Psi(x_i)\Psi(\mathfrak{c}_2)\Psi(\mathfrak{c}_1)\Psi(\mathfrak{c}_2)\\ &= \Psi(\mathfrak{c}_2)(\Psi(x_i) + \Psi(\mathfrak{c}_1) - \Psi(x_i)\Psi(\mathfrak{c}_1))\\ &= \Psi(x_i \vee \mathfrak{c}_1)\Psi(\mathfrak{c}_2).\end{aligned}$$

It follows that for arbitrary finite clauses $\mathfrak{c}_1, \mathfrak{c}_2$, $\Psi(\mathfrak{c}_1 \wedge \mathfrak{c}_2) = \Psi(\mathfrak{c}_1)\Psi(\mathfrak{c}_2)$. This further extends by induction to

$$\Psi(\mathfrak{c}_1 \wedge \cdots \wedge \mathfrak{c}_m) = \prod_{j=1}^{m} \Psi(\mathfrak{c}_j).$$

Finally, we show that $\Psi(\mathfrak{c}_1)\Psi(\mathfrak{c}_2) = 0$ if and only if $\mathfrak{c}_1 \wedge \mathfrak{c}_2$ is a contradiction. When $\mathfrak{c}_1 = x_i$ is a literal, we can assume without loss of generality that x_i is positive and write

$$\Psi(x_i)\Psi(\mathfrak{c}_2) = \varepsilon_{\{i\}}\Psi(\mathfrak{c}_2).$$

It follows that $\Psi(x_i)\Psi(\mathfrak{c}_2) = 0$ only if $\Psi(\mathfrak{c}_2) = (1 - \varepsilon_{\{i\}})\xi$ for some $\xi \in \mathcal{C}\ell_n{}^{\text{idem}}$. In other words, $\Psi(\mathfrak{c}_2) = \Psi(\overline{x_i} \wedge s)$ for Boolean expression s satisfying $\Psi(s) = \xi$. Thus $\Psi(x_i)\Psi(\mathfrak{c}_2) = 0$ if and only if $x_i \wedge \mathfrak{c}_2$ is a contradiction.

Extending to arbitrary clauses by induction, assume now that for clause $\mathfrak{c}_1$ satisfying $1 \leq |\mathfrak{c}_1| < k$ and arbitrary finite clause $\mathfrak{c}_2$, the result holds; i.e., assume $\Psi(\mathfrak{c}_1 \wedge \mathfrak{c}_2) = 0$ if and only if $\mathfrak{c}_1 \wedge \mathfrak{c}_2$ is a contradiction. Then, for literal $x_i \notin \mathfrak{c}_1$, the clause $x_i \vee \mathfrak{c}_1$ is a clause containing k literals and, writing $\mathcal{F} = (x_i \vee \mathfrak{c}_1) \wedge \mathfrak{c}_2$, we have

$$\begin{aligned}\Psi(\mathcal{F}) &= \Psi((x_i \vee \mathfrak{c}_1) \wedge \mathfrak{c}_2)\\ &= \Psi((x_i \wedge \mathfrak{c}_2) \vee (\mathfrak{c}_1 \wedge \mathfrak{c}_2))\\ &= \Psi(x_i \wedge \mathfrak{c}_2) + \Psi(\mathfrak{c}_1 \wedge \mathfrak{c}_2) - \Psi(x_i \wedge \mathfrak{c}_2)\Psi(\mathfrak{c}_1 \wedge \mathfrak{c}_2).\end{aligned}$$

Clearly, if $x_i \wedge \mathfrak{c}_2 \equiv F$ and $\mathfrak{c}_1 \wedge \mathfrak{c}_2 \equiv F$, then $\Psi(\mathcal{F}) = 0$. On the other hand, if exactly one of $x_i \wedge \mathfrak{c}_2$, $\mathfrak{c}_1 \wedge \mathfrak{c}_2$ is a contradiction, the result is nonzero. To see this, assume without loss of generality that $x_i \wedge \mathfrak{c}_2$ is the contradiction. A quick examination shows that $\Psi(\mathcal{F}) = \Psi(\mathfrak{c}_1 \wedge \mathfrak{c}_2) \neq 0$.

Finally, suppose neither $x_i \wedge \mathfrak{c}_2$ nor $\mathfrak{c}_1 \wedge \mathfrak{c}_2$ is a contradiction. In this case, $\Psi(x_i \wedge \mathfrak{c}_2), \Psi(\mathfrak{c}_1 \wedge \mathfrak{c}_2) \neq 0$. Assuming (without loss of generality) that x_i is a positive literal,

$$\begin{aligned}&\Psi(x_i \wedge \mathfrak{c}_2) + \Psi(\mathfrak{c}_1 \wedge \mathfrak{c}_2) - \Psi(x_i \wedge \mathfrak{c}_2)\Psi(\mathfrak{c}_1 \wedge \mathfrak{c}_2)\\ &\quad= \varepsilon_{\{i\}}\Psi(\mathfrak{c}_2) + \Psi(\mathfrak{c}_1 \wedge \mathfrak{c}_2) - \varepsilon_{\{i\}}\Psi(\mathfrak{c}_2)\Psi(\mathfrak{c}_1 \wedge \mathfrak{c}_2)\\ &\quad= \varepsilon_{\{i\}}\Psi(\mathfrak{c}_2) + \Psi(\mathfrak{c}_1)\Psi(\mathfrak{c}_2) - \varepsilon_{\{i\}}\Psi(\mathfrak{c}_1)\Psi(\mathfrak{c}_2)\\ &\quad= \Psi(\mathfrak{c}_1)\Psi(\mathfrak{c}_2)(1 - \varepsilon_{\{i\}}) + \varepsilon_{\{i\}}\Psi(\mathfrak{c}_2).\end{aligned}$$

By hypothesis, $x_i \notin \mathfrak{c}_1$, so $(1 - \varepsilon_{\{i\}})\Psi(\mathfrak{c}_1) \neq 0$, and $x_i \wedge \mathfrak{c}_2$ is not a contradiction, so $\varepsilon_{\{i\}}\Psi(\mathfrak{c}_2) \neq 0$. Rewriting in the canonical basis of $\mathcal{C}\ell_n{}^{\text{idem}}$, it follows that

$$\varepsilon_{\{i\}}\Psi(\mathfrak{c}_2) = \sum_I a_I \varepsilon_{\{i\} \cup I}.$$

On the other hand, the canonical expansion of $\Psi(\mathfrak{c}_1)\Psi(\mathfrak{c}_2)(1-\varepsilon_{\{i\}})$ cannot contain any nonzero terms whose multi-indices contain $\{i\}$ since $\varepsilon_{\{i\}}\widetilde{\varepsilon_{\{i\}}} = 0$. It follows that $\Psi((x_i \vee \mathfrak{c}_1) \wedge \mathfrak{c}_2)$ is nonzero when neither $x_i \wedge \mathfrak{c}_2$ nor $\mathfrak{c}_1 \wedge \mathfrak{c}_2$ is a contradiction.

It now follows that for a general Boolean CNF formula $\mathcal{F} = \mathfrak{c}_1 \wedge \cdots \wedge \mathfrak{c}_m$, $\Psi(\mathcal{F}) = \prod_{j=1}^m \psi(\mathfrak{c}_j) = 0$ if and only if $\mathcal{F}$ is unsatisfiable, and hence the result. □

Example 17.2.3. Consider the Boolean 3-SAT formula

$$\mathcal{F} = (x_1 \vee \overline{x_2} \vee x_3) \wedge (x_2 \vee \overline{x_3}) \wedge (\overline{x_1} \vee x_3) \wedge x_1 \wedge \overline{x_3},$$

which has idem-Clifford representation

$$\begin{aligned}\Psi &= (\varepsilon_{\{1\}} + (1-\varepsilon_{\{2\}}) + \varepsilon_{\{3\}})(\varepsilon_{\{2\}} + (1-\varepsilon_{\{3\}}))\\ &\quad((1-\varepsilon_{\{1\}}) + \varepsilon_{\{3\}})\varepsilon_{\{1\}}(1-\varepsilon_{\{3\}})\\ &= 0.\end{aligned}$$

It is clear that $\mathcal{F}$ is not satisfiable since the fourth and fifth clauses, taken together, form the negation of the third clause; i.e.,

$$\overline{(\overline{x_1} \vee x_3)} = (x_1 \wedge \overline{x_3}).$$

On the other hand, deleting the last two clauses gives the following:

$$\begin{aligned}&(\varepsilon_{\{1\}} + (1-\varepsilon_{\{2\}}) + \varepsilon_{\{3\}})(\varepsilon_{\{2\}} + (1-\varepsilon_{\{3\}}))((1-\varepsilon_{\{1\}}) + \varepsilon_{\{3\}})\\ &\quad= 1 - \varepsilon_{\{1\}} - \varepsilon_{\{2\}} - \varepsilon_{\{3\}} + \varepsilon_{\{1,2\}} + \varepsilon_{\{1,3\}} + 3\varepsilon_{\{2,3\}} - \varepsilon_{\{1,2,3\}},\end{aligned}$$

indicating that the reduced formula

$$\mathcal{F}_1 = (x_1 \vee \overline{x_2} \vee x_3) \wedge (x_2 \vee \overline{x_3}) \wedge (\overline{x_1} \vee x_3)$$

is satisfiable. The solution sets, however, are not easily determined.

While the idem-Clifford algebra is well suited for determining whether or not a given satisfiability problem has solutions, it is not efficient for determining what those solutions are. For example, expanding a product of negations in terms of idempotent generators gives

$$\widetilde{\varepsilon_{\{i\}}}\widetilde{\varepsilon_{\{j\}}}\widetilde{\varepsilon_{\{k\}}} = 1 - \varepsilon_{\{i\}} - \varepsilon_{\{j\}} - \varepsilon_{\{k\}} + \varepsilon_{\{i,j\}} + \varepsilon_{\{j,k\}} + \varepsilon_{\{i,k\}} - \varepsilon_{\{i,j,k\}}.$$

From this, factorization is required to retrieve the solution $(x_i, x_j, x_k) = (F, F, F)$.

In the next section, zeon formulations of graph problems are utilized in coordination with k-SAT problems to recover explicit solution sets for satisfiable CNF formulas.

17.3 Boolean k-SAT Problems as Clique Problems

The basic method of converting a SAT problem in CNF to a clique problem is well known. It was first put forth by Karp [59]. Given a k-SAT formula with m clauses and n total literals, a graph on n vertices is constructed as follows.

(1) Vertices of G are partitioned into m clusters, where m is the number of clauses.
(2) Each cluster is composed of vertices corresponding to the literals from the corresponding clause.
(3) Vertices in each cluster are labeled with their corresponding literals.
(4) Edges are constructed between all pairs of vertices in different clusters except for pairs of the form $\{x_i, \overline{x_i}\}$.

Note that in the above graph construction, there are no edges between vertices in the same cluster. (In the parlance of graph theory, G is m-partite, but not necessarily complete.) Once the graph is constructed, the corresponding truth assignments are given by the m-cliques of the graph; again, m is the number of clauses. Once an m-clique is found, assigning all vertices of the clique with their respective truth values gives a solution to the SAT problem. Thus, literals in the clique that are not negated are assigned TRUE, and any variable associated with a negated literal is assigned FALSE.

Null square properties of zeons make them particularly useful for recovering cliques in finite graphs.

Zeon Approach via Cliques and Independent Sets

Given a k-SAT formula $\mathcal{F}$ with m clauses $\{\mathfrak{c}_i : 1 \leq i \leq m\}$, let S denote the collection of all literals (including negations) appearing in $\mathcal{F}$. For $i = 1, \ldots, m$, let $|\mathfrak{c}_i|$ denote the number of literals appearing in the ith clause[2], and let $n = \sum_{i=1}^{m} |\mathfrak{c}_i|$. The formula $\mathcal{F}$ can be written as

$$\begin{aligned}\mathcal{F} &= \bigwedge_{i=1}^{m} \mathfrak{c}_i \\ &= \bigwedge_{i=1}^{m} \left(\bigvee_{j=1}^{|\mathfrak{c}_i|} u_{ij} \right),\end{aligned}$$

[2]Note that $1 \leq |\mathfrak{c}_i| \leq k$ for each i.

where $u_{ij} \in S$ for each i, j.

Let $G = (V, E)$ be a graph on n vertices constructed as in the previous section. The graph's vertices are partitioned into m independent subsets identified with the clauses of $\mathcal{F}$. Identifying vertices of G with literals u_{ij} of $\mathcal{F}$, adjacency is determined by

$$\{u_{ij}, u_{i_1 j_1}\} \in E \Leftrightarrow (i \neq i_1) \wedge (u_{ij} \neq \overline{u_{i_1 j_1}}).$$

Utilizing the vertex-literal identification above, let $\lambda : V \to {\mathcal{C}\ell_{2n}}^{\mathrm{idem}}$ be the vertex labeling of non-negated literals to generators $x_i \mapsto \varepsilon_{\{i\}}$ and negations to generators $\overline{x_i} \mapsto \varepsilon_{\{n+i\}}$. For notational convenience, we denote $\varepsilon_{\{n+i\}}$ by $\overline{\varepsilon_{\{i\}}}$, so that $\overline{x_i} \mapsto \overline{\varepsilon_{\{i\}}}$. It is important to note that the labeling λ is not assumed to be one-to-one; the same literals may appear in multiple clauses. In any case, ${\mathcal{C}\ell_{2n}}^{\mathrm{idem}}$ will be sufficiently large to accommodate the labeling, even though it is likely to have more generators than necessary.

Let edges of $G' = (V, E')$ be labeled with zeon generators by a bijection $\phi : E' \to \{\zeta_1, \ldots, \zeta_{|E'|}\}$. For each $v \in V$, define

$$\begin{aligned}\psi(v) &= \lambda(v) \prod_{\substack{u \in E' \\ u \text{ is incident to } v}} \phi(u) \\ &= \lambda(v)\zeta_{\mathcal{N}(v)},\end{aligned}$$

where $\zeta_{\mathcal{N}(v)}$ is a zeon basis blade indexed by the edges incident[3] to v in G'. Define $\Phi_{G'} \in {\mathcal{C}\ell_{2n}}^{\mathrm{idem}} \otimes {\mathcal{C}\ell_{|E'|}}^{\mathrm{nil}}$ by

$$\Phi_{G'} = \sum_{v \in V} \psi(v) = \sum_{v \in V} \lambda(v)\zeta_{\mathcal{N}(v)}.$$

Definition 17.3.1. The element $\Phi_{G'} \in {\mathcal{C}\ell_{2n}}^{\mathrm{idem}} \otimes {\mathcal{C}\ell_{|E'|}}^{\mathrm{nil}}$ is called the *zeon clique formulation* of $\mathcal{F}$.

Theorem 17.3.2. *Let G be the graph of a k-SAT formula $\mathcal{F}$ with m clauses and n literals. Let $\Phi_{G'} \in {\mathcal{C}\ell_{2n}}^{\mathrm{idem}} \otimes {\mathcal{C}\ell_{|E'|}}^{\mathrm{nil}}$ be the zeon clique formulation of $\mathcal{F}$. Then,*

$$ {\Phi_{G'}}^m = m! \sum_{\substack{\{I \subseteq V, |I| = m\} \\ I \text{ independent in } G'}} \phi(I)\zeta_{\mathcal{N}(I)},$$

where, for each independent set I in G', $\phi(I) \in {\mathcal{C}\ell_{2n}}^{\mathrm{idem}}$ is an idem-Clifford blade of the form $\varepsilon_J \widetilde{\varepsilon_K}$ representing the solution set of the satisfiability problem by assigning T to literals indexed by J and assigning F to literals indexed by K.

[3]In other words, $\mathcal{N}(v)$ represents the "edge-neighborhood" of v.

Proof. The result is a corollary of Theorem 15.2.6. Details are left as an exercise. □

Example 17.3.3. Consider the following SAT formula:

$$\mathcal{F} = (x_1 \vee \overline{x_2} \vee x_3) \wedge (x_2 \vee \overline{x_3}) \wedge (\overline{x_1} \vee x_3).$$

Graphically, any solutions of $\mathcal{F}$ correspond to 3-cliques in the graph seen in Figure 17.1.

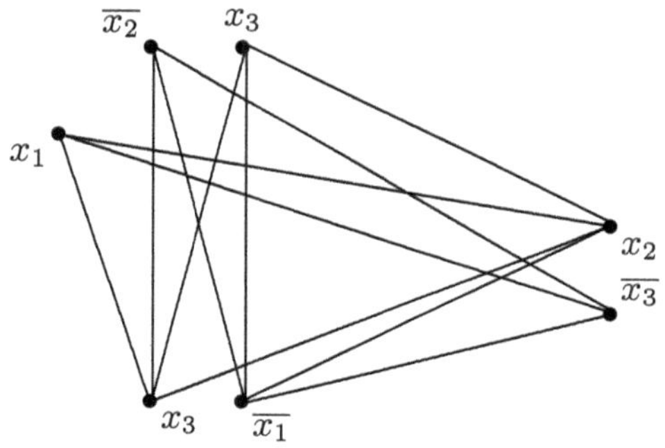

Figure 17.1 Graph of $\mathcal{F} = (x_1 \vee \overline{x_2} \vee x_3) \wedge (x_2 \vee \overline{x_3}) \wedge (\overline{x_1} \vee x_3)$.

The 3-SAT formula $\mathcal{F}$ has zeon-clique formulation

$$\begin{aligned}\Phi = &\ \zeta_{\{1,3\}}\widetilde{\varepsilon_{\{1\}}} + \zeta_{\{1,2\}}\varepsilon_{\{3\}} + \zeta_{\{4,6\}}\varepsilon_{\{2\}} + \zeta_{\{2,5,6\}}\widetilde{\varepsilon_{\{3\}}} \\ &+ \zeta_{\{5,7,9\}}\varepsilon_{\{3\}} + \zeta_{\{4,8,9\}}\widetilde{\varepsilon_{\{2\}}} + \zeta_{\{3,7,8\}}\varepsilon_{\{1\}}.\end{aligned}$$

The solutions are given by

$$\begin{aligned}\Phi^3 = &\ 6(\zeta_{\{1,3,4,5,6,7,9\}}\widetilde{\varepsilon_{\{1\}}}\varepsilon_{\{2,3\}} + \zeta_{\{1,2,3,4,5,6,8,9\}}\widetilde{\varepsilon_{\{1,2,3\}}}) \\ &+ 6(\zeta_{\{1,2,4,5,6,7,9\}}\varepsilon_{\{2,3\}} + \zeta_{\{1,2,3,4,6,7,8\}}\varepsilon_{\{1,2,3\}}.\end{aligned}$$

In particular, the following truth assignments satisfy the problem:

$$\begin{aligned}\widetilde{\varepsilon_{\{1\}}}\varepsilon_{\{2,3\}} &\leftrightarrow (F, T, T) \\ \widetilde{\varepsilon_{\{1\}}}\widetilde{\varepsilon_{\{2\}}}\widetilde{\varepsilon_{\{3\}}} &\leftrightarrow (F, F, F) \\ \varepsilon_{\{2,3\}} &\leftrightarrow (*, T, T) \\ \varepsilon_{\{1,2,3\}} &\leftrightarrow (T, T, T).\end{aligned}$$

17.4 The "Ortho-Idem" Approach to Boolean Satisfiability

In the previous section, the idempotents ε_I and $\widetilde{\varepsilon_I}$ were used as convenient labels. Any family of commuting variables would have worked as well, although valid solution sets would have included higher powers of variables. For example, $x_1{}^3\,\overline{x_4}{}^2 \leftrightarrow (T, F) \leftrightarrow \varepsilon_{\{1\}}\widetilde{\varepsilon_{\{4\}}}$.

The multiplicative properties of idem-Clifford elements and their negations make symbolic computations convenient. We now extend the idem-Clifford algebra $\mathcal{C}\ell_n{}^{\text{idem}}$ to a "new" algebra $\mathcal{C}\ell_n{}^{\mp\text{idem}}$ generated by idempotents $\{\varepsilon_{\{i\}}, \widetilde{\varepsilon_{\{i\}}} : 1 \leq i \leq n\}$, whose multiplication satisfies

$$\varepsilon_{\{i\}}\varepsilon_{\{j\}} = \begin{cases} \varepsilon_{\{j\}}\varepsilon_{\{i\}} & i \neq j \\ \varepsilon_{\{i\}} & i = j; \end{cases}$$

$$\widetilde{\varepsilon_{\{i\}}}\widetilde{\varepsilon_{\{j\}}} = \begin{cases} \widetilde{\varepsilon_{\{j\}}}\widetilde{\varepsilon_{\{i\}}} & i \neq j \\ \widetilde{\varepsilon_{\{i\}}} & i = j; \end{cases}$$

$$\varepsilon_{\{i\}}\widetilde{\varepsilon_{\{j\}}} = \begin{cases} \widetilde{\varepsilon_{\{j\}}}\varepsilon_{\{i\}} & i \neq j \\ 0 & i = j. \end{cases}$$

Utilizing multi-index notation,

$$\mathcal{C}\ell_n{}^{\mp\text{idem}} = \operatorname{span}\left(\{\varepsilon_I\widetilde{\varepsilon_J} : I, J \in 2^{[n]}, I \cap J = \varnothing\}\right).$$

A more straightforward approach to Boolean satisfiability can now be obtained using properties of $\mathcal{C}\ell_n{}^{\mp\text{idem}}$ without needing to construct graphs.

Definition 17.4.1. Suppose $\mathcal{F} = \mathfrak{c}_1 \wedge \mathfrak{c}_2 \wedge \cdots \wedge \mathfrak{c}_m$ is a Boolean k-SAT formula with n literals $X = \{x_1, \ldots, x_n\}$. Define the *ortho-idem clause representation* of $\mathcal{F}$ by

$$\varphi(\mathcal{F}) = \prod_{j=1}^{m} \varphi(\mathfrak{c}_j),$$

where $\varphi(\mathfrak{c}_j)$ is defined by extension of the following for $x_i \in X$ and $a, b \in \{x_i, \overline{x_i} : 1 \leq i \leq n\}$:

$$\begin{aligned} \varphi(x_i) &= \varepsilon_i, \\ \varphi(\overline{x_i}) &= \widetilde{\varepsilon_{\{i\}}}, \\ \varphi(a \vee b) &= \varphi(a) + \varphi(b). \end{aligned}$$

Theorem 17.4.2. *Let* $\mathcal{F} = \mathfrak{c}_1 \wedge \mathfrak{c}_2 \wedge \cdots \wedge \mathfrak{c}_m$ *be a Boolean k-SAT formula with n literals. Then,*

$$\varphi(\mathcal{F}) = \prod_{j=1}^{m} \varphi(\mathfrak{c}_j) = \sum_{\substack{(K,J)\subseteq[n]\times[n] \\ K\cap J=\varnothing}} a_{(K,J)}\varepsilon_K\widetilde{\varepsilon_J},$$

where $a_{(K,J)} \neq 0$ *if and only if* $\mathcal{F}$ *is satisfied by the following assignments:*

$$x_i = \begin{cases} T & i \in K, \\ F & i \in J. \end{cases}$$

Proof. Suppose $\mathcal{F} = \bigwedge_{i=1}^{m} \mathfrak{c}_i$, and let each clause $\mathfrak{c}_i$ be represented in the form

$$\mathfrak{c}_i = \left(\bigvee_{j\in P_i} x_j\right) \vee \left(\bigvee_{j'\in N_i} \overline{x_{j'}}\right),$$

where P_i represents the indices of positive literals in the ith clause and N_i represents the indices of negated literals in the ith clause.

Proof is by induction on m, the number of clauses in $\mathcal{F}$ by expanding the product of idem-Clifford clause representations. In particular, when $m = 1$, the result is clear since

$$\bigvee_{j\in P_1} x_j \vee \bigvee_{j'\in N_1} \overline{x_{j'}} \mapsto \sum_{j\in P_1} \varepsilon_{\{j\}} + \sum_{j'\in N_1} \widetilde{\varepsilon_{\{j'\}}},$$

and $\mathcal{F}$ is satisfied by setting x_j to T for any $j \in P_1$ or by setting $x_{j'}$ to F for any $j' \in N_1$. Assuming now that the result holds for the product of m clauses (some $m \geq 1$), write

$$\prod_{i=1}^{m} \varphi(\mathfrak{c}_i) = \sum_{(K,J)\in S_m} \varepsilon_K\widetilde{\varepsilon_J},$$

where S_m denotes the pairs of multi-indices of solutions to the m-clause CNF problem. Using this notation, consider now the product of $m + 1$

clauses:

$$\begin{aligned}
\prod_{i=1}^{m+1}\varphi(\mathfrak{c}_i) &= \varphi(\mathfrak{c}_{m+1})\prod_{i=1}^{m}\varphi(\mathfrak{c}_i)\\
&= \left(\sum_{j\in P_{m+1}}\varepsilon_{\{j\}} + \sum_{j'\in N_{m+1}}\widetilde{\varepsilon_{\{j'\}}}\right)\prod_{i=1}^{m}\varphi(\mathfrak{c}_i)\\
&= \left(\sum_{j\in P_{m+1}}\varepsilon_{\{j\}} + \sum_{j'\in N_{m+1}}\widetilde{\varepsilon_{\{j'\}}}\right)\sum_{(K,J)\in S_m}\varepsilon_K\widetilde{\varepsilon_J}\\
&= \sum_{(K,J)\in S_m}\sum_{\substack{j\in P_{m+1}\\ j\notin J}}\varepsilon_{K\cup\{j\}}\widetilde{\varepsilon_J} + \sum_{(K,J)\in S_m}\sum_{\substack{j'\in N_{m+1}\\ j'\notin K}}\varepsilon_K\widetilde{\varepsilon_{J\cup\{j'\}}}\\
&= \sum_{\substack{(K,J)\subseteq[n]\times[n]\\ K\cap J=\varnothing}} a_{(K,J)}\varepsilon_K\widetilde{\varepsilon_J},
\end{aligned}$$

where $a_{(K,J)} \neq 0$ if and only if the $(m+1)$-clause formula $\mathcal{F}$ is satisfied by the following assignments:

$$x_i = \begin{cases} T & i\in K,\\ F & i\in J.\end{cases}$$

□

Example 17.4.3. Recall the 3-SAT formula of Example 17.3.3:

$$\mathcal{F} = (x_1\vee\overline{x_2}\vee x_3)\wedge(x_2\vee\overline{x_3})\wedge(\overline{x_1}\vee x_3),$$

which has idem-Clifford clause representations

$$\begin{aligned}
\varphi(\mathfrak{c}_1) &= \varepsilon_{\{1\}} + \widetilde{\varepsilon_{\{2\}}} + \varepsilon_{\{3\}}\\
\varphi(\mathfrak{c}_2) &= \varepsilon_{\{2\}} + \widetilde{\varepsilon_{\{3\}}}\\
\varphi(\mathfrak{c}_3) &= \widetilde{\varepsilon_{\{1\}}} + \varepsilon_{\{3\}}.
\end{aligned}$$

The truth assignments satisfying the problem are revealed by computing $\prod_{j=1}^{3}\varphi(\mathfrak{c}_j)$:

$$\prod_{j=1}^{3}\varphi(\mathfrak{c}_j) = \varepsilon_{\{1,2,3\}} + \varepsilon_{\{2,3\}} + \widetilde{\varepsilon_{\{1,2,3\}}} + \widetilde{\varepsilon_{\{1\}}}\varepsilon_{\{2,3\}}.$$

Representations of $\mathcal{C}\ell_n{}^{\text{idem}}$

Let $\mathcal{J}_n$ denote the Abelian group generated by $\varepsilon_\varnothing$ along with commutative generators $\{\varepsilon_1, \ldots, \varepsilon_n\}$ satisfying ${\varepsilon_i}^2 = \varepsilon_i$ for each $i \in \{1, \ldots, n\}$. The Cayley graph of $\mathcal{J}_4$ (for example) is then readily seen to be the four-dimensional looped hypercube obtained by appending a loop to each vertex of a 4-cube like the one seen in Figure 8.1.

The *idem-Clifford algebra*, $\mathcal{C}\ell_n{}^{\text{idem}}$, is canonically isomorphic to the idempotent-generated semigroup algebra $\mathbb{R}\mathcal{J}_n$. The number of irreducible representations is given by the next theorem.

Theorem 17.4.4. *For any natural number n, there are* 2^n *irreducible representations of* $\mathcal{J}_n$.

Proof. The proof method follows the same format as those of Chapter 8. Details are left as an exercise. □

Faithful representations of $\mathcal{J}_n$ are easily obtained. Let $\tau : \mathcal{J}_n \to \text{End}(\mathbb{C}^{n+1})$ be defined on the set $\{\varepsilon_i\}$ by

$$\tau(\varepsilon_i) = (a^i_{jk}),$$

where

$$a^i_{jk} = \begin{cases} 1 & j = k \neq i, \\ 0 & \text{otherwise.} \end{cases} \tag{17.16}$$

In other words, (a^i_{jk}) is the matrix with ones on the diagonal except in the ith position and zeros elsewhere. These matrices are all idempotent and commute pairwise. Extending multiplicatively to all of $\mathcal{J}_n$, we obtain

$$\tau(\varepsilon_I) = (a^I_{jk}),$$

where

$$a^I_{jk} = \begin{cases} 1 & j = k \notin I, \\ 0 & \text{otherwise.} \end{cases}$$

For each $i = 1, \ldots, n$, the matrix defined in (17.16) represents a *hyperplane projection* in $\mathbb{C}^{n+1}$. In particular, the ith matrix represents a projection onto the hyperplane orthogonal to the ith unit coordinate vector of $\mathbb{C}^{n+1}$.

Exercises

Exercise 17.1: For literals a, b, c, verify the following equivalence:

$$(\overline{a \wedge b}) \to c \equiv (a \vee c) \wedge (b \vee c).$$

Exercise 17.2: Convert the following logical formula to CNF:

$$\psi = ((a \vee b) \to c) \wedge \overline{(d \wedge e)}.$$

Exercise 17.3: Obtain a solution for the logical formula:

$$\psi = ((A \vee B) \to C) \wedge \overline{(D \vee E)}.$$

Exercise 17.4: Let Ψ be the mapping defined in Proposition 17.2.2. Directly verify that the idempotent property of literals extends to general clauses: $\Psi(\mathfrak{c} \wedge \mathfrak{c}) = \Psi(\mathfrak{c})$.

Exercise 17.5: Let Ψ be the mapping defined in Proposition 17.2.2. Show that De Morgan's Laws

$$\begin{aligned} \overline{x_i} \vee \overline{x_j} &\equiv \overline{x_i \wedge x_j}, \text{ and} \\ \overline{x_i \vee x_j} &\equiv \overline{x_i} \wedge \overline{x_j} \end{aligned}$$

are preserved under Ψ by showing that

$$\begin{aligned} \Psi(\overline{x_i} \vee \overline{x_j}) &= \Psi(\overline{x_i \wedge x_j}), \text{ and} \\ \Psi(\overline{x_i \vee x_j}) &= \Psi(\overline{x_i} \wedge \overline{x_j}). \end{aligned}$$

Exercise 17.6: For each $i = 1, \ldots, n$, let $\widetilde{\varepsilon_{\{i\}}} = 1 - \varepsilon_{\{i\}}$. Verify the following:

$$\begin{aligned} \widetilde{\varepsilon_{\{i\}}}^{\,2} &= \widetilde{\varepsilon_{\{i\}}} \\ \varepsilon_{\{i\}}\widetilde{\varepsilon_{\{i\}}} &= 0 \\ \varepsilon_{\{i\}} + \widetilde{\varepsilon_{\{i\}}} - \varepsilon_{\{i\}}\widetilde{\varepsilon_{\{i\}}} &= 1. \end{aligned}$$

Exercise 17.7: Let $n \geq 3$, and suppose $i, j, k \in [n]$. Verify the following:

$$\varepsilon_{\{i\}}(\varepsilon_{\{j\}} + \varepsilon_{\{k\}} - \varepsilon_{\{j,k\}}) = \varepsilon_{\{i,j\}} + \varepsilon_{\{i,k\}} - \varepsilon_{\{i,j,k\}}.$$

Exercise 17.8: Let $n \geq 3$, and suppose $i, j, k \in [n]$. Verify the following identities[4]:

$$1 - (\varepsilon_{\{j\}} + \varepsilon_{\{k\}} - \varepsilon_{\{j,k\}}) = (1 - \varepsilon_{\{j\}})(1 - \varepsilon_{\{k\}})$$

and

$$1 - \varepsilon_{\{j,k\}} = (1 - \varepsilon_{\{j\}}) + (1 - \varepsilon_{\{k\}}) - (1 - \varepsilon_{\{j\}} - \varepsilon_{\{k\}} + \varepsilon_{\{j,k\}}).$$

Exercise 17.9: Prove Theorem 17.4.4.

Exercise 17.10: Prove Theorem 17.3.2.

[4]These identities represent De Morgan's Laws.

PART IV
Induced Operators

Chapter 18

Induced Operators and Kravchuk Polynomials

In this chapter, based on [99], the term *induced operator* will generally refer to an operator on a Clifford algebra $\mathcal{C}\ell_Q(V)$ obtained from an operator on the underlying vector space V spanned by the algebra's generators. The term *reduced operator* will generally refer to an operator on the paravector space $V_* := \mathbb{R} \oplus V$ obtained from an operator on the Clifford algebra. Finally, a *deduced operator* will be an operator on V obtained by restricting an operator on the full algebra, provided the operator on V also induces the operator on the full algebra.

The relationship between Kravchuk polynomials and Clifford algebras is multifaceted. As first seen in [99], Kravchuk polynomials appear as traces of conjugation operators in Clifford algebras and in Clifford Berezin integrals of Clifford polynomials. Regarding Kravchuk matrices as linear operators on a vector space V, the action induced on the Clifford algebra over V is equivalent to blade conjugation (i.e., reflections across sets of orthogonal hyperplanes). Such operators also have a natural interpretation in terms of raising and lowering operators on the algebra.

Table 18.1 Kravchuk Notation

Notation	Meaning
$K_\ell(x;n)$	ℓth Kravchuk polynomial of order n
$\mathfrak{K}_n$	nth Kravchuk matrix
${\mathfrak{K}_n}^{\mathrm{sym}}$	nth symmetric Kravchuk matrix
$\mathcal{K}_n$	nth *normalized* Kravchuk matrix

18.1 Kravchuk Polynomials and Kravchuk Matrices

Kravchuk polynomials arise as discrete orthogonal polynomials with respect to the binomial distribution. They have numerous applications in harmonic analysis, statistics, coding theory, and quantum probability. Kravchuk polynomials appear in Szëgo's classic work on orthogonal polynomials [106].

Kravchuk polynomials have been studied from the point of view of harmonic analysis and special functions, as in the work of Dunkl [34, 35], and they can be regarded as the discrete version of Hermite polynomials [4]. In combinatorics and coding theory, Kravchuk polynomials are essential in MacWilliams' theorem on weight enumerators [62], [69]. They also provide fundamental examples in association schemes [29], [30], [31].

In quantum theory, Kravchuk matrices interpreted as operators give rise to two new interpretations in the context of both classical and quantum random walks [41]. The latter interpretation underlies the basis of quantum computing.

The three-term recurrence relation for the Kravchuk polynomials of order n is as follows. Define $K_0(x;n) := 1$ and $K_1(x;n) := x$. For $\ell \geq 2$, the ℓth Kravchuk polynomial is given by

$$K_\ell(x;n) := xK_{\ell-1}(x;n) + (\ell-1)(n-\ell+2)K_{\ell-2}(x;n).$$

An explicit formula for the nth Kravchuk polynomial is given by

$$K_\ell(x,n) := \sum_{i=0}^{n}(-1)^i\binom{x}{i}\binom{n-x}{\ell-i}.$$

Consider a Bernoulli random walk starting at the origin, jumping to the left with probability q and to the right with probability p, where $p+q=1$ and $pq \neq 0$. After n steps, the position is $x = n-2j$, where j denotes the number of jumps to the left. A generating function for Kravchuk polynomials is then given by

$$G(v) = (1+v)^{(n+x)/2}(1-v)^{(n-x)/2} = \sum_{\ell=0}^{n}\frac{v^\ell}{\ell!}K_\ell(x,n).$$

As functions of j, the generating function can be written

$$G(v) = (1+v)^{n-j}(1-v)^j = \sum_{\ell=0}^{n}\frac{v^\ell}{\ell!}K_\ell(j,n).$$

For nonnegative integers i and j, the Kravchuk polynomials satisfy the orthogonality relation

$$\sum_{\ell=0}^{n}\binom{n}{\ell}K_i(\ell,n)K_j(\ell,n) = 2^n\binom{n}{i}\delta_{i,j},$$

where $\delta_{i,j}$ denotes the Kronecker delta. Moreover, the Kravchuk polynomials exhibit the symmetry relation

$$\binom{n}{i} K_j(i,n) = \binom{n}{j} K_i(j,n).$$

Let $\{e_i : 1 \leq i \leq n+1\}$ denote the standard basis for $\mathbb{R}^{n+1}$. As defined in the work of Feinsilver and Fitzgerald [38], the nth *Kravchuk matrix*, $\mathfrak{K}_n$, is the $(n+1)\times(n+1)$ matrix defined via the Kravchuk polynomial generating function according to

$$(1+x)^{n-j}(1-x)^j = \sum_{i=0}^{n} x^i \langle e_i|\mathfrak{K}_n|e_j\rangle,$$

with $\langle e_i|\mathfrak{K}_n|e_j\rangle = K_i(j;n)$, the ith Kravchuk polynomial evaluated at j.

Beginning with the Hadamard matrix

$$\mathfrak{K}_1 = \begin{pmatrix} 1 & 1 \\ 1 & -1 \end{pmatrix},$$

the next few Kravchuk matrices are seen to be

$$\mathfrak{K}_2 = \begin{pmatrix} 1 & 1 & 1 \\ 2 & 0 & -2 \\ 1 & -1 & 1 \end{pmatrix}, \quad \mathfrak{K}_3 = \begin{pmatrix} 1 & 1 & 1 & 1 \\ 3 & 1 & -1 & -3 \\ 3 & -1 & -1 & 3 \\ 1 & -1 & 1 & -1 \end{pmatrix},$$

$$\mathfrak{K}_4 = \begin{pmatrix} 1 & 1 & 1 & 1 & 1 \\ 4 & 2 & 0 & -2 & -4 \\ 6 & 0 & -2 & 0 & 6 \\ 4 & -2 & 0 & 2 & -4 \\ 1 & -1 & 1 & -1 & 1 \end{pmatrix}.$$

The *symmetric Kravchuk matrices* ${\mathfrak{K}_n}^{\mathrm{sym}}$ are defined by ${\mathfrak{K}_n}^{\mathrm{sym}} := \mathfrak{K}_n B$, where $B := (b_{ii})$ is the $(n+1)\times(n+1)$ diagonal matrix of binomial coefficients; i.e., $b_{ii} = \binom{n}{i}$ for $i = 0, \ldots, n$. The first few symmetric Kravchuk matrices are

$${\mathfrak{K}_2}^{\mathrm{sym}} := \begin{pmatrix} 1 & 2 & 1 \\ 2 & 0 & -2 \\ 1 & -2 & 1 \end{pmatrix}, \quad {\mathfrak{K}_3}^{\mathrm{sym}} := \begin{pmatrix} 1 & 3 & 3 & 1 \\ 3 & 3 & -3 & -3 \\ 3 & -3 & -3 & 3 \\ 1 & -3 & 3 & -1 \end{pmatrix},$$

$${\mathfrak{K}_4}^{\mathrm{sym}} := \begin{pmatrix} 1 & 4 & 6 & 4 & 1 \\ 4 & 8 & 0 & -4 & -4 \\ 6 & 0 & -12 & 0 & 6 \\ 4 & -8 & 0 & 8 & -4 \\ 1 & -4 & 6 & -4 & 1 \end{pmatrix}.$$

18.2 Endomorphisms of V and $\mathcal{C}\ell_Q(V)$

Let V be an n-dimensional vector space over $\mathbb{R}$ equipped with a nondegenerate quadratic form Q. Recall that the *Clifford algebra* $\mathcal{C}\ell_Q(V)$ is the real algebra obtained from associative linear extension of the Clifford vector product

$$\mathbf{x}\mathbf{y} := \langle \mathbf{x}, \mathbf{y}\rangle_Q + \mathbf{x} \wedge \mathbf{y},\ \forall \mathbf{x}, \mathbf{y} \in V.$$

Further, when Q is nondegenerate, the mapping $\|\cdot\|_Q : V \to \mathbb{R}$ defined by

$$\|\mathbf{x}\|_Q = |\langle \mathbf{x}, \mathbf{x}\rangle_Q|^{1/2},\ \ (\mathbf{x} \in V)$$

is the *Q-seminorm* on V.

Definition 18.2.1. Given an arbitrary blade $\mathfrak{u} \in \mathcal{C}\ell_Q(V)$, a collection of vectors $\mathfrak{U} = \{\mathbf{u}_1, \ldots, \mathbf{u}_{\sharp\mathfrak{u}}\}$ is said to constitute an *exterior factorization* of $\mathfrak{u}$ if

$$\mathfrak{u} = \beta \mathbf{u}_1 \wedge \cdots \wedge \mathbf{u}_{\sharp\mathfrak{u}}$$

for some nonzero scalar β. If, utilizing the Clifford product,

$$\mathfrak{u} = \alpha \mathbf{u}_1 \cdots \mathbf{u}_{\sharp\mathfrak{u}}$$

for some scalar α, then the collection $\mathfrak{U}$ is said to constitute a *Clifford factorization* of $\mathfrak{u}$.

Beginning with a Q-orthogonal set of vectors $\{\mathbf{u}_1, \ldots, \mathbf{u}_\ell\} \subset V$, their product is an ℓ-blade, $\mathfrak{u} \in \mathcal{C}\ell_Q(V)$. Define the mapping $\varphi_\mathfrak{u} : \mathcal{C}\ell_Q(V) \to \mathcal{C}\ell_Q(V)$ by

$$\varphi_\mathfrak{u}(x) := \mathfrak{u}\, x\, \frac{\tilde{\mathfrak{u}}}{\mathfrak{u}\tilde{\mathfrak{u}}} = \mathfrak{u} x \mathfrak{u}^{-1}.$$

For a fixed blade $\mathfrak{u}$, the linear map $x \mapsto \mathfrak{u} x \mathfrak{u}^{-1}$ is an endomorphism on $\mathcal{C}\ell_Q(V)$ referred to as the *conjugation of x by the blade* $\mathfrak{u}$, or simply as a *blade conjugation.*

A thorough study of the operator $\varphi_\mathfrak{u}$ begins with the restriction $\Phi_\mathfrak{u} = \varphi_\mathfrak{u}\Big|_V$. When $\Phi_\mathfrak{u}$ is an endomorphism on V having eigenvalue λ, let $\mathcal{E}_\lambda$ denote the corresponding eigenspace. A *blade test* is now given by the following theorem.

Theorem 18.2.2. *A homogeneous, grade-k multivector $\mathfrak{u} \in \mathcal{C}\ell_Q(V)$ is a blade if and only if $\Phi_\mathfrak{u}$ is an endomorphism on V with eigenvalues $\lambda_1 = (-1)^{k-1}$ and $\lambda_2 = -\lambda_1 = (-1)^k$ such that*

$$\dim(\mathcal{E}_{\lambda_1}) = k,$$
$$\dim(\mathcal{E}_{\lambda_2}) = n - k.$$

Proof. First, if $\mathfrak{u} = \mathbf{u}_1 \cdots \mathbf{u}_k$ is a blade, invertibility of $\mathfrak{u}$ follows from

$$\mathfrak{u}\tilde{\mathfrak{u}} = \mathbf{u}_1 \cdots \mathbf{u}_k \mathbf{u}_k \cdots \mathbf{u}_1 = \prod_{i=1}^{k} \langle \mathbf{u}_i, \mathbf{u}_i \rangle_Q \in \mathbb{R} \neq 0.$$

Whence, $\mathfrak{u}^{-1} = \dfrac{\tilde{\mathfrak{u}}}{\mathfrak{u}\tilde{\mathfrak{u}}}$. Decomposing an arbitrary vector $\mathbf{x} \in V$ into components parallel and orthogonal to $\mathbf{u}_k$,

$$\begin{aligned} \mathbf{u}_k \mathbf{x} \mathbf{u}_k = \mathbf{u}_k(\mathbf{x}_{\|} + \mathbf{x}_{\perp})\mathbf{u}_k &= \langle \mathbf{x}_{\|}, \mathbf{u}_k \rangle_Q \mathbf{u}_k - \langle \mathbf{u}_k, \mathbf{u}_k \rangle_Q \mathbf{x}_{\perp} \\ &= \alpha \mathbf{x}_{\|} - \beta \mathbf{x}_{\perp} \in V \end{aligned} \tag{18.1}$$

for scalars α and β. Associative extension gives $\mathfrak{u}\mathbf{x}\tilde{\mathfrak{u}} \in V$, so that $\Phi_{\mathfrak{u}} : V \to V$ is a well-defined linear operator. Invertibility of $\Phi_{\mathfrak{u}}$ follows from invertibility of $\mathfrak{u}$. Hence, ${\Phi_{\mathfrak{u}}}^{-1} = \Phi_{\mathfrak{u}^{-1}}$.

Keep in mind that the factors of $\mathfrak{u}$ are unique only up to isomorphism, i.e., change of basis. Writing $u = \mathbf{x}_1 \cdots \mathbf{x}_k$ for Q-orthogonal anisotropic vectors $\{\mathbf{x}_i : 1 \leq i \leq n\}$ and choosing any factor $\mathbf{x}_i$, it follows immediately that

$$\begin{aligned} \Phi_{\mathfrak{u}}(\mathbf{x}_i) &= \frac{1}{\mathfrak{u}\tilde{\mathfrak{u}}} \mathbf{x}_1 \cdots \mathbf{x}_k \mathbf{x}_i \mathbf{x}_k \cdots \mathbf{x}_1 \\ &= \frac{(-1)^{k-i} Q(\mathbf{x}_i)(-1)^{i-1}}{\mathfrak{u}\tilde{\mathfrak{u}}} \mathbf{x}_i \prod_{j \neq i} Q(\mathbf{x}_k) \\ &= \frac{(-1)^{k-1}}{\mathfrak{u}\tilde{\mathfrak{u}}} \mathbf{x}_i \prod_{1 \leq j \leq k} Q(\mathbf{x}_j) \\ &= \frac{(-1)^{k-1}}{\mathfrak{u}\tilde{\mathfrak{u}}} \mathbf{x}_i \, \mathfrak{u}\tilde{\mathfrak{u}} = (-1)^{k-1} \mathbf{x}_i. \end{aligned} \tag{18.2}$$

Hence, $\{\mathbf{x}_i : 1 \leq i \leq k\}$ is a basis for the eigenspace $\mathcal{E}_{\lambda_1}$.

Further, if $\mathbf{v}$ is Q-orthogonal to $\mathcal{E}_{\lambda_1}$, then

$$\begin{aligned} \Phi_{\mathfrak{u}}(\mathbf{v}) &= \frac{1}{\mathfrak{u}\tilde{\mathfrak{u}}} \mathbf{x}_1 \cdots \mathbf{x}_k \mathbf{v} \mathbf{x}_k \cdots \mathbf{x}_1 \\ &= \frac{(-1)^k}{\mathfrak{u}\tilde{\mathfrak{u}}} \mathbf{v} \prod_{1 \leq j \leq k} Q(\mathbf{x}_j) \\ &= \frac{(-1)^k \mathfrak{u}\tilde{\mathfrak{u}}}{\mathfrak{u}\tilde{\mathfrak{u}}} \mathbf{v} = (-1)^k \mathbf{v}. \end{aligned} \tag{18.3}$$

Hence, $\mathbf{v}$ is an eigenvector of $\varphi_{\mathfrak{u}}$ associated with eigenvalue $\lambda_2 = (-1)^k$. Since $\mathbf{v}$ was arbitrarily chosen from the orthogonal complement of $\mathcal{E}_{\lambda_1}$, it follows that $\dim(\mathcal{E}_{\lambda_2}) = n - k$.

Conversely, suppose Φ is a linear operator on V with the prescribed eigenvalues and eigenspaces, and let $\{\mathbf{u}_j : 1 \leq j \leq k\}$ be an arbitrary Q-orthogonal basis of $\mathcal{E}_{\lambda_1}$. Setting $\mathfrak{u} = \prod_{1\leq j\leq k} \mathbf{u}_j$, it is clear that $\mathfrak{u}$ is an invertible product of Q-orthogonal vectors. Hence, $\mathfrak{u}$ is a k-blade. By the previous arguments, the action on V of conjugation by $\mathfrak{u}$ (i.e., $\mathbf{x} \mapsto \mathfrak{u}\mathbf{x}\mathfrak{u}^{-1}$) is exactly the action of Φ on V. Therefore, $\Phi = \Phi_{\mathfrak{u}}$. □

Considering conjugation by a blade $\mathfrak{u}$, we adopt the notational convention $\lambda_1 := (-1)^{\sharp\mathfrak{u}-1}$ and $\lambda_2 := (-1)^{\sharp\mathfrak{u}}$. Observing that $\lambda_1 = \pm 1$ and $\lambda_2 = -\lambda_1$, it is clear that $\Phi_{\mathfrak{u}}$ acts as the identity on $\mathcal{E}_{+1}$ and as a reflection (or reversal) on $\mathcal{E}_{-1}$. It follows immediately that $\Phi_{\mathfrak{u}}$ is an involution; i.e., ${\Phi_{\mathfrak{u}}}^2 = \mathbb{I}$.

Recall from Chapter 7 that the collection of all Q-orthogonal transformations on V forms a group called the *orthogonal group* of Q, denoted $\mathrm{O}_Q(V)$. Further recall that the *conformal orthogonal group*, denoted $\mathrm{CO}_Q(V)$, is the direct product of the orthogonal group with the group of dilations. It is not difficult to see that blade conjugation is an orthogonal transformation on V. The next lemma makes this formal.

Lemma 18.2.3. *Let* $\mathfrak{u} \in \mathcal{C}\ell_Q(V)$ *be a blade. Then,* $\Phi_{\mathfrak{u}} \in \mathrm{O}_Q(V)$.

Proof. The result is established by showing that the eigenspaces $\mathcal{E}_{\lambda_1}$ and $\mathcal{E}_{\lambda_2}$ are orthogonal with respect to the quadratic form Q. Begin by letting $\mathbf{v} \in \mathcal{E}_{\lambda_1}$ and $\mathbf{w} \in \mathcal{E}_{\lambda_2}$. Then,

$$\mathfrak{u}\mathbf{v}u^{-1}\mathfrak{u}\mathbf{w}u^{-1} = \lambda_1\lambda_2\mathbf{v}\mathbf{w} = \lambda_1\lambda_2\left(\langle \mathbf{v}, \mathbf{w}\rangle_Q + \mathbf{v}\wedge\mathbf{w}\right).$$

Letting $\alpha = \langle \mathbf{v}, \mathbf{w}\rangle_Q$ and noting that $\lambda_1\lambda_2 = -1$, this implies

$$\mathfrak{u}(\mathbf{v}\mathbf{w})\mathfrak{u}^{-1} = -\alpha - \mathbf{v}\wedge\mathbf{w}. \tag{18.4}$$

On the other hand, noting that $\mathbf{v}\mathbf{w} = \alpha + \mathbf{v}\wedge\mathbf{w}$, one also finds

$$\mathfrak{u}(\mathbf{v}\mathbf{w})\mathfrak{u}^{-1} = \mathfrak{u}\left(\alpha + \mathbf{v}\wedge\mathbf{w}\right)\mathfrak{u}^{-1} = \alpha - \mathbf{v}\wedge\mathbf{w}. \tag{18.5}$$

Equality of (18.4) and (18.5) then implies $\alpha = 0$. □

Given an arbitrary k-blade $\mathfrak{u}$, one now readily defines the *level-ℓ induced map* ${\varphi_{\mathfrak{u}}}^{(\ell)}$ on blades of grade $\ell > 1$ by

$${\varphi_{\mathfrak{u}}}^{(\ell)}(\mathbf{w}_1\dots\mathbf{w}_\ell) := \mathfrak{u}\prod_{j=1}^{\ell}\mathbf{w}_j\mathfrak{u}^{-1} = \prod_{j=1}^{\ell}\mathfrak{u}\mathbf{w}_j\mathfrak{u}^{-1} = \prod_{j=1}^{\ell}\Phi_{\mathfrak{u}}(\mathbf{w}_j).$$

As an immediate consequence of Lemma 18.2.3, the induced map is a well-defined invertible linear transformation on the grade-ℓ subspace $\bigwedge^{\ell} V$ of $C\ell_Q(V)$; that is, the image of an ℓ-blade is also an ℓ-blade. In fact, the induced map is the restriction of $\varphi_{\mathfrak{u}}$ to the grade-ℓ subspace of $C\ell_Q(V)$:

$$\varphi_{\mathfrak{u}}{}^{(\ell)} = \varphi_{\mathfrak{u}}\Big|_{\bigwedge^{\ell} V}.$$

Given a blade $\mathfrak{u}$, a useful generalization of the conjugation operator is defined by setting

$$\psi_{\mathfrak{u}}(x) := \mathfrak{u}\, x\, \tilde{\mathfrak{u}} \tag{18.6}$$

for $x \in C\ell_Q(V)$. This linear map $x \mapsto \mathfrak{u}x\tilde{\mathfrak{u}}$ is an endomorphism on $C\ell_Q(V)$ referred to as *generalized conjugation of x by the blade $\mathfrak{u}$*, or simply as a *generalized blade conjugation.* When $\mathfrak{u}\tilde{\mathfrak{u}} = 1$, one sees that $\varphi_{\mathfrak{u}} = \psi_{\mathfrak{u}}$.

For convenience, the restriction of $\psi_{\mathfrak{u}}$ to V is denoted by $\Psi_{\mathfrak{u}}$. Properties of $\Psi_{\mathfrak{u}}$ differ from those of $\Phi_{\mathfrak{u}}$ in scaling. In particular, $\Psi_{\mathfrak{u}}$ is an element of the conformal orthogonal group $\mathrm{CO}_Q(V)$. Applying the reasoning used in the proof of Theorem 18.2.2 reveals that

$$\Psi_{\mathfrak{u}}(\mathbf{v}) = \begin{cases} (-1)^{\sharp\mathfrak{u}-1}\mathfrak{u}\tilde{\mathfrak{u}}\,\mathbf{v} & \text{if } \mathbf{v}|\mathfrak{u}, \\ (-1)^{\sharp\mathfrak{u}}\mathfrak{u}\tilde{\mathfrak{u}}\,\mathbf{v} & \text{if } \mathbf{v}|\mathfrak{u}^\star. \end{cases} \tag{18.7}$$

Consequently,

$$\Psi_{\mathfrak{u}} = (\mathfrak{u}\tilde{\mathfrak{u}})\Phi_{\mathfrak{u}}.$$

The geometric significance of the conjugation $\Phi_{\mathfrak{u}}(x) = \mathfrak{u}\mathbf{x}\mathfrak{u}^{-1}$ is well explained in the paper by Lounesto and Latvamaa [68]. For example, when Q is the quadratic form of *signature* (p,q) corresponding to the Clifford algebra $C\ell_{p,q}$ and $\mathfrak{u}\tilde{\mathfrak{u}} = 1$, the following exact sequences exist:

$$1 \to \mathbb{Z}_2 \to \mathrm{Pin}(p,q) \stackrel{\Phi_{\mathfrak{u}}}{\to} \mathrm{O}(p,q) \to 1,$$

$$1 \to \mathbb{Z}_2 \to \mathrm{Spin}(p,q) \stackrel{\Phi_{\mathfrak{u}}}{\to} \mathrm{SO}(p,q) \to 1.$$

Here, $\mathrm{Pin}(p,q)$ and $\mathrm{Spin}(p,q)$ are the *Pin* and *Spin* groups, which constitute double coverings of the corresponding orthogonal groups. The irreducible representations of these groups are *pinors* and *spinors.*

Corollary 18.2.4. *Given a unit blade $\mathfrak{u} \in C\ell_Q(V)$, where Q is positive definite, the map $\mathbf{x} \mapsto \mathfrak{u}\mathbf{x}\mathfrak{u}^{-1}$ represents a composition of hyperplane reflections across pairwise-orthogonal hyperplanes. In particular, when $\sharp\mathfrak{u}$ is even, the mapping is a composition of rotations by π radians in each of*

$\natural\mathfrak{u}/2$ orthogonal planes. When $n \equiv \natural\mathfrak{u} \equiv 1 \pmod 2$, the mapping is a composition of π radian rotations in $(n - \natural\mathfrak{u})/2$ orthogonal planes. When $n \equiv 0 \pmod 2$ and $\natural\mathfrak{u} \equiv 1 \pmod 2$, the mapping is a composition of π radian rotations in $(n - \natural\mathfrak{u} - 1)/2$ orthogonal planes along with one reflection through an additional orthogonal hyperplane.

Proof. Recall that when $\mathbf{u}, \mathbf{v}$ are unit vectors separated by angle α (measured from $\mathbf{u}$ to $\mathbf{v}$), and $\mathbf{x} \in V$, the mapping $\mathbf{x} \mapsto \mathbf{x}''$ defined by the composition of reflections across hyperplanes orthogonal to $\mathbf{u}$ and $\mathbf{v}$, respectively, corresponds to rotation by angle 2α in the $\mathbf{uv}$-plane (as pictured in Figure 4.1).

Letting $\mathfrak{u}$ be a blade of grade k, the proof of Theorem 18.2.2 shows that for any unit vector $\mathbf{v} \in \mathcal{E}_{(-1)^{1+(\natural\mathfrak{u} \pmod 2)}}$, the product $-\mathbf{vxv}$ is the reflection of $\mathbf{x}$ across the hyperplane orthogonal to $\mathbf{v}$. Orthogonality of the constituent vectors of $\mathfrak{u}$ then guarantees that the map $\mathbf{x} \mapsto \mathfrak{u}\mathbf{x}\mathfrak{u}^{-1}$ represents a composition of reflections across pairwise orthogonal hyperplanes. The composition of any two such reflections thereby results in rotation by π radians. $\square$

The next definition addresses how a vector can be said to "divide" a blade.

Definition 18.2.5. Let $\mathfrak{u}$ be a blade in $\mathcal{C}\ell_Q(V)$. A vector $\mathbf{w} \in V$ is said to *divide* $\mathfrak{u}$ if and only if there exists a blade $\mathfrak{u}' \in \mathcal{C}\ell_Q(V)$ of grade $\natural\mathfrak{u} - 1$ such that $\mathfrak{u} = \pm\mathbf{w}\mathfrak{u}'$. In this case, one writes $\mathbf{w}\big|\mathfrak{u}$.

An important interpretation of Theorem 18.2.2 is that any Q-orthogonal basis of the eigenspace $\mathcal{E}_{\lambda_1}$ of $\Phi_\mathfrak{u}$ determines a *Clifford factorization* of $\mathfrak{u}$. Moreover, any Q-orthogonal basis of $\mathcal{E}_{\lambda_2}$ determines a Clifford factorization of the dual blade $\mathfrak{u}^\star$.

Quantum Random Variables and States of the Clifford Algebra

In quantum probability, self-adjoint operators are analogous to random variables, with the expectation being given by the operator's normalized trace. Good references for the underlying theory include books by Gudder [52], Meyer [70], and Parthasarathy [77].

The role played by Kravchuk polynomials and Kravchuk matrices in quantum probability has been studied in a number of papers by Feinsilver *et al.* [39, 41]. Spectral properties of symmetric Kravchuk matrices, which are clearly self-adjoint, have also been considered in work by Feinsilver and Fitzgerald [38].

When $\mathcal{A}$ is a complex algebra with involution $*$ and unit 1, a positive linear $*$-functional on $\mathcal{A}$ satisfying $\varphi(1) = 1$ is called a *state* on $\mathcal{A}$. The pair $(\mathcal{A}, \varphi)$ is called an *algebraic probability space.*

As seen in Lemma 18.2.3, $\Phi_{\mathfrak{u}}$ is a Q-orthogonal transformation on V. An immediate consequence is that $\Phi_{\mathfrak{u}}$ is self-adjoint with respect to the Q-inner product, as made formal in the following lemma.

Lemma 18.2.6. *The blade conjugation operator $\Phi_{\mathfrak{u}}$ is self-adjoint with respect to $\langle \cdot, \cdot \rangle_Q$.*

Proof. The proof is left as an exercise. □

Hence, $\Phi_{\mathfrak{u}}$ is correctly regarded as a *quantum random variable.* From the spectral properties uncovered in Theorem 18.2.2, one now interprets $\Phi_{\mathfrak{u}}$ as a quantum random variable ξ taking values ± 1 and satisfying

$$\mathbb{P}(\xi = \lambda_1) = \frac{\sharp\mathfrak{u}}{n},$$

$$\mathbb{P}(\xi = \lambda_2) = \frac{n - \sharp\mathfrak{u}}{n}.$$

The expectation, $\langle \xi \rangle = \mathrm{Tr}(\Phi_{\mathfrak{u}})$, is thus seen to be

$$\langle \xi \rangle = \frac{1}{n}\left((-1)^{\sharp\mathfrak{u}-1}\sharp\mathfrak{u} + (-1)^{\sharp\mathfrak{u}}(n - \sharp\mathfrak{u})\right) = \begin{cases} 1 - \dfrac{2\sharp\mathfrak{u}}{n} & \sharp\mathfrak{u} \equiv 0 \pmod 2, \\ \dfrac{2\sharp\mathfrak{u}}{n} - 1 & \sharp\mathfrak{u} \equiv 1 \pmod 2. \end{cases}$$

As Lemma 18.2.8 will show, the scalar projection $\langle \cdot \rangle_0 : \mathcal{C}\ell_Q(V) \to \mathbb{R}$ defines a state on the Clifford algebra. For convenience, define the notation $\pi_0(u) := \langle u \rangle_0$. Further, when Q is a definite quadratic form on V, one can ensure $u^* u \geq 0$ for arbitrary $u \in \mathcal{C}\ell_Q(V)$ and appropriate choice of involution. In this way one can formulate the additional requirement $\rho(u^* u) \geq 0$ for states of the probability space.

On the other hand, when Q is indefinite, π_0 is still a state on $\mathcal{C}\ell_Q(V)$, but one cannot guarantee $u^* u \geq 0$ for all $u \in \mathcal{C}\ell_Q(V)$ using any of the canonical Clifford involutions. To address this potential shortcoming, the blade adjoint involution is defined.

Definition 18.2.7. Let $\mathfrak{u} \in \mathcal{C}\ell_Q(V)$ be a blade, and write $\mathfrak{u} = \prod_{i=1}^{\sharp\mathfrak{u}} \mathbf{u}_i$ for appropriately ordered vectors of V. Define the *adjoint* $\mathfrak{u}^\dagger \in \mathcal{C}\ell_Q(V)$ by

$$\mathfrak{u}^\dagger := \frac{\mathfrak{u}\tilde{\mathfrak{u}}}{|\mathfrak{u}\tilde{\mathfrak{u}}|}\tilde{\mathfrak{u}}. \tag{18.8}$$

Clearly, $(\mathrm{u}^\dagger)^\dagger = \mathrm{u}$; i.e., $\dagger$ is an involution on blades of $\mathcal{C}\ell_Q(V)$ and extends linearly to the entire algebra.

Lemma 18.2.8. *A* state *of the Clifford algebra $\mathcal{C}\ell_Q(V)$ is defined by the functional $\pi_0(u) := \langle u \rangle_0$, $u \in \mathcal{C}\ell_Q(V)$. For any nondegenerate quadratic form Q, the functional satisfies $\pi_0(u^\dagger u) \geq 0$ for all $u \in \mathcal{C}\ell_Q(V)$, where $\dagger$ denotes the blade adjoint defined by* (18.8).

Proof. Direct computation shows that π_0 is a linear functional satisfying the required conditions. □

Remark 18.2.9. In fact, when Q is a definite quadratic form on V,

$$u^\dagger = \begin{cases} \tilde{u} & \text{when } Q \text{ is positive definite,} \\ \overline{u} & \text{when } Q \text{ is negative definite.} \end{cases}$$

Another formulation of combinatorial integrals, attributed to Berezin [11], proves useful. For our purposes, the following definition is established.

Definition 18.2.10. The *Clifford Berezin integral* of arbitrary $u \in \mathcal{C}\ell_Q(V)$ is defined for anisotropic $\mathbf{x} \in V$ by

$$\int_{\mathfrak{B}} u \, d\mathbf{x} := u \lfloor \mathbf{x}. \tag{18.9}$$

Berezin integrals are defined for blades via iteration; i.e., given blade $\mathfrak{x} = \mathbf{x}_1 \cdots \mathbf{x}_{\sharp \mathfrak{x}}$,

$$\int_{\mathfrak{B}} u \, d\mathfrak{x} := ((u \lfloor \mathbf{x}_1) \lfloor \cdots) \lfloor \mathbf{x}_{\sharp \mathfrak{x}}.$$

Considering the iterated Clifford Berezin integral associated with the pseudo scalar ω, one obtains the following simplification:

$$\int_{\mathfrak{B}} u \, d\omega := \langle u\omega \rangle_0.$$

One can now define a *pseudo*-state by making use of the pseudo scalar, ω. The following lemma is obvious.

Lemma 18.2.11. *The mapping $\rho_{\mathfrak{B}}(u) := \int_B u \, d\omega : \mathcal{C}\ell_Q(V) \to \mathbb{R}$ is a positive linear functional on $\mathcal{C}\ell_Q(V)$ satisfying $\rho_{\mathfrak{B}}(\omega) = 1$.*

In quantum probability terms, blades represent *pure states* in $\mathcal{C}\ell_Q(V)$. Following the notation of Shale and Stinespring in their work on states of the Clifford algebra, one associates left and right operators with blades of the Clifford algebra [94] .

Given a blade $\mathfrak{u} \in \mathcal{C}\ell_Q(V)$, the operators $\mathfrak{L}_\mathfrak{u}$ and $\mathfrak{R}_\mathfrak{u}$ are defined on $\mathcal{C}\ell_Q(V)$ by $\mathfrak{L}_\mathfrak{u} v = \mathfrak{u}v$ and $\mathfrak{R}_\mathfrak{u} v = v\mathfrak{u}$, respectively. The map $\varphi_\mathfrak{u}$ thus satisfies

$$\varphi_\mathfrak{u} = \mathfrak{L}_\mathfrak{u}\mathfrak{R}_{\mathfrak{u}^{-1}} = \mathfrak{R}_{\mathfrak{u}^{-1}}\mathfrak{L}_\mathfrak{u}.$$

It is easily verified that

$$\varphi_\mathfrak{u} = \left[\frac{1}{2}\left(\mathfrak{L}_\mathfrak{u} + \mathfrak{R}_{\mathfrak{u}^{-1}}\right)\right]^2,$$

so that $\frac{1}{2}\left(\mathfrak{L}_\mathfrak{u} + \mathfrak{R}_{\mathfrak{u}^{-1}}\right)$ represents a "square root" of the conjugation operator $\varphi_\mathfrak{u}$, and hence a "fourth root of unity."

Kravchuk Polynomials from Blade Conjugation

Discussion of the level-1 operator $\Phi_\mathfrak{u}$ now concludes with the recovery of Kravchuk polynomials from the operator's characteristic polynomial. Moreover, this characteristic polynomial is recovered from an iterated Berezin integral.

Theorem 18.2.12. *Let $\mathfrak{u} \in \mathcal{C}\ell_Q(V)$ be a blade, let $\{\mathbf{e}_\ell : 1 \leq \ell \leq n\}$ be a fixed Q-orthonormal basis for V, and for each $\ell = 1, \ldots, n$, define the Clifford polynomial $\xi_\ell(t)$ by*

$$\xi_\ell(t) := \Phi_\mathfrak{u}(\mathbf{e}_\ell) - t\mathbf{e}_\ell.$$

If the collection $\{\xi_\ell(t) : 1 \leq \ell \leq n\}$ is orthogonal, then the corresponding iterated Berezin integral is

$$\int_{\mathfrak{B}} \omega \, \mathrm{d}\xi_n(t) \cdots \mathrm{d}\xi_1(t) = \begin{cases} \pm \sum_{\ell=0}^{n} K_\ell(n - \sharp\mathfrak{u}, n) t^\ell & \text{if } \sharp\mathfrak{u} \text{ even}, \\ \pm \sum_{\ell=0}^{n} K_\ell(\sharp\mathfrak{u}, n) t^\ell & \text{if } \sharp\mathfrak{u} \text{ odd}. \end{cases} \tag{18.10}$$

Proof. With due consideration to Theorem 5.1.4, the exterior product satisfies

$$\bigwedge_{\ell=1}^{n} \xi_\ell(t) = \bigwedge_{\ell=1}^{n} \left(\Phi_\mathfrak{u}(\mathbf{e}_\ell) - t\mathbf{e}_\ell\right) = \det(\Phi_\mathfrak{u} - t\mathbb{I})\,\omega, \tag{18.11}$$

where $\mathbb{I}$ is the identity on V. Reversing the order of the differentials reverses both the pseudo scalar and the computation of the determinant, resulting in an identity. Moreover, from the definition of the Berezin integral and the fact that $\omega_{\llcorner \mathfrak{z}} = \omega_{\mathfrak{z}}$ for any blade $\mathfrak{z}$, one can rewrite the integral in the form

$$\begin{aligned}\int_{\mathfrak{B}} \omega \, \mathrm{d}\xi_n(t) \cdots \mathrm{d}\xi_1(t) &= \langle \omega(\xi_n(t) \wedge \cdots \wedge \xi_1(t)) \rangle_0 \\ &= \langle \omega \det(\Phi_{\mathfrak{u}} - t\mathbb{I})\omega \rangle_0 = \omega^2 \det(\Phi_{\mathfrak{u}} - t\mathbb{I}).\end{aligned}$$

As a consequence of Theorem 18.2.2, when $\mathfrak{u} \in \mathcal{C}\ell_Q(V)$ is a blade, the characteristic polynomial $\chi_{\Phi_{\mathfrak{u}}}(t)$ of $\Phi_{\mathfrak{u}}$ is of the form

$$\chi_{\Phi_{\mathfrak{u}}}(t) = \begin{cases} \pm(t-1)^{n-\sharp\mathfrak{u}}(t+1)^{\sharp\mathfrak{u}} & \text{if } \sharp\mathfrak{u} \equiv 0 \pmod 2, \\ \pm(t-1)^{\sharp\mathfrak{u}}(t+1)^{n-\sharp\mathfrak{u}} & \text{if } \sharp\mathfrak{u} \equiv 1 \pmod 2. \end{cases} \tag{18.12}$$

Recalling that (18.12) is the generating function for Kravchuk polynomials and seeing that the characteristic polynomial of $\Phi_{\mathfrak{u}}$ is given by (18.11), (18.10) follows immediately. □

Theorem 18.2.12 indicates that the Berezin integral is a polynomial $F(t) := \int_{\mathfrak{B}} \omega \, d\xi_n(t) \cdots d\xi_1(t)$ having the following property:

$$\left.\frac{d^\ell}{dt^\ell} F(t)\right|_{t=0} = \begin{cases} \ell! K_\ell(n - \sharp\mathfrak{u}, n) & \text{if } \sharp\mathfrak{u} \equiv 0 \pmod 2, \\ \ell! K_\ell(\sharp\mathfrak{u}, n) & \text{if } \sharp\mathfrak{u} \equiv 1 \pmod 2. \end{cases}$$

Corollary 18.2.13. *Let $\mathfrak{u} \in \mathcal{C}\ell_Q(V)$ be a blade, define $\bar{t} := t/\mathfrak{u}\tilde{\mathfrak{u}}$, let $\{\mathbf{e}_\ell : 1 \le \ell \le n\}$ be a fixed Q-orthonormal basis for V, and for each $\ell = 1, \ldots, n$, define the Clifford polynomial $\upsilon_\ell(t)$ by*

$$\upsilon_\ell(t) := \Psi_{\mathfrak{u}}(\mathbf{e}_\ell) - t\mathbf{e}_\ell.$$

If the collection $\{\upsilon_\ell(t) : 1 \le \ell \le n\}$ is orthogonal, then the corresponding iterated Berezin integral is

$$\int_{\mathfrak{B}} \omega \, \mathrm{d}\upsilon_n(t) \cdots \mathrm{d}\upsilon_1(t) = \begin{cases} \pm(\mathfrak{u}\tilde{\mathfrak{u}})^n \displaystyle\sum_{\ell=0}^{n} K_\ell(n - \sharp\mathfrak{u}, n)\,(t/\mathfrak{u}\tilde{\mathfrak{u}})^\ell & \text{if } \sharp\mathfrak{u} \text{ even}, \\ \pm(\mathfrak{u}\tilde{\mathfrak{u}})^n \displaystyle\sum_{\ell=0}^{n} K_\ell(\sharp\mathfrak{u}, n)(t/\mathfrak{u}\tilde{\mathfrak{u}})^\ell & \text{if } \sharp\mathfrak{u} \text{ odd}, \end{cases}$$

*where $K_\ell(j, n)$ denotes the ℓ*th Kravchuk polynomial of order n.

Proof. Noting that for each $\mathbf{x} \in V$, $\Psi_{\mathfrak{u}}(\mathbf{x}) = (\mathfrak{u}\tilde{\mathfrak{u}})\Phi_{\mathfrak{u}}(\mathbf{x})$, the result follows from (18.12) and the proof of Theorem 18.2.12. □

The appearance of Kravchuk polynomials in the Berezin integral is only the first hint at the intricate relationship between Kravchuk polynomials and Clifford algebras.

18.3 Kravchuk Matrices as Operators on Clifford Algebras

Realizations of endomorphisms of V are developed as $|V| \times |V|$ real matrices. The *dual* of an endomorphism $A \in \mathrm{End}(V)$, denoted $A^\dagger$, is realized as the matrix transpose of the right-regular representation $\breve{A}$; i.e., $A^\dagger \mapsto \breve{A}^\dagger$.

Given a Q-orthogonal basis $\{\mathbf{u}_i : 1 \leq i \leq n\}$ for V, a matrix representation of $A \in \mathrm{End}(V)$ is determined by

$$\breve{A}_{ij} = \langle A(\mathbf{u}_i), \mathbf{u}_j \rangle_Q.$$

The relationships among Q, the Q-inner product, and the right-regular representation of Q are understood by $Q(\mathbf{x}) = \langle \mathbf{x}, \mathbf{x} \rangle_Q = \langle \mathbf{x} | \breve{Q} | \mathbf{x} \rangle$.

Letting $\mathfrak{K}_n$ denote the nth Kravchuk matrix, which is a square matrix of order $n+1$, a number of properties are known [38, 40]:

- The eigenvalues of $\mathfrak{K}_n$ are $\lambda_1 = 2^{n/2}$ of multiplicity $\lceil (n+1)/2 \rceil$, and $\lambda_2 = -2^{n/2}$ of multiplicity $\lfloor (n+1)/2 \rfloor$.
- ${\mathfrak{K}_n}^2 = 2^n \mathbb{I}$.
- The rows of $\mathfrak{K}_n$ are orthogonal with respect to the order-n binomial matrix $B = (b_{ij})$ defined by

$$b_{ij} := \begin{cases} 0 & i \neq j, \\ \binom{n}{j} & i = j. \end{cases} \tag{18.13}$$

  In particular, $\mathfrak{K}_n B {\mathfrak{K}_n}^\dagger = 2^n B$.

Letting $\mathcal{K}_n := 2^{-n/2} \mathfrak{K}_n$ denote the nth *normalized Kravchuk matrix*, the following corresponding properties are easily derived:

- The eigenvalues of $\mathcal{K}_n$ are $\lambda_1 = 1$ of multiplicity $\lceil (n+1)/2 \rceil$, and $\lambda_2 = -1$ of multiplicity $\lfloor (n+1)/2 \rfloor$.
- ${\mathcal{K}_n}^2 = \mathbb{I}$.
- The rows of $\mathcal{K}_n$ are orthogonal with respect to the order-n binomial matrix $B = (b_{ij})$ defined by (18.13). In particular, $\mathcal{K}_n B {\mathcal{K}_n}^\dagger = B$.

In light of the results from Section 18.2, the first property above indicates that $\mathcal{K}_n$ might represent a blade conjugation operator on $\mathcal{C}\ell_Q(V)$. In fact, this is often the case as the next theorem shows.

Theorem 18.3.1. *Let $\mathcal{C}\ell_Q(V)$ be the Clifford algebra of a nondegenerate quadratic form Q over an $(n+1)$-dimensional vector space V. Let $\mathcal{K}_n$ be the nth normalized Kravchuk matrix. Then,*

(1) When $n \equiv 0 \pmod 4$, $\mathcal{K}_n$ represents conjugation by a blade of grade $(n+2)/2$ associated with eigenspace $\mathcal{E}_{+1}$ or conjugation by a blade of grade $n/2$ associated with eigenspace $\mathcal{E}_{-1}$.

(2) When $n \equiv 1 \pmod 4$, $\mathcal{K}_n$ represents conjugation by a blade of grade $(n+1)/2$ associated with eigenspace $\mathcal{E}_{+1}$.

(3) When $n \equiv 3 \pmod 4$, $\mathcal{K}_n$ represents conjugation by a blade of grade $(n+1)/2$ associated with eigenspace $\mathcal{E}_{-1}$.

(4) When $n \equiv 2 \pmod 4$, $\mathcal{K}_n$ has no consistent interpretation as a blade conjugation. However, $\mathcal{K}_n$ does have a consistent interpretation as generalized conjugation by a unit blade $\mathfrak{u}$ of grade $(n+2)/2$ associated with eigenspace $\mathcal{E}_{+1}$ or generalized conjugation by a blade of grade $n/2$ associated with eigenspace $\mathcal{E}_{-1}$, if and only if the following condition is satisfied:

$$\mathfrak{u}\tilde{\mathfrak{u}} = \mathfrak{u}^\star\widetilde{\mathfrak{u}^\star} = -1. \tag{18.14}$$

Proof. Only the exceptional case, $n \equiv 2 \pmod 4$, is treated in detail. Recall that $\mathcal{K}_n$ has eigenvalues $\lambda_1 = 1$ and $\lambda_2 = -1$ of multiplicity $(n+2)/2$ and $n/2$, respectively. Moreover, the eigenspaces $\mathcal{E}_{+1}$ and $\mathcal{E}_{-1}$ are of dimension $(n+2)/2$ and $n/2$, respectively.

In the case $n \equiv 2 \pmod 4$, $\lceil (n+1)/2 \rceil = (n+2)/2$ is even, thereby making -1 the eigenvalue required for consistency with blade conjugation, a contradiction. Similarly, $\lfloor (n+1)/2 \rfloor = n/2$ is odd, making the required eigenvalue $+1$, which is again a contradiction. Hence, $\mathcal{K}_n$ cannot represent blade conjugation.

On the other hand, if $\mathfrak{u} = \prod_{\ell=1}^{\sharp\mathfrak{u}} \mathbf{u}_\ell$ is a unit blade of grade $(n+2)/2$, $\mathbf{x}|\mathfrak{u}$ implies

$$\mathfrak{u}\mathbf{x}\tilde{\mathfrak{u}} = (-1)^{\sharp\mathfrak{u}-1} \prod_{\ell=1}^{\sharp\mathfrak{u}} Q(\mathbf{u}_\ell)\mathbf{x} = (-1)^{n/2}(\mathfrak{u}\tilde{\mathfrak{u}})\mathbf{x}.$$

Hence, $\mathbf{x} \mapsto \mathbf{x}$ if and only if $\mathfrak{u}\tilde{\mathfrak{u}} = -1$.

Similarly, if $\mathbf{x}|\mathfrak{u}^\star$, one finds

$$\mathfrak{u}\mathbf{x}\tilde{\mathfrak{u}} = (-1)^{\sharp\mathfrak{u}} \prod_{\ell=1}^{\sharp\mathfrak{u}} Q(\mathbf{u}_\ell)\mathbf{x} = (-1)^{(n+2)/2}(\mathfrak{u}\tilde{\mathfrak{u}})\mathbf{x},$$

satisfying $\mathbf{x} \mapsto -\mathbf{x}$ if and only if $\mathfrak{u}\tilde{\mathfrak{u}} = -1$. Thus, $\mathcal{K}_n$ represents the specified generalized conjugations in the case $n \equiv 2 \pmod 4$.

Proofs of the remaining cases can be handled in similar fashion, once claims regarding the appropriate eigenspaces have been justified.

Observe first that the cases $n \equiv 1 \pmod 4$ and $n \equiv 3 \pmod 4$ lead to eigenspaces of equal dimension for $\lambda_1 = +1$ and $\lambda_2 = -1$. The blade conjugation requirement $\mathbf{x}|\mathfrak{u} \Rightarrow \mathbf{x} \mapsto (-1)^{\sharp\mathfrak{u}-1}$ thereby determines the correct eigenspace associated with the blade. Specifically, $n \equiv 1 \pmod 4$ implies $\sharp\mathfrak{u} = (n+1)/2$ is odd, so the corresponding eigenvalue must be $(-1)^{\sharp\mathfrak{u}-1} = +1$. Similarly, $n \equiv 3 \pmod 4$ implies $(n+1)/2$ is even, so the required eigenvalue is -1.

In the case $n \equiv 0 \pmod 4$ the eigenspaces are of distinct dimensions. Further, $\lceil (n+1)/2 \rceil = (n+2)/2$ is odd, thereby making $+1$ the eigenvalue required for consistency. Similarly, $\lfloor (n+1)/2 \rfloor = n/2$ is even, making the required eigenvalue -1. □

In light of these results, when $n \not\equiv 2 \pmod 4$, the Kravchuk matrices now satisfy

$$\mathfrak{K}_n \simeq 2^{n/2} \varphi_{\mathfrak{u}} \Big|_V$$

for an appropriate blade $\mathfrak{u} \in \mathcal{C}\ell_Q(V)$. Further, when Q is positive definite, the normalized Kravchuk matrices represent compositions of reflections across orthogonal hyperplanes as detailed in Corollary 18.2.4.

In the case $n \equiv 2 \pmod 4$,

$$\mathfrak{K}_n \simeq 2^{n/2} \psi_{\mathfrak{u}} \Big|_V$$

for appropriate blade $\mathfrak{u}$ satisfying (18.14), provided Q is indefinite.

To better see that the case $n \equiv 2 \pmod 4$ is not vacuous (i.e., that (18.14) has solutions), consider the following. First, let $\mathfrak{S}$ denote an $(n+1) \times (n+1)$ diagonal matrix with $\{\pm 1\}$ along the diagonal such that among the first $n/2+1$ diagonal entries, -1 appears an odd number of times. Similarly, -1 should appear an odd number of times among the remaining $n/2$ diagonal entries.

Let E be a matrix whose rows are unit eigenvectors of $\mathcal{K}_n$, and set $Q_\kappa := E^{-1}\mathfrak{S}E^{\dagger^{-1}}$ noting that Q_κ is then nonsingular and symmetric. It follows that Q_k represents a nondegenerate quadratic form satisfying

$$EQ_\kappa E^\dagger = \mathfrak{S}.$$

Denoting the ith row of E by $\mathcal{k}_i$, define the blade $\mathfrak{u} := \mathcal{k}_1 \cdots \mathcal{k}_{(n/2)+1} \in \mathcal{C}\ell_{Q_\kappa}(V)$. It follows that $\mathfrak{u}^\star = \mathcal{k}_{(n/2)+2} \cdots \mathcal{k}_n$.

By construction, it now follows that $\mathfrak{u}\tilde{\mathfrak{u}} = \mathfrak{u}^\star\tilde{\mathfrak{u}^\star} = -1$, satisfying (18.14). Moreover, one can see that

$$\mathfrak{u}\mathbf{x}\tilde{\mathfrak{u}} = \begin{cases} \mathbf{x} = (\mathfrak{u}\tilde{\mathfrak{u}})(-1)^{(n/2+1)-1}\mathbf{x} & \text{if } \mathbf{x}|\mathfrak{u}, \\ -\mathbf{x} = (\mathfrak{u}\tilde{\mathfrak{u}})(-1)^{(n/2+1)}\mathbf{x} & \text{if } \mathbf{x}|\mathfrak{u}^\star \end{cases}$$

so that $\mathbf{x} \mapsto \mathfrak{u}\mathbf{x}\tilde{\mathfrak{u}}$ represents generalized conjugation by blade $\mathfrak{u}$, in accordance with (18.7). Similarly, one establishes that $\mathcal{K}_n$ also represents generalized conjugation by $\mathfrak{u}^\star$.

18.4 Spectral Properties of Induced Operators

The notation $X \twoheadrightarrow \mathfrak{X}$ will be used to indicate that an operator $\mathfrak{X}$ on $\mathcal{C}\ell_Q(V)$ has been induced by the action of operator X on the underlying vector space V.

Definition 18.4.1. An operator $X \in \mathcal{L}(V)$ is said to be *deduced* from $\mathfrak{X} \in \mathcal{L}(\mathcal{C}\ell_Q(V))$ if $X \twoheadrightarrow \mathfrak{X}$. In this case, it is clear that $X = \mathfrak{X}\Big|_V$.

The *grade function* $\mathrm{gr} : \mathcal{C}\ell_Q(V) \to \mathbb{N}_0$ is defined on an arbitrary multivector $u = \sum_{I\in 2^{[n]}} \alpha_I\, \mathfrak{u}_I \in \mathcal{C}\ell_Q(V)$ by

$$\mathrm{gr}(u) := \max_{\{I\in 2^{[n]}:\alpha_I\neq 0\}} \sharp\mathfrak{u}.$$

It should be clear that the grade function is basis independent.

Definition 18.4.2. An operator φ on $\mathcal{C}\ell_Q(V)$ satisfying $\mathrm{gr}(\varphi(u)) \leq \mathrm{gr}(u)$ for all blades $u \in \mathcal{C}\ell_Q(V)$ is said to be *lowering*. Similarly, if $\mathrm{gr}(\varphi(u)) \geq \mathrm{gr}(u)$ for all blades $u \in \mathcal{C}\ell_Q(V)$, then φ is said to be *raising*. The operator φ is said to be *grade preserving* if it is both lowering and raising.

As a consequence of Q-orthogonality, $\varphi_\mathfrak{u}$ is *blade preserving*; i.e., the image of a blade is a blade. Hence, the level-ℓ induced map ${\varphi_\mathfrak{u}}^{(\ell)}$is a well-defined linear operator on $\bigwedge^\ell V$, the grade-ℓ subspace of $\mathcal{C}\ell_Q(V)$ spanned by the basis ℓ-vectors.

The first result shows that, as in the case of the level-1 map $\Phi_\mathfrak{u}$, the level-ℓ map determines a quantum observable.

Lemma 18.4.3. *The level-ℓ induced map ${\varphi_\mathfrak{u}}^{(\ell)}$ is self-adjoint with respect to $\langle\cdot,\cdot\rangle_Q$. Moreover, ${\varphi_\mathfrak{u}}^{(\ell)}$ is Q-orthogonal.*

Proof. Letting $\mathfrak{x}$ and $\mathfrak{y}$ be arbitrary ℓ-blades in $\mathcal{C}\ell_Q(V)$, it suffices to show that $\langle {\varphi_\mathfrak{u}}^{(\ell)}(\mathfrak{x}), \mathfrak{y}\rangle_Q = \langle \mathfrak{x}, {\varphi_\mathfrak{u}}^{(\ell)}(\mathfrak{y})\rangle_Q$.

First, since $\mathfrak{u}$ is a blade, let $\mathfrak{U} = \{\mathbf{u}_1, \ldots, \mathbf{u}_n\}$ be a Q-orthogonal basis for V such that $\mathfrak{u}$ is a basis blade for the grade $\sharp\mathfrak{u}$ subspace of $\mathcal{C}\ell_Q(V)$. Hence, one can write $\mathfrak{u} = \mathbf{u}_1 \cdots \mathbf{u}_{\sharp\mathfrak{u}}$. Further, $\mathfrak{x}$ and $\mathfrak{y}$ can be written as linear combinations of basis ℓ-blades: $\mathfrak{x} = \sum_{|J|=\ell} \alpha_J \mathbf{u}_J$ and $\mathfrak{y} = \sum_{|I|=\ell} \beta_I \mathbf{u}_I$. It is therefore sufficient to show that $\langle \varphi_{\mathfrak{u}}{}^{(\ell)}(\mathbf{u}_I), \mathbf{u}_J \rangle_Q = \langle \mathbf{u}_I, \varphi_{\mathfrak{u}}{}^{(\ell)}(\mathbf{u}_J) \rangle_Q$ holds for arbitrary basis ℓ-blades $\mathbf{u}_I$ and $\mathbf{u}_J$.

Since each vector of $\mathbf{u}_I$ is either a factor of $\mathfrak{u}$ or of its orthogonal complement $\mathfrak{u}^\star$, $\mathfrak{u}\mathbf{u}_I\mathfrak{u}^{-1} = \prod_{i\in I} \mathfrak{u}\mathbf{u}_i\mathfrak{u}^{-1}$ implies

$$\begin{aligned}
\mathfrak{u}\mathbf{u}_I\mathfrak{u}^{-1}\widetilde{\mathbf{u}_J} &= \left(\prod_{i\in I:1\le i\le \sharp\mathfrak{u}} (-1)^{\sharp\mathfrak{u}-1} \prod_{i\in I:i>\sharp\mathfrak{u}} (-1)^{\sharp\mathfrak{u}} \mathbf{u}_I \right) \widetilde{\mathbf{u}_J} \\
&= \left(\prod_{i\in I:1\le i\le \sharp\mathfrak{u}} (-1)^{\sharp\mathfrak{u}-1} \prod_{i\in I:i>\sharp\mathfrak{u}} (-1)^{\sharp\mathfrak{u}} \mathbf{u}_I \right) (-1)^{\frac{\ell(\ell-1)}{2}} \mathbf{u}_J \\
&= (-1)^{\frac{\ell(\ell-1)}{2}} \left(\prod_{i\in I:1\le i\le \sharp\mathfrak{u}} (-1)^{\sharp\mathfrak{u}-1} \prod_{i\in I:i>\sharp\mathfrak{u}} (-1)^{\sharp\mathfrak{u}} \right) \mathbf{u}_I\mathbf{u}_J.
\end{aligned} \tag{18.15}$$

On the other hand, a similar calculation shows

$$\mathbf{u}_I\widetilde{\mathfrak{u}\mathbf{u}_J\mathfrak{u}^{-1}} = (-1)^{\frac{\ell(\ell-1)}{2}} \left(\prod_{j\in J:1\le j\le \sharp\mathfrak{u}} (-1)^{\sharp\mathfrak{u}-1} \prod_{j\in J:j>\sharp\mathfrak{u}} (-1)^{\sharp\mathfrak{u}} \right) \mathbf{u}_I\mathbf{u}_J. \tag{18.16}$$

Note that the product $\mathbf{u}_I\mathbf{u}_J$ is given by

$$\mathbf{u}_I\mathbf{u}_J = (-1)^{\sum_{j\in J}|\{i\in I: i>j\}|} \prod_{k\in I\cap J} Q(\mathbf{u}_k)\, \mathbf{u}_{I\triangle J},$$

where $I\triangle J$ is the set-symmetric difference of I and J. Hence,

$$\langle \mathbf{u}_I\mathbf{u}_J\rangle_0 = \begin{cases} (-1)^{\sum_{j\in J}|\{i\in I:i>j\}|} \prod_{k\in I\cap J} Q(\mathbf{u}_k) & \text{if } I = J, \\ 0 & \text{otherwise.} \end{cases}$$

It now follows that $\mathfrak{u}\mathbf{u}_I\mathfrak{u}^{-1}\widetilde{\mathbf{u}_J}$ is nonzero only if $I = J$. Note that $I = J$ implies

$$\begin{aligned}
(-1)^{\frac{\ell(\ell-1)}{2}} &\left(\prod_{i\in I:1\le i\le \sharp\mathfrak{u}} (-1)^{\sharp\mathfrak{u}-1} \prod_{i\in I:i>\sharp\mathfrak{u}} (-1)^{\sharp\mathfrak{u}} \right) \\
&= (-1)^{\frac{\ell(\ell-1)}{2}} \left(\prod_{j\in J:1\le j\le \sharp\mathfrak{u}} (-1)^{\sharp\mathfrak{u}-1} \prod_{j\in J:j>\sharp\mathfrak{u}} (-1)^{\sharp\mathfrak{u}} \right),
\end{aligned} \tag{18.17}$$

which shows (18.15) is equal to (18.16).

Orthogonality is similarly established:

$$\langle \mathfrak{u}\mathbf{u}_I\mathfrak{u}^{-1}, \mathfrak{u}\mathbf{u}_J\mathfrak{u}^{-1}\rangle_Q = \langle \mathfrak{u}\mathbf{u}_I\mathfrak{u}^{-1}\widetilde{\mathfrak{u}\mathbf{u}_J\mathfrak{u}^{-1}}\rangle_0 = (-1)^{\frac{\ell(\ell-1)}{2}}\langle \mathfrak{u}\mathbf{u}_I\mathfrak{u}^{-1}\mathfrak{u}\mathbf{u}_J\mathfrak{u}^{-1}\rangle_0$$

$$= \left\langle (-1)^{\frac{\ell(\ell-1)}{2}}\left(\prod_{j\in J:1\leq j\leq \sharp\mathfrak{u}}(-1)^{\sharp\mathfrak{u}-1}\prod_{j\in J:j>\sharp\mathfrak{u}}(-1)^{\sharp\mathfrak{u}}\right)^2 \mathbf{u}_I\mathbf{u}_J\right\rangle_0$$

$$= \left\langle (-1)^{\frac{\ell(\ell-1)}{2}}\mathbf{u}_I\mathbf{u}_J\right\rangle_0 = \langle \mathbf{u}_I\widetilde{\mathbf{u}_J}\rangle_0 = \langle \mathbf{u}_I, \mathbf{u}_J\rangle_Q. \tag{18.18}$$

□

Hence, ${\varphi_\mathfrak{u}}^{(\ell)}$ is a quantum observable for each $\ell = 1, \ldots, n$. The expectation of this quantum observable is given by its trace.

Lemma 18.4.4. *For fixed blade $\mathfrak{u} \in \mathcal{C}\ell_Q(V)$, the level-$\ell$ induced map ${\varphi_\mathfrak{u}}^{(\ell)}$ has eigenvalues ± 1. The dimensions of the corresponding eigenspaces are given by*

$$\dim({\mathcal{E}_{-1}}^{(\ell)}) = \begin{cases} \displaystyle\sum_{j\text{ odd}}\binom{\sharp\mathfrak{u}}{j}\binom{n-\sharp\mathfrak{u}}{\ell-j} & \sharp\mathfrak{u}\equiv 0 \pmod 2,\\ \displaystyle\sum_{j\text{ odd}}\binom{n-\sharp\mathfrak{u}}{j}\binom{\sharp\mathfrak{u}}{\ell-j} & \sharp\mathfrak{u}\equiv 1 \pmod 2,\end{cases} \tag{18.19}$$

and

$$\dim({\mathcal{E}_{+1}}^{(\ell)}) = \begin{cases} \displaystyle\sum_{j\text{ even}}\binom{\sharp\mathfrak{u}}{j}\binom{n-\sharp\mathfrak{u}}{\ell-j} & \sharp\mathfrak{u}\equiv 0 \pmod 2,\\ \displaystyle\sum_{j\text{ even}}\binom{n-\sharp\mathfrak{u}}{j}\binom{\sharp\mathfrak{u}}{\ell-j} & \sharp\mathfrak{u}\equiv 1 \pmod 2,\end{cases} \tag{18.20}$$

with sums being taken over all admissible values of the index.

Proof. Let $\mathfrak{U}_1, \mathfrak{U}_2$ be Q-orthogonal unit bases of the $\varphi_\mathfrak{u}$ eigenspaces $\mathcal{E}_{\lambda_1}$ and $\mathcal{E}_{\lambda_2}$, respectively. Writing $\mathfrak{U}_1 = \{\mathbf{u}_1, \ldots, \mathbf{u}_{\sharp\mathfrak{u}}\}$ and $\mathfrak{U}_2 = \{\mathbf{u}_{\sharp\mathfrak{u}+1}, \ldots, \mathbf{u}_n\}$, one immediately sees that

$$\varphi_\mathfrak{u}(\mathbf{u}_I) = \prod_{|I\cap\mathfrak{U}_1|}(-1)^{\sharp\mathfrak{u}-1}\prod_{|I\cap\mathfrak{U}_2|}(-1)^{\sharp\mathfrak{u}}\mathbf{u}_I.$$

Computing the dimension of ${\mathcal{E}_{-1}}^{(\ell)}$ thereby amounts to counting the multi-indices $I \in 2^{[n]}$ in which an odd number of generators having negative eigenvalues occur. When $\sharp\mathfrak{u}$ is even, an odd number of generators in I must come from $\mathfrak{U}_1$ and an even number from $\mathfrak{U}_2$. When $\sharp\mathfrak{u}$ is odd, an even number of generators must come from $\mathfrak{U}_1$ and an odd number from $\mathfrak{U}_2$. Keeping in mind that $|\mathfrak{U}_1| = \sharp\mathrm{u}$ and $|\mathfrak{U}_2| = n - \sharp\mathrm{u}$, (18.19) is established by counting cases. The dimension of ${\mathcal{E}_{+1}}^{(\ell)}$ is computed by similar counting arguments to reveal (18.20). □

Corollary 18.4.5. *The trace and determinant of the level-ℓ induced operator are given by the following:*

$$\mathrm{tr}({\varphi_{\mathrm{u}}}^{(\ell)}) = \begin{cases} K_\ell(\sharp\mathrm{u}, n) & \sharp\mathrm{u} \equiv 0 \pmod 2, \\ K_\ell(n - \sharp\mathrm{u}, n) & \sharp\mathrm{u} \equiv 1 \pmod 2, \end{cases}$$

and

$$\det({\varphi_{\mathrm{u}}}^{(\ell)}) = (-1)^{\dim({\mathcal{E}_{-1}}^{(\ell)})}.$$

Proof. The trace follows from the explicit formula for the ℓth Kravchuk polynomial of order n seen in (18.1). The determinant follows from Lemma 18.4.4. □

Recall that the level-ℓ induced map is the restriction of φ_{u} to the grade-ℓ subspace of $\mathcal{C}\ell_Q(V)$; i.e.,for each $\ell = 0, \ldots, n$,

$${\varphi_{\mathrm{u}}}^{(\ell)} = \varphi_{\mathrm{u}}\Big|_{\bigwedge^\ell V}.$$

Further observing that these subspaces are Q-orthogonal to each other, φ_{u} is seen to be a direct operator sum:

$$\varphi_{\mathrm{u}} := \bigoplus_{\ell=0}^{n} {\varphi_{\mathrm{u}}}^{(\ell)}.$$

Relative to an ordered Q-orthogonal basis $\{\mathbf{u}_j : 1 \leq j \leq n\}$ of V and induced canonical blade basis $\{\mathbf{u}_I : I \in 2^{[n]}\}$, the matrix representation of the level-ℓ induced map of φ_{u} is an order $\binom{n}{\ell}$ square matrix.

For integer $i \in \{0, \ldots, 2^n - 1\}$, let $\underline{i}$ denote the corresponding subset representation of i; that is, $\underline{i}$ is a subset of the n-set, $[n]$, satisfying $i = \sum_{\ell \in \underline{i}} 2^\ell$. An ordering is then naturally imposed on the basis blades by

$$\mathbf{u}_{\underline{i}} \prec \mathbf{u}_{\underline{j}} \begin{cases} \text{if } |\underline{i}| < |\underline{j}|, \text{ or} \\ \text{if } |\underline{i}| = |\underline{j}| \text{ and } i < j. \end{cases} \tag{18.21}$$

The matrix representation of φ_{u} with respect to the basis ordered by (18.21) is then a $2^n \times 2^n$ block-diagonal matrix $\check{\varphi}_{\mathrm{u}} = (\phi_{IJ})$ defined by

$$\phi_{IJ} := \langle \mathbf{u}_I | \varphi_{\mathrm{u}} | \mathbf{u}_J \rangle$$

for multi-indices $I, J \in 2^{[n]}$.

For integers k, ℓ and matrix M, the notation $\langle k|M|\ell\rangle$ will denote the entry of M in row k, column ℓ. Letting $\langle \mathbf{1}|$ denote the row vector of ones allows the jth column sum to be written $\langle \mathbf{1}|M|j\rangle$.

Corollary 18.4.6. *The trace of the blade conjugation operator on $\mathcal{C}\ell_Q(V)$ is obtained by summing a column of the nth Kravchuk matrix, $\mathfrak{K}_n$. More specifically, letting* $\mathbf{1}$ *denote the vector of all ones,*

$$\operatorname{tr}(\varphi_{\mathfrak{u}}) = \begin{cases} \displaystyle\sum_{\ell=0}^{n} K_\ell(\sharp\mathfrak{u}, n) = \langle \mathbf{1}|\mathfrak{K}_n|\sharp\mathfrak{u}\rangle & \text{if } \sharp\mathfrak{u} \equiv 0 \pmod 2, \\ \displaystyle\sum_{\ell=0}^{n} K_\ell(n-\sharp\mathfrak{u}, n) = \langle \mathbf{1}|\mathfrak{K}_n|n-\sharp\mathfrak{u}\rangle & \text{if } \sharp\mathfrak{u} \equiv 1 \pmod 2. \end{cases}$$

Proof. Since $\varphi_{\mathfrak{u}}$ is blade-preserving, its matrix is block-diagonal (under appropriate ordering of basis). The rest follows from summing over levels. □

Given an orthogonal collection $\mathfrak{U} = \{\mathfrak{u}_j : 1 \leq j \leq |\mathfrak{U}|\}$ of blades in $\mathcal{C}\ell_Q(V)$, attention now turns to linear combinations of operators $\xi = \sum_{j=1}^{|\mathfrak{U}|} a_j\, \varphi_{\mathfrak{u}_j}$. Since $\varphi_{\mathfrak{u}}$ is self-adjoint for each $\mathfrak{u}$, bilinearity of $\langle\cdot,\cdot\rangle_Q$ guarantees that ξ is self-adjoint with respect to the inner product $\langle\cdot,\cdot\rangle_Q$. Hence, ξ also represents a quantum random variable.

Further, the restriction of ξ to the grade-ℓ subspace $\bigwedge^\ell V$ is once again a well-defined linear operator. The *level-ℓ induced operator* is defined accordingly:

$$\xi^{(\ell)} := \xi\Big|_{\bigwedge^\ell V}.$$

The next result concerns traces of linear combinations of conjugation operators of a specified grade. It is here that the symmetric Kravchuk matrices ${\mathfrak{K}_n}^{\mathrm{sym}}$ come into play.

Proposition 18.4.7. *Let $\mathfrak{U}$ denote a fixed orthogonal basis of unit blades for $\mathcal{C}\ell_Q(V)$. For fixed $g \in \{0, 1, \ldots, n\}$, define the operator $\xi_g := \sum_{\{\mathfrak{u}\in\mathfrak{U}:\sharp\mathfrak{u}=g\}} a_{\mathfrak{u}}\varphi_{\mathfrak{u}}$, for scalar coefficients $a_{\mathfrak{u}}$. Then,*

$$\operatorname{tr}(\xi_g{}^{(\ell)}) = \begin{cases} K_\ell(g,n) \sum_{\mathfrak{u}\in\mathfrak{U}} a_{\mathfrak{u}} & \text{if } g \equiv 0 \pmod 2, \\ K_\ell(n-g,n) \sum_{\mathfrak{u}\in\mathfrak{U}} a_{\mathfrak{u}} & \text{if } g \equiv 1 \pmod 2. \end{cases} \tag{18.22}$$

In particular, when $a_{\mathfrak{u}} = 1$ *for all* $\mathfrak{u}$,

$$\operatorname{tr}(\xi_g) = \begin{cases} \langle \mathbf{1} | \mathfrak{K}_n{}^{\text{sym}} | g \rangle & \text{if } g \equiv 0 \pmod 2, \\ \langle \mathbf{1} | \mathfrak{K}_n{}^{\text{sym}} | n-g \rangle & \text{if } g \equiv 1 \pmod 2. \end{cases}$$

In other words, the trace of ξ_g *is obtained by summing the appropriate column of the symmetric Kravchuk matrix* $\mathfrak{K}_n{}^{\text{sym}}$.

Proof. Linearity of the trace functional together with Corollary 18.4.6 establishes (18.22). If $a_{\mathfrak{u}} = 1$ for all $\mathfrak{u}$, these sums of coefficients are reduced to binomial coefficients:

$$\operatorname{tr}(\xi_g{}^{(\ell)}) = \begin{cases} \binom{n}{g} K_\ell(g,n) & \text{if } g \equiv 0 \pmod 2, \\ \binom{n}{n-g} K_\ell(n-g,n) & \text{if } g \equiv 1 \pmod 2. \end{cases}$$

Summing over levels then gives the sum of the gth column of the symmetric Kravchuk matrix $\mathfrak{K}_n{}^{\text{sym}}$. □

18.5 Reductions and Other Inductions

As already seen, a linear operator A on an n-dimensional vector space naturally induces an operator on the 2^n-dimensional Clifford algebra $\mathcal{C}\ell_Q(V)$. Dual to this notion is the idea of reducing an operator on $\mathcal{C}\ell_Q(V)$ to an operator on the paravectors of V, that is, $\mathbb{R} \oplus V$, which we denote by V_*.

Definition 18.5.1. Let $\mathfrak{A}$ be an endomorphism on $\mathcal{C}\ell_Q(V)$, defined by its action on basis blades. An operator A on V_* is said to be *grade-reduced* from $\mathfrak{A}$ if its action on the ordered generators $\{\mathbf{e}_0, \ldots, \mathbf{e}_n\}$ of V_* satisfies

$$\langle \mathbf{e}_i | A | \mathbf{e}_j \rangle := \sum_{\substack{\sharp\mathfrak{a}=i \\ \sharp\mathfrak{b}=j}} \langle \mathfrak{a} | \mathfrak{A} | \mathfrak{b} \rangle, \tag{18.23}$$

where the sum is taken over blades in some fixed basis of $\mathcal{C}\ell_Q(V)$.

When (18.23) holds, it is convenient to write $\mathfrak{A} \searrow A$ and refer to A as an operator *grade-reduced* from $\mathfrak{A}$. Similarly, if X is a linear operator on

V_* having ordered basis $\{\mathbf{e}_0, \mathbf{e}_1, \ldots, \mathbf{e}_n\}$, the operator $\mathfrak{X}$ on $\mathcal{C}\ell_Q(V)$ whose action is defined on basis blades by

$$\mathfrak{X}(\mathfrak{u}) := \sum_{\mathfrak{v}\in\mathcal{B}} \langle \mathbf{e}_{\sharp\mathfrak{u}}|X|\mathbf{e}_{\sharp\mathfrak{v}}\rangle \mathfrak{v} \tag{18.24}$$

is said to be *grade-induced* on $\mathcal{C}\ell_Q(V)$ by X. When (18.24) holds, it is convenient to write $X \nearrow \mathfrak{X}$.

Considering spaces of linear operators, one sees

$$\mathcal{L}(V_*) \nearrow \mathcal{L}(\mathcal{C}\ell_Q(V)) \searrow \mathcal{L}(V_*).$$

More specifically, one has the next sequence of lemmas and corollaries.

Lemma 18.5.2. *Regarding the diagonal matrix B of binomial coefficients as an operator on V_*, let $X \in \mathcal{L}(V_*)$. Then, $X \nearrow \mathfrak{X} \searrow BXB$.*

Proof. Suppose $X \nearrow \mathfrak{X}$. Then, letting A denote the grade reduction of $\mathfrak{X}$,

$$\langle \mathbf{e}_i|A|\mathbf{e}_j\rangle := \sum_{\substack{\sharp\mathfrak{a}=i\\ \sharp\mathfrak{b}=j}} \langle \mathfrak{a}|\mathfrak{X}|\mathfrak{b}\rangle = \sum_{\substack{\sharp\mathfrak{a}=i\\ \sharp\mathfrak{b}=j}} \langle \mathbf{e}_{\sharp\mathfrak{a}}|X|\mathbf{e}_{\sharp\mathfrak{b}}\rangle = \binom{n}{i}\langle \mathbf{e}_i|X|\mathbf{e}_j\rangle\binom{n}{j}.$$

□

For convenience, the notation $A \twoheadrightarrow \mathfrak{A}$ will be used to indicate that the operator $\mathfrak{A}$ on $\mathcal{C}\ell_Q(V)$ is induced by the (conjugation) action of A on V. Considering spaces of linear operators on V, its Clifford algebra $\mathcal{C}\ell_Q(V)$, and the space of *paravectors* $V_* := \mathbb{R} \oplus V$, induced and reduced operators satisfy the following:

$$\mathcal{L}(V) \twoheadrightarrow \mathcal{L}(\mathcal{C}\ell_Q(V)) \searrow \mathcal{L}(V_*).$$

Lemma 18.5.3. *Let $X \in \mathcal{L}(V_*)$, suppose $X \nearrow \mathfrak{X}$, and let B be the diagonal matrix of binomial coefficients regarded as an operator on V_*. Then,*

$$\operatorname{tr}(\mathfrak{X}) = \operatorname{tr}(XB).$$

Proof. Suppose $X \nearrow \mathfrak{X}$. Then,

$$\begin{aligned}\operatorname{tr}(\mathfrak{X}) &= \sum_{\mathfrak{a}\in\mathfrak{B}} \langle \mathfrak{a}|\mathfrak{X}|\mathfrak{a}\rangle = \sum_{i=0}^{n}\sum_{\sharp\mathfrak{a}=i} \langle \mathfrak{a}|\mathfrak{X}|\mathfrak{a}\rangle\\ &= \sum_{i=0}^{n}\sum_{\sharp\mathfrak{a}=i} \langle \mathbf{e}_i|X|\mathbf{e}_i\rangle = \sum_{i=0}^{n}\binom{n}{i}\langle \mathbf{e}_i|X|\mathbf{e}_i\rangle = \operatorname{tr}(XB).\end{aligned} \tag{18.25}$$

□

Lemma 18.5.4. *Let $X, Y \in \mathcal{L}(V_*)$, suppose $X \nearrow \mathfrak{X}$ and $Y \nearrow \mathfrak{Y}$, and let $\mathfrak{X} \circ \mathfrak{Y}$ denote the operator on $\mathcal{C}\ell_Q(V)$ represented by the* Hadamard product *of matrix representations of the grade-induced operators. Then,*

$$\operatorname{tr}(\mathfrak{X} \circ \mathfrak{Y}) = \operatorname{tr}((X \circ Y)B),$$

where B is the diagonal matrix of binomial coefficients regarded as an operator on V_*.

Proof. First, observe that

$$\begin{aligned}(\mathfrak{X} \circ \mathfrak{Y})(\mathfrak{u}) &= \sum_{\mathfrak{v} \in \mathfrak{B}} \langle \mathfrak{u} | \mathfrak{X} \circ \mathfrak{Y} | \mathfrak{v} \rangle \mathfrak{v} \\ &= \sum_{\mathfrak{v} \in \mathfrak{B}} \langle \mathfrak{u} | \mathfrak{X} | \mathfrak{v} \rangle \langle \mathfrak{u} | \mathfrak{Y} | \mathfrak{v} \rangle \mathfrak{v} = \sum_{\mathfrak{v} \in \mathfrak{B}} \langle \mathbf{e}_{\sharp \mathfrak{u}} | X | \mathbf{e}_{\sharp \mathfrak{v}} \rangle \langle \mathbf{e}_{\sharp \mathfrak{u}} | Y | \mathbf{e}_{\sharp \mathfrak{v}} \rangle \mathfrak{v} \\ &= \sum_{\mathfrak{v} \in \mathfrak{B}} \langle \mathbf{e}_{\sharp \mathfrak{u}} | X \circ Y | \mathbf{e}_{\sharp \mathfrak{v}} \rangle \mathfrak{v}. \end{aligned} \tag{18.26}$$

Hence,

$$\begin{aligned}\operatorname{tr}(\mathfrak{X} \circ \mathfrak{Y}) &= \sum_{\mathfrak{u} \in \mathfrak{B}} \sum_{\mathfrak{v} \in \mathfrak{B}} \langle \mathfrak{u} | \mathfrak{X} \circ \mathfrak{Y} | \mathfrak{u} \rangle \\ &= \sum_{\mathfrak{v} \in \mathfrak{B}} \langle \mathfrak{u} | \mathfrak{X} | \mathfrak{v} \rangle \langle \mathfrak{u} | \mathfrak{Y} | \mathfrak{v} \rangle \mathfrak{v} = \sum_{\mathfrak{u} \in \mathfrak{B}} \langle \mathbf{e}_{\sharp \mathfrak{u}} | X | \mathbf{e}_{\sharp \mathfrak{u}} \rangle \langle \mathbf{e}_{\sharp \mathfrak{u}} | Y | \mathbf{e}_{\sharp \mathfrak{u}} \rangle \\ &= \sum_{\ell=0}^{n} \binom{n}{\ell} \langle \mathbf{e}_\ell | X \circ Y | \mathbf{e}_\ell \rangle = \operatorname{tr}((X \circ Y)B). \end{aligned} \tag{18.27}$$

□

Proposition 18.5.5. *Regarding the Kravchuk matrix $\mathfrak{K}_n$ and the symmetric Kravchuk matrix ${\mathfrak{K}_n}^{\mathrm{sym}}$ as operators on V_* and letting $\mathcal{K}$ denote the appropriate grade induced operator on $\mathcal{C}\ell_Q(V)$, one immediately sees*

$$\mathfrak{K}_n \nearrow \mathcal{K} \searrow B{\mathfrak{K}_n}^{\mathrm{sym}}.$$

Proof. The result follows immediately from Lemma 18.5.2 by observing $B\mathfrak{K}_n B = B(\mathfrak{K}_n B) = B{\mathfrak{K}_n}^{\mathrm{sym}}$.

□

Remark 18.5.6. In fact, letting $\mathfrak{B}$ denote the operator on $\mathcal{C}\ell_Q(V)$ grade-induced by B^{-1} on V_*, one finds $B^{-1} \nearrow \mathfrak{B} \searrow B$. Moreover,

$$B^{-2} \nearrow \mathfrak{B}^{\circ 2} \searrow \mathbb{I},$$

where $\mathfrak{B}^{\circ 2} = \mathfrak{B} \circ \mathfrak{B}$ denotes the operator on $\mathcal{C}\ell_Q(V)$ represented by the *Hadamard product* of matrix representations of $\mathfrak{B}$.

Proposition 18.5.7. *If Ψ is Q-orthogonal on V, then there exists a diagonal operator Δ on V_* satisfying*

$$\Psi \twoheadrightarrow \psi \searrow \Delta.$$

Proof. If Ψ is Q-orthogonal on V, then the induced operator ψ is grade-preserving on $\mathcal{C}\ell_Q(V)$. In other words, ℓ-blades map only to ℓ-blades under the induced action of ψ. The result then follows from the definition of grade reduction. □

Corollary 18.5.8. *If Φ represents blade conjugation on V, then $\Phi \twoheadrightarrow \varphi \searrow \Delta$ for some diagonal operator Δ on V_*.*

Remark 18.5.9. Using a suitably ordered basis, lowering operators are represented by lower triangular matrices, raising operators by upper triangular matrices, and grade preserving operators by block-diagonal matrices.

Form-Induced Operators

The quadratic form Q naturally induces a linear operator $\mathcal{Q}$ on V defined so that the Q-inner product coincides with the Euclidean inner product. In other words,

$$\langle \mathbf{x}, \mathbf{y} \rangle_Q = \langle \mathcal{Q}(\mathbf{x}) | \mathbf{y} \rangle. \tag{18.28}$$

Definition 18.5.10. Given the operator $\mathcal{Q}$ on V induced by the quadratic form Q on V, the linear operator $\mathfrak{Q}$ on $\mathcal{C}\ell_Q(V)$ induced by $\mathcal{Q}$ is said to be *form-induced* by Q. In particular, given a canonical basis $\{\mathbf{e}_1, \ldots, \mathbf{e}_n\}$ for V, one defines $\mathfrak{Q}$ by

$$\mathfrak{Q}(\mathbf{e}_I) := \sum_{J \in 2^{[n]}} \left(\prod_{\ell \in I \cap J} Q(\mathbf{e}_\ell) \right) \mathbf{e}_J.$$

For convenience, write $Q \rightsquigarrow \mathfrak{Q}$ when (18.28) holds and $\mathcal{Q} \twoheadrightarrow \mathfrak{Q}$.

As will be seen shortly, symmetric Kravchuk matrices are related to operators form-induced by negative definite quadratic forms. First, another well-known family of matrices is recalled.

The family $\mathfrak{H}$ of *Sylvester-Hadamard matrices* is defined as the collection of tensor (Kronecker) powers of the initial matrix $H = \begin{pmatrix} 1 & 1 \\ 1 & -1 \end{pmatrix}$. In particular, $\mathfrak{H} := \{H^{\otimes n} : n \in \mathbb{N}\}$, where

$$H^{\otimes n} = \underbrace{H \otimes H \otimes \cdots \otimes H}_{n \text{ times}}.$$

Remark 18.5.11. In terms of the Pauli matrices, $H = \sigma_x + \sigma_z$.

Writing $\{\bullet, \circ\}$ in place of $\{1, -1\}$, the matrices of $\mathfrak{H}$ for $n = 2, 3$ are

$$H^{\otimes 2} = \begin{pmatrix} \bullet & \bullet & \bullet & \bullet \\ \bullet & \circ & \bullet & \circ \\ \bullet & \bullet & \circ & \circ \\ \bullet & \circ & \circ & \bullet \end{pmatrix}, \quad H^{\otimes 3} = \begin{pmatrix} \bullet & \bullet & \bullet & \bullet & \bullet & \bullet & \bullet & \bullet \\ \bullet & \circ & \bullet & \circ & \bullet & \circ & \bullet & \circ \\ \bullet & \bullet & \circ & \circ & \bullet & \bullet & \circ & \circ \\ \bullet & \circ & \circ & \bullet & \bullet & \circ & \circ & \bullet \\ \bullet & \bullet & \bullet & \bullet & \circ & \circ & \circ & \circ \\ \bullet & \circ & \bullet & \circ & \circ & \bullet & \circ & \bullet \\ \bullet & \bullet & \circ & \circ & \circ & \circ & \bullet & \bullet \\ \bullet & \circ & \circ & \bullet & \circ & \bullet & \bullet & \circ \end{pmatrix}.$$

Obviously $H^{\otimes n}$ represents a linear operator on a 2^n-dimensional vector space. Since the eigenvalues of the nth Hadamard matrix are $\pm 2^{n/2}$, each with multiplicity 2^{n-1}, it follows immediately that $H^{\otimes n}$ represents generalized conjugation by a blade of grade 2^{n-1}.

In the Clifford algebra of negative definite signature, that is, $\mathcal{C}\ell_Q(V) \cong \mathcal{C}\ell_{0,n}$, the Hadamard matrix represents the operator form-induced by Q. In particular, let $\mathfrak{H}$ denote the operator represented by $H^{\otimes n}$, with rows and columns indexed by integers $\{0, \ldots, 2^n - 1\}$. Using the binary representation of integers, one sees that $\mathfrak{H}$ acts on basis blade $\mathbf{e}_I$ according to

$$\mathbf{e}_I \mapsto \sum_{J \in 2^{[n]}} \left(\prod_{\ell \in I \cap J} Q(\mathbf{e}_\ell) \right) \mathbf{e}_J = \sum_{J \in 2^{[n]}} (-1)^{I \cap J} \mathbf{e}_J,$$

so that in terms of the Euclidean inner product,

$$\langle \mathfrak{H}(\mathbf{e}_I) | \mathbf{e}_J \rangle = (-1)^{|I \cap J|} = \prod_{\ell \in I \cap J} Q(\mathbf{e}_\ell).$$

Relative to the Q-inner product, one finds

$$\begin{aligned} \langle \mathfrak{H}(\mathbf{e}_I), \mathbf{e}_J \rangle_Q &:= \langle (-1)^{|I \cap J|} \tilde{\mathbf{e}_J} \mathbf{e}_J \rangle_0 \\ &= (-1)^{|I \cap J| + |J|} \\ &= (-1)^{|J \setminus I|} = \prod_{\{\ell \in J : \ell \notin I\}} Q(\mathbf{e}_\ell). \end{aligned}$$

Regarding the symmetric Kravchuk matrix ${\mathfrak{K}_n}^{\mathrm{sym}}$ as a linear operator on V_*, the preceding discussion leads to the following theorem.

Lemma 18.5.12. *Let $\mathcal{S}$ denote the operator on V_* represented by the symmetric Kravchuk matrix ${\mathfrak{K}_n}^{\mathrm{sym}}$. Let $\mathcal{H}$ denote the operator on $\mathcal{C}\ell_Q(V)$ represented by the nth Sylvester-Hadamard matrix $H^{\otimes n}$ as described previously, and let Q be negative definite on V. Then,*

$$Q \rightsquigarrow \mathcal{H} \searrow \mathcal{S}.$$

Proof. It is known [40] that the nth symmetric Kravchuk matrix is obtained by summing entries of the nth Hadamard matrix by selecting rows and columns indexed by integers whose binary representations are of Hamming weights corresponding to the row and column index of the symmetric Kravchuk matrix. These Hamming weights correspond precisely to the grades of canonical blades in $\mathcal{C}\ell_Q(V)$, so that this summing is accomplished by grade deduction of the nth Hadamard matrix; i.e.,

$$\langle \mathbf{e}_i | \mathcal{S} | \mathbf{e}_j \rangle = \sum_{\substack{\sharp \mathfrak{a}=i \\ \sharp \mathfrak{b}=j}} \langle \mathfrak{a} | \mathcal{H} | \mathfrak{b} \rangle.$$

Hence, $\mathcal{H} \searrow \mathcal{S}$. □

Along these lines, define the *binomial grade operator* Γ on $\mathcal{C}\ell_Q(V)$ by

$$\Gamma(\mathfrak{u}) := \binom{|V|}{\sharp \mathfrak{u}} \mathfrak{u}.$$

The inverse is given by $\Gamma^{-1}(\mathfrak{u}) := \binom{|V|}{\sharp \mathfrak{u}}^{-1} \mathfrak{u}$.

Remark 18.5.13. Let $\mathcal{H}'$ be defined on $\mathcal{C}\ell_Q(V)$ by $\mathcal{H}' := \Gamma^{-1}\mathcal{H}$, and let $\mathfrak{x}$ denote a basis blade chosen randomly (with all blades having equal probability) from the canonical basis $\{\mathbf{e}_I : I \in 2^{[n]}\}$ of $\mathcal{C}\ell_Q(V)$, where Q is negative definite. The operator $\mathcal{H}'$ then has the following *probabilistic* interpretation:

$$\langle \mathcal{H}'(\mathbf{e}_I), \mathbf{e}_J \rangle_Q = (-1)^{|I \cap J|} \mathbb{P}\left(\mathfrak{x} = \mathbf{e}_I \mid \sharp \mathfrak{x} = |I|\right).$$

Lemma 18.5.14. *Regarding the nth Kravchuk matrix $\mathfrak{K}_n$ as an operator on V_*, let $\mathcal{H}'$ be defined on $\mathcal{C}\ell_Q(V)$ by $\mathcal{H}' := \Gamma^{-1}\mathcal{H}$. Then,*

$$\mathcal{H}' \searrow \mathfrak{K}_n.$$

Proof. In terms of canonical basis blades,

$$\mathcal{H}'(\mathbf{e}_I) = \sum_{J \in 2^{[n]}} (-1)^{|I \cap J|} \binom{n}{|J|}^{-1} \mathbf{e}_J. \tag{18.29}$$

Denote the grade-reduction of $\mathcal{H}'$ by X. Then,

$$\begin{aligned}
\langle \mathbf{e}_i | X | \mathbf{e}_j \rangle &= \sum_{|I|=i, |J|=j} (-1)^{|I \cap J|} \binom{n}{|J|}^{-1} \\
&= \binom{n}{|J|}^{-1} \langle \mathbf{e}_i | \mathfrak{K}_n{}^{\mathrm{sym}} | \mathbf{e}_j \rangle \\
&= \langle \mathbf{e}_i | \mathfrak{K}_n | \mathbf{e}_j \rangle.
\end{aligned}$$

□

18.6 Kravchuk Matrices from Operator Reduction

In light of the interpretation of symmetric Kravchuk matrices as operators on V_* obtained by grade-reduction of operators on Clifford algebras of negative definite signature, the goal now is to uncover a similar relationship between the Kravchuk matrices $\mathfrak{K}_n$ and operators on $\mathcal{C}\ell_Q(V)$.

When Q is positive definite, the corresponding Clifford algebra is denoted more simply by $\mathcal{C}\ell_n$, where $n = \dim(V)$. Note that for each $j = 1, \ldots, n$, the left contraction operator defines a lowering operator on $\mathcal{C}\ell_n$ by linear extension of the action on blades; i.e., $\Lambda_j(\mathfrak{u}) := \mathbf{e}_j \lrcorner \mathfrak{u}$. The *grade-reduction* of this operator is an operator L_j on the paravector space $V_* := \text{span}(\{\mathbf{e}_\varnothing, \mathbf{e}_1, \ldots, \mathbf{e}_n\})$, whose action is given by

$$\begin{aligned} L_j(\mathbf{e}_k) &= \sum_{|I|=k} \langle (\mathbf{e}_j \lrcorner \mathbf{e}_I), \mathbf{e}_{I\setminus\{j\}} \rangle \mathbf{e}_{k-1} \\ &= \sum_{|I|=k} (-1)^{\sharp\{i \in I : i < j\}} \mathbf{e}_{k-1}. \end{aligned}$$

Recall that under the action of a lowering operator, the nonzero image of a grade-k blade is a blade of grade $k-1$. Hence, $\Lambda_j \searrow L_j$, where the matrix representation of $L_j = (\ell_{im})$ is determined by

$$\ell_{im} = \begin{cases} K_i(m; n) & m = i+1, \\ 0 & \text{otherwise.} \end{cases}$$

In other words, the jth column of the nth Kravchuk matrix appears as the super-diagonal of the jth lowering operator, L_j. Similarly, letting Ξ_j denote the jth raising operator, defined by action on blades according to $\Xi_j(\mathfrak{u}) := \mathbf{e}_j \wedge \mathfrak{u}$, one finds $\Xi_j \searrow R_j$, where the matrix representation of $R_j \in \mathcal{L}(V_*)$ satisfies

$$r_{im} = \begin{cases} K_i(m; n) & m = i-1, \\ 0 & \text{otherwise.} \end{cases}$$

The next proposition provides the basis for a broader interpretation of $\mathfrak{K}_n$ as an operator on V_*.

Proposition 18.6.1. *For $\ell = 0, \ldots, n$, let ψ_ℓ denote the sum of homogeneous grade-ℓ basis elements of the Euclidean Clifford algebra $\mathcal{C}\ell_{n,0}$. That is,*

$$\psi_\ell := \sum_{\{I : |I| = \ell\}} \mathbf{e}_I.$$

Then, for $\ell = 0, \ldots, n-1$, *the left contraction* $\psi_\ell \lrcorner \widetilde{\psi_{\ell+1}}$ *is*

$$\psi_\ell \lrcorner \widetilde{\psi_{\ell+1}} = \sum_{j=1}^{n} K_\ell(j-1; n-1)\,\mathbf{e}_j.$$

Proof. Note that by expanding ψ_ℓ in terms of canonical basis blades of grade ℓ, one finds

$$\begin{aligned}\psi_\ell \lrcorner \widetilde{\psi_{\ell+1}} &= \left(\sum_{\{I:|I|=\ell\}} \mathbf{e}_I\right) \lrcorner \left(\sum_{\{J:|I|=\ell+1\}} \widetilde{\mathbf{e}_J}\right)\\ &= \sum_{\{J:|J|=\ell+1\}} \sum_{j\in J} \mathbf{e}_{J\setminus\{j\}} \widetilde{\mathbf{e}_J}\\ &= \sum_{\{J:|J|=\ell+1\}} \sum_{j\in J} (-1)^{\sharp\{m\in J: m<j\}} \mathbf{e}_j.\end{aligned}$$

It follows that the coefficient of $\mathbf{e}_j$ in $\psi_\ell \lrcorner \widetilde{\psi_{\ell+1}}$ is given by

$$\langle \psi_\ell \lrcorner \widetilde{\psi_{\ell+1}}, \mathbf{e}_j\rangle = \sum_{\{J:|J|=\ell, j\notin J\}} (-1)^{\sharp\{m\in J: m<j\}}.$$

Regarding i as $\sharp\{m \in J : m < j\}$, note that each multi-index J appearing in the sum can be constructed by choosing i elements of $[j-1]$ and choosing $\ell - i$ elements of $\{j+1, \ldots, n\}$. It follows immediately that

$$\begin{aligned}\sum_{\{J:|J|=\ell, j\notin J\}} (-1)^{\sharp\{m\in J: m<j\}} &= \sum_{i=0}^{n-1} (-1)^i \binom{j-1}{i}\binom{n-j}{\ell-i}\\ &= K_\ell(j-1; n-1).\end{aligned}$$

□

Corollary 18.6.2. *For* $\ell = 0, \ldots, n+1$, *let* ψ_ℓ *denote the sum of homogeneous grade-*ℓ *basis elements of the Euclidean Clifford algebra* $\mathcal{C}\ell_{n+1,0}$. *Then, the* n*th Kravchuk matrix is the* $(n+1) \times (n+1)$ *square matrix*

$$\mathfrak{K}_n = \begin{pmatrix} \psi_0 \lrcorner \widetilde{\psi_1} \\ \psi_1 \lrcorner \widetilde{\psi_2} \\ \vdots \\ \psi_n \lrcorner \widetilde{\psi_{n+1}} \end{pmatrix}.$$

Proof. Recall that the entries of the nth Kravchuk matrix are given by $\langle i|\mathfrak{K}_n|j\rangle = K_i(j; n)$, where $K_i(j; n)$ denotes the ith Kravchuk polynomial evaluated at j. Let $\{\mathbf{e}_j' : 0 \leq j \leq n\}$ be the standard orthonormal basis of row vectors for $\mathbb{R}^n$. By Proposition 18.6.1, working in $\mathcal{C}\ell_{n+1,0}$ gives

$$\begin{aligned}\psi_\ell \lrcorner \widetilde{\psi_{\ell+1}} &= \sum_{j=1}^{n+1} K_\ell(j-1;n)\,\mathbf{e}_j \\ &= \sum_{j=0}^{n} K_\ell(j;n)\,\mathbf{e}_j',\end{aligned}$$

for each $\ell = 0, \ldots, n$. It is now clear that the row vector $\psi_\ell \lrcorner \widetilde{\psi_{\ell+1}}$ is the ℓth row of $\mathfrak{K}_n$. □

The probabilistic/combinatorial interpretation of the nth Kravchuk matrix now comes into focus by considering the implications of the theorem. Thinking of $\{\mathbf{e}_1, \ldots, \mathbf{e}_n\}$ as a collection of n coins, the ℓth row of the matrix represents flipping ℓ of the coins simultaneously. All possible $\binom{n}{\ell}$ subsets are flipped and their outcomes summed. The random variable associated with a subset takes value ± 1, determined in the following way, based on column. In the jth column, coins indexed by numbers less than j are given value -1, so that the value of random variable X_I associated with subset I is

$$X_I = (-1)^{\sharp\{i \in I : i < j\}}.$$

The row ℓ, column j entry of $\mathfrak{K}_n$ is then

$$\langle \mathbf{e}_\ell | \mathfrak{K}_n | \mathbf{e}_j \rangle = \sum_{|I|=\ell} X_I = \sum_{|I|=\ell} (-1)^{\sharp\{i \in I : i \leq j\}}.$$

A Clifford operator calculus interpretation of the nth Kravchuk matrix is found by considering the following. Let β denote the canonical blade basis for $\mathcal{C}\ell_n$ with orthonormal generators $\{\mathbf{e}_1, \ldots, \mathbf{e}_n\}$. Define $\Lambda_\star$ as the lowering operator sum

$$\Lambda_\star := \sum_{\mathfrak{u} \in \beta} \Lambda_{\mathfrak{u}}.$$

It is not difficult to see that the matrix representation of $\Lambda_\star$ relative to β is given by

$$\langle \mathfrak{u} | \Lambda_\star | \mathfrak{w} \rangle = \begin{cases} \alpha & \text{if } \exists \mathfrak{b} \in \beta \text{ such that } \mathfrak{b} \lrcorner \mathfrak{u} = \alpha \mathfrak{w}, \\ 0 & \text{otherwise.} \end{cases}$$

In fact, when there exists $\mathfrak{b} \in \beta$ such that $\mathfrak{b} \lrcorner \mathfrak{u} = \alpha \mathfrak{w}$, the blade $\mathfrak{b}$ is unique. Hence, the matrix representation is more simply defined by $\langle \mathfrak{u} | \Lambda_\star | \mathfrak{w} \rangle = \alpha$, where α is the unique solution of $\mathfrak{b} \lrcorner \mathfrak{u} - \alpha \mathfrak{w} = 0$ when a solution exists.

The nth Kravchuk matrix is now obtained by an operator reduction from the $2^n \times 2^n$ matrix representation of $\Lambda_\star$.

Theorem 18.6.3. *For $0 \leq \ell, j \leq n$, the entries of the nth Kravchuk matrix satisfy*

$$\langle \mathbf{e}_\ell | \mathfrak{K}_n | \mathbf{e}_j \rangle = \sum_{\sharp \mathfrak{w} = \ell + 1} \langle \mathfrak{w} | \Lambda_\star | \mathbf{e}_j \rangle.$$

18.7 Kravchuk Classification of Clifford Algebras

Fixing canonical unit vector basis $\{\mathbf{e}_j : 0 \leq j \leq n\}$ for $\mathbb{R}^{n+1}$, define the vector $\mathfrak{h}_n \in \mathbb{R}^{n+1}$ by

$$\mathfrak{h}_n = \sum_{j=0}^{n} (-1)^{j(j-1)/2} \mathbf{e}_j.$$

Theorem 18.7.1. *Clifford algebras $\mathcal{C}\ell_{p,q}$ and $\mathcal{C}\ell_{p',q'}$ are isomorphic if and only if $\langle \mathfrak{h}_n | \mathfrak{K}_n | \mathbf{e}_q - \mathbf{e}_{q'} \rangle = 0$, where $n = p + q = p' + q'$.*

Proof. Clearly, the dimension n of the underlying vector spaces must be $p + q = p' + q'$. Let $\{\mathbf{e}_I : I \in 2^{[n]}\}$ and $\{\mathbf{f}_I : I \in 2^{[n]}\}$ be Grassmann bases for $\mathcal{C}\ell_{p,q}$ and $\mathcal{C}\ell_{p',q'}$, respectively. Isomorphism can then be determined by counting the basis blades squaring to 1. Letting n_+ and n_- denote the number of basis blades of $\mathcal{C}\ell_{p,q}$ squaring to 1 and -1, respectively, the sum of basis blade squares is given by

$$\begin{aligned} s = n_+ - n_- &= n_+ - (2^n - n_+) \\ &= 2n_+ - 2^n. \end{aligned}$$

Hence, isomorphism is determined just as well by summing squares of basis blades. Further, considering the relationship ${\mathbf{e}_I}^2 = (-1)^{(|I|(|I|-1))/2} \mathbf{e}_I \widetilde{\mathbf{e}_I}$ among basis blades, isomorphism is determined just as well by summing products of the form $\mathbf{e}_I \widetilde{\mathbf{e}_I}$.

In light of Proposition 18.6.1, isomorphism is thus determined by summing columns of Kravchuk matrices since the qth column sum represents the sum of basis blade products $\mathbf{e}_I \widetilde{\mathbf{e}_I}$ in $\mathcal{C}\ell_{p,q}$. □

Exercise

Exercise 18.1: Prove Lemma 18.2.6.

Chapter 19

Graph-Induced Operators

In this chapter, based on [100], operators are induced on fermion and zeon algebras by the action of adjacency matrices and combinatorial Laplacians on the vector spaces spanned by the graph's vertices. Properties of the algebras automatically give information about the graph's spanning trees and vertex coverings by cycles & matchings. Combining the properties of operators induced on fermions and zeons gives a fermion-zeon convolution that recovers the number of Hamiltonian cycles in an arbitrary graph.

19.1 Operators on Fermions and Zeons

Denote by $\mathfrak{F}_n$ the associative algebra over $\mathbb{R}$ generated by the collection $\{\mathfrak{f}_1, \ldots, \mathfrak{f}_n, \mathfrak{f}_1{}^\dagger, \ldots, \mathfrak{f}_n{}^\dagger\}$ satisfying the canonical anticommutation relations (CAR):

$$\{\mathfrak{f}_i, \mathfrak{f}_j\}_+ = \{\mathfrak{f}_i{}^\dagger, \mathfrak{f}_j{}^\dagger\}_+ = 0, \tag{19.1}$$

$$\mathfrak{f}_j{}^2 = \mathfrak{f}_j{}^{\dagger 2} = 0, \tag{19.2}$$

$$\{\mathfrak{f}_i, \mathfrak{f}_j{}^\dagger\}_+ = \delta_{ij}. \tag{19.3}$$

The algebra $\mathfrak{F}_n$ is called the *n-particle fermion algebra*. The generators $\mathfrak{f}_i$ and $\mathfrak{f}_i{}^\dagger$ are referred to as the ith *annihilation operator* and ith *creation operator*, respectively.

Zeon algebras can be regarded as commutative subalgebras of fermion algebras. It is not difficult to see that (19.1) implies commutativity of disjoint pairs of fermions; i.e., letting $\zeta_{\{j\}} = \mathfrak{f}_{2j-1}\mathfrak{f}_{2j} \in \mathfrak{F}_{2n}$ for $j = 1, \ldots, n$ ensures that the zeons satisfy the zeon canonical commutation relations (ZCR):

$$[\zeta_{\{i\}}, \zeta_{\{j\}}] = 0,$$

$$\zeta_{\{j\}}{}^2 = 0.$$

Here, $\zeta_{\{j\}}$ is defined as a pair of fermion annihilators, although creators could have been used as easily.

Assume $G = (V, E)$ is a graph on n vertices. For $i = 1, \ldots, n$ associate vertex $\mathbf{v}_i \in V$ with the ith fermion creation/annihilation pair, $\gamma_i = \frac{1}{\sqrt{2}}(\mathfrak{f}_i + \mathfrak{f}_i{}^\dagger) \in \mathfrak{F}_n$. Note that the fermion CAR imply $\{\gamma_i, \gamma_j\}_+ = 0$ and $\gamma_i{}^2 = 1$ for $i = 1, \ldots, n$. Hence, there is no cause for concern in defining multi-index notation by writing the ordered product

$$\gamma_I := \prod_{j\in I} \gamma_j = \frac{1}{2^{|I|/2}} \prod_{j\in I} (\mathfrak{f}_j + \mathfrak{f}_j{}^\dagger).$$

Remark 19.1.1. The collection $\{\gamma_i : 1 \leq i \leq n\}$ generates the 2^n-dimensional Euclidean Clifford algebra, commonly denoted $\mathcal{C}\ell_n$. This algebra is isomorphic to *fermion toy Fock space*.

For convenience, V is regarded as both the vertex set of G and as the vector space generated by the vertices of G. Let $\mathcal{L}(V)$ denote the space of linear operators on V. Beginning with an operator $X \in \mathcal{L}(V)$, the corresponding operator $\Psi \in \mathcal{L}(\mathfrak{F}_n)$ induced by X is defined naturally by linear extension of the following action on basis blades:

$$\Psi(\gamma_I) := \prod_{j\in I} X(\gamma_j).$$

Operators are induced in similar manner on $\mathcal{C}\ell_n{}^{\text{nil}}$. For convenience, when Ψ is an operator on the algebra $\mathcal{A}$ induced by the operator X acting on the vector space of generators of $\mathcal{A}$, it will be convenient to write $X \twoheadrightarrow \Psi$.

Because the algebras $\mathfrak{F}_n$ and $\mathcal{C}\ell_n{}^{\text{nil}}$ have natural grade decompositions, an operator X on V similarly induces operators $\Psi^{(\ell)}$ and $\Xi^{(\ell)}$, respectively, on the grade-ℓ subspaces of the algebras. More concisely,

$$\Psi^{(\ell)}(\gamma_I) = \begin{cases} \prod_{j\in I} X(\gamma_j) & |I| = \ell, \\ 0 & \text{otherwise.} \end{cases}$$

Using Dirac notation, one writes $\Psi^{(\ell)}(\gamma_I) = \langle \gamma_I | \Psi^{(\ell)}$. The use of Dirac notation in matrix representations is made clear by the following: Given $n \times n$ matrix A, the ith row and jth column of A are given by $\langle i|A|j\rangle$. In this context, the matrix A acts on row vectors by *right* multiplication. While this may be less common in linear algebra, it makes sense when dealing with adjacency matrices, combinatorial Laplacians, and transition matrices associated with Markov chains.

Once again, let $G = (V, E)$ be a graph, where V is regarded both as the vertex set of G and also the vector space spanned by vertices of G.

Let A denote an operator on V, suppose $A \twoheadrightarrow \Phi \in \mathcal{L}(\mathfrak{F}_n)$, and suppose $A \twoheadrightarrow \Xi \in \mathcal{L}({\mathcal{C}\ell_n}^{\mathrm{nil}})$. Under the vertex-fermion and vertex-zeon associations $\mathbf{v}_i \mapsto \gamma_i$ and $\mathbf{v}_i \mapsto \zeta_i$, respectively, there should be no ambiguity in adopting the conventions $\langle \mathbf{v}_I|\Phi|\mathbf{v}_J\rangle = \langle \gamma_I|\Phi|\gamma_J\rangle$ and $\langle \mathbf{v}_I|\Xi|\mathbf{v}_J\rangle = \langle \zeta_I|\Xi|\zeta_J\rangle$.

Spanning Trees via Fermion Trace

When $G = (V, E)$ is a simple graph on n vertices, the *combinatorial Laplacian* of G is the $n \times n$ matrix $L = (\ell_{ij})$ defined by

$$\ell_{ij} = \begin{cases} \deg(v_i) & \text{if } i = j, \\ -1 & \text{if } \{v_i, v_j\} \in E, \\ 0 & \text{otherwise.} \end{cases}$$

Equivalently, if D is the diagonal matrix of vertex degrees and A is the adjacency matrix of G, then $L = D - A$.

Let $\mathrm{Tr}(\cdot)$ denote *normalized* operator trace. In particular, if A acts on a finite-dimensional vector space V, then

$$\mathrm{Tr}(A) := \frac{1}{\dim(V)} \sum_{i=1}^{\dim(V)} \langle \mathbf{v}_i|A|\mathbf{v}_i\rangle$$

for any orthonormal basis $\{\mathbf{v}_i : 1 \le i \le n\}$ of V.

The following well known result of Kirchhoff [60] gives meaning to the normalized trace of operators on $\mathfrak{F}_V$ induced by the combinatorial Laplacian. It is recalled here without proof.

Theorem 19.1.2 (Kirchhoff). *For a given connected graph G with n labeled vertices, let $\lambda_1, \ldots, \lambda_{n-1}$ be the non-zero eigenvalues of its Laplacian matrix. Then the number of spanning trees of G is*

$$t_G = \frac{1}{n}\lambda_1\lambda_2 \cdots \lambda_{n-1}.$$

Equivalently the number of spanning trees is equal to any cofactor of the Laplacian matrix of G.

Viewing the Laplacian as an operator on the vector space V spanned by vertices of G, the combinatorial properties of fermions now allow numbers of spanning trees to be recovered from the trace of the induced operator.

Proposition 19.1.3. *Let L denote the combinatorial Laplacian associated with a finite graph, $G = (V, E)$, and let Φ be the operator on the fermion* *algebra $\mathfrak{F}_n$ induced by L; i.e., $L \twoheadrightarrow \Phi$. Then, the normalized trace of the induced map at level $n-1$ satisfies the following:*

$$\mathrm{Tr}\left(\Phi^{(n-1)}\right) = \sharp\{\text{spanning trees of } G\}.$$

Proof. First, for any $k \times n$ matrix $B = (b_{ij})$, associate the ith row of B with the fermion vector $\mathbf{b}_i := \sum_{j=1}^{n} \frac{b_{ij}}{\sqrt{2}} (\mathfrak{f}_j + \mathfrak{f}_j{}^{\dagger})$. For multi-index I of cardinality k, let B_I denote the coefficient submatrix of B whose columns are indexed by elements of I. The fermion CAR guarantee that the coefficient of γ_I in the expansion of $\mathfrak{z} = \prod_{\ell=1}^{k} \mathbf{b}_\ell$ is then $\langle \mathfrak{z}, \gamma_I \rangle = \det(B_I)$.

By definition of $L \twoheadrightarrow \Phi$, one immediately finds $\langle \mathbf{v}_I | \Phi^{(n-1)} | \mathbf{v}_I \rangle = \det(L_I)$, which is a cofactor of the combinatorial Laplacian of G. By Kirchhoff's theorem, the number of spanning trees of a connected graph G on n vertices is equal to any cofactor of the Laplacian of G. Summing over the n diagonal elements of the level $n-1$ induced operator then gives the stated result. □

When $k < |V|$, the trace of the level-k operator $\Phi^{(k)}$ can be interpreted by using the All Minors Matrix Tree Theorem [20]. In this interpretation, diagonal elements of $\Phi^{(k)}$ enumerate forests in G having $|V| - k$ trees, with each tree satisfying a number of additional properties. The formulation of interest here proceeds as follows.

Given a set F of weighted edges in a directed graph with no loops, let a_F denote the product of the weights. In particular, when $A = (a_{ij})$ is the graph's weighted adjacency matrix, one has

$$a_F = \prod_{(i,j)\in F} a_{ij}.$$

Observe that adopting similar notation for the combinatorial Laplacian $L = (\ell_{ij})$, one finds

$$\ell_F := \prod_{(i,j)\in F} \ell_{ij} = (-1)^{|F|} a_F.$$

Here, the sign is determined by the parity of the cardinality of F, since each edge weight is negative in L.

Theorem 19.1.4 (Matrix Tree Theorem). *Suppose L is the combinatorial Laplacian of a finite graph G, and let $U \subset V$ where $k = |U|$. Then,*

$$\det(L_U) = \sum_F \varepsilon(\varpi^*) \ell_F,$$

where the sum is over all forests F such that

i. F contains exactly $|U|$ trees;
ii. each tree in F contains exactly one vertex in U; and
iii. each arc in F is directed away from the vertex in U of the tree containing that arc.

F defines a matching $\varpi^* : W \to U$ so $\varpi^*(j) = i$ if and only if i and j are in the same tree of F.

Corollary 19.1.5 (Forest enumeration via fermion traces). *Let $L = (\ell_{ij})$ denote the combinatorial Laplacian associated with a finite graph, $G = (V, E)$, and let Φ be the operator on the fermion algebra $\mathfrak{F}_n$ induced by L; i.e., $L \twoheadrightarrow \Phi$. Then, the normalized trace of the induced map at level k satisfies the following:*

$$\operatorname{Tr}\left(\Phi^{(k)}\right) = \sum_F \varepsilon(\varpi^*)\ell_F.$$

Proof. Let J be a multi index of cardinality k. Let L_J denote the submatrix of L whose rows and columns are indexed by J. By definition, $L \twoheadrightarrow \Phi$ implies

$$\begin{aligned}\langle v_J|\Phi|v_J\rangle &= \det(L_J)\\ &= \sum_{\pi\in\mathcal{S}_{|J|}} (-1)^{\operatorname{sgn}\pi} \prod_{j=1}^{|J|} \ell_{j\,\pi(j)}.\end{aligned}$$

□

Remark 19.1.6. By the Matrix Tree Theorem [20],

$$\langle v_J|\Phi|v_J\rangle = \det(L(J'|J')) = \sum_F \varepsilon(\varpi^*)\ell_F.$$

The sum is over all forests F such that

(1) F contains exactly $n-k$ trees;
(2) each tree in F contains exactly one vertex in J'; and
(3) each arc in F is directed away from the vertex in J' of the tree containing that arc.

Tree-Induced Operators on Fermions

Let $\mathcal{G}(V)$ denote the collection of graphs on vertex set V. Using Kirchhoff's theorem as motivation, the goal now is to define a mapping $\mathcal{G}(V) \to \mathcal{L}(\mathfrak{F}_n)$ such that information about spanning trees of subgraphs is obtained from fermion traces.

Letting G be a graph on n vertices, define the *tree-induced operator* $\Gamma \in \mathcal{L}(\mathfrak{F}_n)$ as follows. For multi-index I of order k, let A_I denote the $k \times n$ submatrix of the graph's adjacency matrix consisting of rows indexed by I. Regarding these rows as vectors in V, A_I is of the form $A_I = \begin{pmatrix} \mathbf{a}_{i_1} \\ \vdots \\ \mathbf{a}_{i_k} \end{pmatrix}$. Now define the $k \times n$ matrix $D_I = (d_{ij})$ by

$$d_{ij} = \begin{cases} \sum_{\ell \in I}^{n} a_{i\ell} & \text{if } i = j, \\ 0 & \text{otherwise.} \end{cases}$$

Letting $T_I = D_I - A_I$, delete the first row of T_I to obtain the matrix T_{I^*}, and let γ_* denote the fermion generator associated with this deleted row under the vertex-fermion correspondence. Regarding rows $\{\mathbf{t}_j\}$ of T_{I^*} as vectors of $\mathfrak{F}_n$, define

$$\Gamma(\gamma_I) = \gamma_* \prod_{j \in I^*} \mathbf{t}_j.$$

The consistent choice of row for deletion makes this construction well-defined. The action of Γ on each γ_I is defined this way and extends linearly to the algebra. For convenience, the notation $G \rightsquigarrow \Gamma$ will be used when Γ is tree-induced by G. The next result is an immediate corollary.

Proposition 19.1.7. *Let G be a graph on n vertices and let Γ be the operator on the fermion algebra $\mathfrak{F}_n$ tree-induced by G; i.e., $G \rightsquigarrow \Gamma$. Then, for positive integer $k \leq n$, the trace of the graph-induced map at level k satisfies the following:*

$$\operatorname{tr}\left(\Gamma^{(k)}\right) = \sum_{\substack{W \subseteq V \\ |W|=k}} \sharp\{\text{spanning trees of } G_W\}.$$

Here, G_W denotes the subgraph of G induced by vertex subset W. In particular, for each multi-index I,

$$\langle \mathbf{v}_I | \Gamma | \mathbf{v}_I \rangle = \sharp\{\text{spanning trees of } G_{V_I}\}.$$

Proof. Note that by letting L_I be the submatrix of T_I whose columns are indexed by I, one obtains the combinatorial Laplacian of the subgraph G_{V_I} of G whose vertices are indexed by I. In the construction of $G \rightsquigarrow \Gamma$, premultiplication by γ_* has the effect of deleting the first column of ${T_I}^*$ so that $\langle \mathbf{v}_I|\Gamma|\mathbf{v}_I\rangle = \det(L_{I^*})$, where L_{I^*} is a cofactor of the combinatorial Laplacian of the subgraph G_{V_I}. The rest follows from Kirchhoff's theorem. □

Cycle-Matching Covers via Zeon Trace

Let A be the adjacency matrix of a graph with vertex set V of cardinality n. Regarding A as a linear operator on span(V), an operator Ξ is induced on ${\mathcal{C}\ell_n}^{\text{nil}}$ by multiplication. Graph-theoretic properties of Ξ are immediately seen.

Proposition 19.1.8. *Let A denote the adjacency matrix of a simple graph with vertex set V, viewed as a linear transformation on the vector space generated by V. Let $\Xi^{(k)}$ denote the corresponding operator induced on the grade-k subspace of the zeon algebra $\mathfrak{Z}_V$. For fixed subset $I \subseteq V$, let X_I denote the number of disjoint cycle covers of the subgraph induced by I. Similarly, let M_J denote the number of perfect matchings on the subgraph induced by $J \subseteq V$ (nonzero only for J of even cardinality). Then,*

$$\operatorname{tr}(\Xi^{(k)}) = \sum_{\substack{I\subset V\\ |I|=k}} \sum_{J\subseteq I} X_{I\setminus J} M_J.$$

Proof. Given any $k \times n$ matrix $B = (b_{ij})$, associate the ith row of B with the zeon vector $\mathbf{b}_i := \sum_{j=1}^{n} b_{ij}\,\zeta_j$. For multi-index I of cardinality k, let B_I denote the coefficient submatrix of B whose columns are indexed by elements of I. The coefficient of ζ_I in the expansion of $\mathfrak{z} = \prod_{\ell=1}^{k} \mathbf{b}_\ell$ is then $\langle \mathfrak{z}, \zeta_I\rangle = \operatorname{per}(B_I)$.

For $n \times n$ adjacency matrix A, one immediately finds $\langle \mathbf{v}_I|\Xi^{(k)}|\mathbf{v}_I\rangle = \operatorname{per}(A_I)$, where A_I is the $k\times k$ submatrix of A whose rows and columns are indexed by elements of I. Writing $A_I = (\Xi_{ij})$ and applying the definition of the matrix permanent,

$$\operatorname{per}(A_I) = \sum_{\pi\in S_k} \prod_{\ell=1}^{k} \Xi_{\ell\pi(\ell)}.$$

Recall that by an elementary result of group theory, every permutation $\pi \in S_k$ has a unique (up to order) factorization as a product of disjoint cycles. Further recall that A_I is the adjacency matrix of the subgraph induced by vertex set $\mathbf{v}_I$. Observing that disjoint 2-cycles (i.e., transpositions) represent edges with no shared vertices (i.e. matchings of subgraphs), it follows that

$$\mathrm{per}(A_I) = \sum_{J \subseteq I} X_{I \setminus J} M_J,$$

where $X_{I\setminus J}$ and M_J are defined as in the statement of the theorem. Summing over all multi-indices of size k then gives the trace of the induced operator. □

The following corollary is immediate.

Corollary 19.1.9 (Cycle-matching convolution). *Let A denote the adjacency matrix of a simple graph with vertex set V, and suppose $A \twoheadrightarrow \Xi \in \mathcal{L}({\mathcal{C}\ell_n}^{\mathrm{nil}})$. Let X_I and $M_{I'}$ be defined as in the statement of Proposition 19.1.8, where I' is the complement of I in $[n]$. Then,*

$$\langle \mathbf{v}_{[n]} | \Xi | \mathbf{v}_{[n]} \rangle = \mathrm{tr}(\Xi^{(n)}) = \sum_{I \subseteq V} X_I M_{I'}.$$

Hamiltonian Cycles via Fermion-Zeon Convolution

By combining properties of induced operators on fermions and zeons, one is able to count the Hamiltonian cycles in an arbitrary graph. The following result of Goulden and Jackson [51] lies at the heart of the main result. The statement of the theorem has been adapted to the notation developed herein.

Theorem 19.1.10 (Goulden-Jackson). *Let H_c be the number of directed Hamiltonian circuits in a digraph on n vertices with adjacency matrix A. Then*

$$H_c = \sum_{I} (-1)^{|I|} \det(A_I) \mathrm{per}(A_{I'})$$

where the sum is over all $I \subseteq [n] \setminus \{c\}$ for any $c \in [n]$, and $\det(A_\varnothing) := 1$.

Liu [66] generalized Goulden and Jackson's result to obtain a formula somewhat better suited for consideration as fermion-zeon convolution. In particular, the requirement of summing over subsets omitting one generator can be avoided. In the following theorem, the graph is assumed to be undirected, and all parameters of Liu's original formulation are assumed to be zero. Division by 2 is seen as correction for cycles appearing in two orientations.

Theorem 19.1.11 (Liu). *Let G be a finite graph with adjacency matrix A, and let H_c denote the number of Hamiltonian cycles in G. Then,*

$$H_c = \frac{1}{2n} \sum_{I \subseteq [n]} (-1)^{n-|I|} |I| \mathrm{per}(A_I) \det(A_{I'}).$$

To make the concept of fermion-zeon convolution rigorous, let $\varphi \in \mathcal{L}(\mathfrak{F}_n)$, let $\xi \in \mathcal{L}(\mathcal{C}\ell_n{}^{\mathrm{nil}})$, and define combinatorial integrals on $\mathcal{L}(\mathfrak{F}_n) \otimes \mathcal{L}(\mathcal{C}\ell_n{}^{\mathrm{nil}})$ by

$$\begin{aligned} \int (\varphi * \xi)(\mathbf{v}_I)\, d\gamma_I\, d\zeta_{I'} &:= \int (\varphi(\gamma_I) \otimes \xi(\zeta_{I'}))\, d\gamma_I\, d\zeta_{I'} \\ &= \langle \varphi(\gamma_I) \otimes \xi(\zeta_{I'}), \gamma_I \otimes \zeta_{I'} \rangle, \end{aligned}$$

where $I' = [n] \setminus I$ is the complement of I in the n-set. In other words, the value of the integral is the coefficient of $\gamma_I \otimes \zeta_{I'}$ in the canonical expansion of $(\varphi(\gamma_I) \otimes \xi(\zeta_{I'})) \in \mathfrak{F}_n \otimes \mathcal{C}\ell_n{}^{\mathrm{nil}}$. The *fermion-zeon convolution* of φ and ξ is then defined by

$$\begin{aligned} \int_{\mathfrak{F}\mathfrak{Z}} (\varphi * \xi)\, d\gamma\, d\zeta &:= \sum_{I \subseteq [n]} \int (\varphi * \xi)(\mathbf{v}_I)\, d\gamma_I\, d\zeta_{I'} \\ &= \sum_{I \subseteq [n]} \langle (\varphi(\gamma_I) \otimes \xi(\zeta_{I'})), \gamma_I \otimes \zeta_{I'} \rangle. \end{aligned} \tag{19.4}$$

To see how Liu's result naturally appears as fermion-zeon convolution, one first defines the *star dual* of $\Xi \in \mathcal{L}(\mathcal{C}\ell_n{}^{\mathrm{nil}})$ by

$$\langle \zeta_I | \Xi^\star | \zeta_J \rangle := \langle \zeta_{I'} | \Xi | \zeta_{J'} \rangle$$

for $I, J \subseteq [n]$, where I' and J' denote the set complements of I and J, respectively, in the n-set. Second, for operators X, Y on spaces of equal finite dimension, denote the componentwise product of operators X, Y by $X \odot Y$. In this way, the $\odot$ product of $\Psi \in \mathcal{L}(\mathcal{C}\ell_Q(V))$ and $\Xi^\star \in \mathcal{L}(\mathcal{C}\ell_n{}^{\mathrm{nil}})$ satisfies

$$\langle \mathbf{v}_I | \Psi \odot \Xi^\star | \mathbf{v}_J \rangle = \langle \mathbf{v}_I | \Psi | \mathbf{v}_J \rangle \langle \mathbf{v}_I | \Xi^\star | \mathbf{v}_J \rangle$$

for all $I, J \subseteq [n]$.

Letting σ denote the diagonal operator on the power set 2^V defined by

$$\langle \mathbf{v}_I | \sigma | \mathbf{v}_I \rangle := (-1)^{|I|} |I'|,$$

the unifying theorem can now be stated.

Theorem 19.1.12 (FZ Convolution). *Let G be a simple graph on n vertices having adjacency matrix A. Suppose $A \twoheadrightarrow \Phi \in \mathcal{L}(\mathfrak{F}_n)$ and $A \twoheadrightarrow \Xi \in \mathcal{L}(\mathcal{C}\ell_n{}^{\text{nil}})$. Then, the number H_c of Hamiltonian cycles in G is given by the following:*

$$H_c = \frac{1}{2n}\operatorname{tr}(\sigma\Phi \odot \Xi^\star).$$

Proof. The proof of Theorem 19.1.12 is now an easy corollary of Theorem 19.1.11.

$$\begin{aligned}
H_c &= \frac{1}{2n}\sum_{I\subseteq[n]}(-1)^{n-|I|}|I|\operatorname{per}(A_I)\det(A_{I'})\\
&= \frac{1}{2n}\sum_{I\in 2^{[n]}}(-1)^{n-|I|}|I|\langle \mathbf{v}_{I'}|\Phi|\mathbf{v}_{I'}\rangle\langle \mathbf{v}_I|\Xi|\mathbf{v}_I\rangle\\
&= \frac{1}{2n}\sum_{J\in 2^{[n]}}(-1)^{|J|}|J'|\langle \mathbf{v}_J|\Phi|\mathbf{v}_J\rangle\langle \mathbf{v}_{J'}|\Xi|\mathbf{v}_{J'}\rangle\\
&= \frac{1}{2n}\sum_{J\in 2^{[n]}}(-1)^{|J|}|J'|\langle \mathbf{v}_J|\Phi|\mathbf{v}_J\rangle\langle \mathbf{v}_J|\Xi^\star|\mathbf{v}_J\rangle\\
&= \frac{1}{2n}\operatorname{tr}(\sigma\Phi \odot \Xi^\star).
\end{aligned}$$

□

The fermion-zeon convolution interpretation is made clear by observing the following:

$$\begin{aligned}
H_c &= \frac{1}{2n}\sum_{J\in 2^{[n]}}(-1)^{|J|}|J'|\langle \mathbf{v}_J|\Phi|\mathbf{v}_J\rangle\langle \mathbf{v}_J|\Xi^\star|\mathbf{v}_J\rangle\\
&= \frac{1}{2n}\sum_{J\in 2^{[n]}}\langle\sigma(\mathbf{v}_J),\mathbf{v}_J\rangle\int(\Phi(\mathbf{v}_J)\otimes\Xi(\mathbf{v}_{J'}))(d\gamma_J\otimes d\zeta_{J'})\\
&= \frac{1}{2n}\sum_{J\in 2^{[n]}}\int(\sigma\Phi(\mathbf{v}_J)\otimes\Xi(\mathbf{v}_{J'}))(d\gamma_J\otimes d\zeta_{J'})\\
&= \frac{1}{2n}\int_{\mathfrak{F}3}(\sigma\Phi * \Xi)\,d\gamma\,d\zeta.
\end{aligned}$$

Observing that the operator trace is given by the sum

$$\operatorname{tr}(\sigma\Phi\odot\Xi^\star) = \sum_{I\subseteq[n]}\langle \mathbf{v}_I|\sigma\Phi\odot\Xi^\star|\mathbf{v}_I\rangle,$$

it follows immediately that the Hamiltonian cycles of the subgraph G_J induced by vertex set $J \subseteq V$ are enumerated by a similar sum; i.e.,

$$H_c(G_J) = \sum_{I \subseteq J} \langle \mathbf{v}_I | \sigma\Phi \odot \Xi^\star | \mathbf{v}_I \rangle.$$

Enumerating cycles of length k in a graph is then done by summing over k-vertex subgraphs of G. The next corollary makes this more formal.

Corollary 19.1.13. *Let G be a simple graph on n vertices having adjacency matrix A, and let $k \leq n$. Suppose $A \twoheadrightarrow \Phi \in \mathcal{L}(\mathfrak{F}_n)$ and $A \twoheadrightarrow \Xi \in \mathcal{L}(\mathcal{C}\ell_n{}^{\text{nil}})$. Then, the number c_k of k-cycles in G is given by*

$$c_k = \frac{1}{2k} \sum_{\substack{J \in 2^{[n]} \\ |J|=k}} \sum_{I \subseteq J} \langle \mathbf{v}_I | \sigma\Phi \odot \Xi^\star | \mathbf{v}_I \rangle.$$

The Middle-Levels Theorem

For odd $n \in \mathbb{N}$, let $\mathcal{Q}_n$ denote the n-dimensional hypercube, and let $\mathcal{M}_n$ denote the subgraph induced by vertices of Hamming weight $\lfloor n/2 \rfloor$ and $\lceil n/2 \rceil$. Note that $\mathcal{M}_n$ is a connected graph on $2\binom{n}{\lfloor n/2 \rfloor}$ vertices, referred to herein as the *middle-levels subgraph* of $\mathcal{Q}_n$.

The *middle-levels theorem* asserts that for every odd $n \geq 1$, the graph $\mathcal{M}_n$ has a Hamiltonian cycle. This theorem originated as a conjecture attributed largely to Havel, although more general formulations have been attributed to Dejter, Erdös, Trotter and others. It remained open for 30 years before being solved by Mütze [72].

Theorem 19.1.14 (Middle-Levels Theorem)**.** *For any odd integer n, the middle-levels subgraph $\mathcal{M}_n$ of the hypercube $\mathcal{Q}_n$ contains a Hamiltonian cycle.*

Reformulating the statement of the Middle-Levels Theorem in terms of fermion-zeon convolution is left as an exercise for the reader.

Exercises

Exercise 19.1: Letting $\lambda_0 \leq \lambda_1 \leq \cdots \leq \lambda_{n-1}$ denote the eigenvalues of the combinatorial Laplacian L, show that the minimum eigenvalue λ_0 is always zero.

Exercise 19.2: Show that the number of times zero appears as an eigenvalue of the combinatorial Laplacian of G is equal to the number of connected components of G.

Exercise 19.3: Show that when G has multiple connected components, its combinatorial Laplacian is block-diagonal.

Exercise 19.4: Let G be a simple graph on n vertices having adjacency matrix A. Suppose $A \twoheadrightarrow \Phi \in \mathcal{L}(\mathfrak{F}_n)$ and $A \twoheadrightarrow \Xi \in \mathcal{L}(\mathcal{C}\ell_n{}^{\text{nil}})$. Let $\mathfrak{A}$ be the nilpotent adjacency matrix of G, defined in Definition 15.1.1. Show that the coefficent of $\zeta_{[n]}$ in $\text{tr}(\mathfrak{A}^n)$ is given by fermion-zeon convolution as

$$\langle \text{tr}(\mathfrak{A}^n), \zeta_{[n]} \rangle = \text{tr}(\sigma\Phi \odot \Xi^\star).$$

Exercise 19.5: Give a fermion-zeon convolution formulation of the Middle-Levels Theorem.

Chapter 20

Solutions and Hints to Selected Exercises

Solution 1.1: *Hint:* Show that for all $g \in G$, $g^2 = g \Rightarrow g = e$. Then show inverses are two sided by showing $(gg')(gg') = gg'$. Finally, show that the identity is two-sided by considering $ge = g(g'g)$.

Solution 1.7: The projection of $\mathbf{x}$ onto $\mathbf{u}$ is

$$\begin{aligned}\mathbf{x}_{\|} &= \pi_{\mathbf{u}}(\mathbf{x}) \\ &= \left\langle \left(\frac{1}{\sqrt{6}}, -\frac{2}{\sqrt{6}}, \frac{1}{\sqrt{6}}\right), (-3,0,2) \right\rangle \left(\frac{1}{\sqrt{6}}, -\frac{2}{\sqrt{6}}, \frac{1}{\sqrt{6}}\right) \\ &= \left(-\frac{1}{6}, \frac{1}{3}, -\frac{1}{6}\right).\end{aligned}$$

Then, the vector orthogonal to $\mathbf{u}$ is

$$\begin{aligned}\mathbf{x}_{\perp} &= \mathbf{x} - \pi_{\mathbf{u}}(\mathbf{x}) \\ &= \left(-\frac{17}{6}, -\frac{1}{3}, \frac{13}{6}\right).\end{aligned}$$

It is left to verify that $\mathbf{x}_{\perp}$ is orthogonal to $\mathbf{u}$ and that $\mathbf{x} = \mathbf{x}_{\|} + \mathbf{x}_{\perp}$.

Solution 1.8: Setting

$$X = \mathbb{I} - 2\mathbf{v}\mathbf{v}^{\dagger} - 2\mathbf{u}\mathbf{u}^{\dagger} + 4\langle \mathbf{v}, \mathbf{u}\rangle \mathbf{v}\mathbf{u}^{\dagger},$$

verify directly that $X^{\dagger}X = I$ and that $|X| = 1$.

Solution 2.1:

$$\begin{aligned}(x+1)^6 &= \sum_{k=0}^{6} \binom{6}{k} x^{n-k} 1^k \\ &= x^6 + 6x^5 + 15x^4 + 20x^3 + 15x^2 + 6x + 1.\end{aligned}$$

Solution 2.4: Let A be the adjacency matrix of a graph G on n vertices. Proof is by induction on the exponent k. When $k = 1$, the result holds by definition of the adjacency matrix. In the inductive step, recall that for $n \times n$ matrices $A = (a_{ij})$ and $B = (b_{ij})$, writing $AB = (c_{ij})$ yields the following:

$$c_{ij} = \sum_{\ell=1}^{n} a_{i\ell} b_{\ell j}.$$

Suitable choices for A and B above, along with an appropriate inductive hypothesis, will lead to the result.

Solution 3.2:

$$\begin{aligned}(1+i)^5 &= \sum_{\ell=0}^{5} \binom{5}{\ell} i^\ell \\ &= \sum_{\substack{0\le\ell\le5 \\ \ell\equiv0 \pmod 2}} \binom{5}{\ell} i^\ell + \sum_{\substack{0\le\ell\le5 \\ \ell\equiv1 \pmod 2}} \binom{5}{\ell} 1^{n-\ell} i^\ell \\ &= 1 - \binom{5}{2} + \binom{5}{4} + i\left(\binom{5}{1} - \binom{5}{3} + \binom{5}{5}\right) \\ &= 1 - 10 + 5 + i(5 - 10 + 1) \\ &= -4 - 4i.\end{aligned}$$

Solution 3.3: In polar form, $w = (1 - i\sqrt{3}) = 2e^{i5\pi/3}$. By De Moivre's formula, $w^6 = 2^6 e^{6(5\pi/3)} = 64e^{10\pi} = 64$.

Solution 3.4:

$$\begin{aligned}f(z) &= e^{i\pi/6} z \\ &= \left(\frac{\sqrt{3}}{2} + \frac{1}{2} i\right) z.\end{aligned}$$

Solution 3.5:

$$\begin{aligned}w_0 &= 2 \\ w_1 &= 2e^{i\pi/4} \\ w_2 &= 2e^{i\pi/2} = 2i\end{aligned}$$

$$\begin{aligned}
w_3 &= 2e^{i3\pi/4} \\
w_4 &= 2e^{i\pi} = -1 \\
w_5 &= 2e^{i5\pi/4} \\
w_6 &= 2e^{i3\pi/2} = -2i \\
w_7 &= 2e^{i7\pi/4}.
\end{aligned}$$

The points are evenly distributed along a circle of radius 2, centered at the origin.

Solution 3.11: Rewriting $z = (1 - i\sqrt{3})$ in exponential form, we obtain $z = 2e^{i\pi/3}$. Thus, $z^6 = 2^6 e^{i6\pi/3} = 64$.

Solution 3.16:

a. $f(z) = (\frac{1}{\sqrt{2}} + \frac{1}{\sqrt{2}}i)z$
b. $f(z) = -\overline{z}$
c. $f(z) = iz$

Solution 4.2: Write $\mathbf{v} = v_0 + \mathbf{v}'$ and $\mathbf{u} = u_0 + \mathbf{u}'$. Then,

$$\begin{aligned}
\overline{\mathbf{v}}\mathbf{u} &= (v_0 - \mathbf{v}')(u_0 + \mathbf{u}') \\
&= v_0 u_0 - u_0 \mathbf{v}' + v_0 \mathbf{u}' - \mathbf{v}'\mathbf{u}' \\
&= (v_0 + \mathbf{v}')(u_0 - \mathbf{u}')
\end{aligned}$$

Solution 4.3: Recalling that the volume of the parallelepiped generated by $\mathbf{u}, \mathbf{v}, \mathbf{w}$ is given by $V = |\mathbf{u} \cdot (\mathbf{v} \times \mathbf{w})|$, one finds that

$$\begin{aligned}
V &= \left|\frac{1}{2}\left(u\,(vw - wv) + (vw - wv)\,u\right)\right| \\
&= \left|\frac{1}{2}\left(uvw - uwv + vwu - wvu\right)\right|.
\end{aligned}$$

Solution 4.4: Hint: Write $\mathbf{x} = \mathbf{x}_1 + \mathbf{x}_2$, where $\mathbf{x}_1$ is parallel to $\mathbf{u}$ and $\mathbf{x}_2$ is orthogonal to $\mathbf{u}$.

Solution 4.5: Observing that $\langle \mathbf{u}|\mathbf{u}\rangle = 1$,

$$\begin{aligned}
(|\mathbf{u}\rangle\langle\mathbf{u}|)(|\mathbf{u}\rangle\langle\mathbf{u}|) &= |\mathbf{u}\rangle \langle\mathbf{u}|\mathbf{u}\rangle \langle\mathbf{u}| \\
&= \langle\mathbf{u}|\mathbf{u}\rangle \, |\mathbf{u}\rangle\langle\mathbf{u}| \\
&= |\mathbf{u}\rangle\langle\mathbf{u}|.
\end{aligned}$$

Solution 4.8: Let I, J, K be defined as follows:

$$I = \sigma_{\mathrm{y}}\sigma_{\mathrm{z}}$$
$$J = \sigma_{\mathrm{x}}\sigma_{\mathrm{z}}$$
$$K = \sigma_{\mathrm{x}}\sigma_{\mathrm{y}}.$$

It follows immediately that $I^2 = J^2 = K^2 = \begin{pmatrix} -1 & 0 \\ 0 & -1 \end{pmatrix}$ and that

$$IJ = K = -JI$$
$$JK = I = -KJ$$
$$KI = J = -IK.$$

Let $\mathbf{w} = (w_1, w_2, w_3) \in \mathbb{R}^3$ be a unit vector, let $\theta \in \mathbb{R}$, and let $\mathbf{x} = (x_1, x_2, x_3) \in \mathbb{R}^3$ be arbitrary. Set

$$Q = \begin{pmatrix} \cos\frac{\theta}{2} & 0 \\ 0 & \cos\frac{\theta}{2} \end{pmatrix} + \sin\left(\frac{\theta}{2}\right)\begin{pmatrix} w_3 & w_1 - iw_2 \\ w_1 + iw_2 & -w_3 \end{pmatrix}$$

and

$$X = \begin{pmatrix} x_3 & x_1 - ix_2 \\ x_1 + ix_2 & -x_3 \end{pmatrix}.$$

The mapping $\mathbf{x} \mapsto X$ is readily seen to be a vector space isomorphism $\mathbb{R}^3 \to \mathrm{SU}(2)$. With respect to this isomorphism, the mapping $X \mapsto QXQ^{-1}$ corresponds to rotation of $\mathbf{x}$ about the unit vector $\mathbf{w}$ through angle θ with right-hand orientation in $\mathbb{R}^3$.

Solution 5.6:

(1) $u \wedge v = 17e_{\{1,2\}} - 24e_{\{1,3\}} - e_{\{2,3\}}$.
(2) $u \wedge w = 7e_{\{1,2\}} + 3e_{\{1,3\}} - 9e_{\{2,3\}}$.
(3) $u \wedge v \wedge w = 73e_{\{1,2,3\}}$.

Solution 5.8: Let $\mathbf{u}_1 = \mathbf{e}_{\{2\}} + 5\mathbf{e}_{\{3\}}$, $\mathbf{u}_2 = \frac{3\mathbf{e}_{\{3\}}}{2} - \mathbf{e}_{\{1\}}$, and $\alpha = 2$.

Solution 6.4: Writing $u = \sum_I u_I\mathbf{e}_I$ and $v = \sum_I v_I\mathbf{e}_I$,

$$\hat{u}\hat{v} = \left(\sum_I (-1)^{|I|} u_I \mathbf{e}_I\right)\left(\sum_J (-1)^{|J|} v_J \mathbf{e}_J\right)$$
$$= \sum_I \sum_{K \subseteq I} (-1)^{|k|}(-1)^{|I|-|K|} u_k v_{I\setminus K}\mathbf{e}_\mathbf{I}.$$

Writing $\varphi(u) = \hat{u}$, it is straightforward to verify that $\varphi\hat{u} = u$.

Solution 6.9: First, observe that $\left(\sum_{i=1}^{p+q} \mathbf{e}_i\right)^2 = p - q$. Proceeding by induction for $k \geq 2$, it follows that

$$\left(\sum_{i=1}^{p+q} \mathbf{e}_i\right)^k \begin{cases} (p-q)^{k/2} & k \equiv 0 \pmod 2, \\ (p-q)^{(k-1)/2} \sum_{i=1}^{p+q} \mathbf{e}_i & k \equiv 1 \pmod 2. \end{cases}$$

Solution 7.1: Let $\mathbf{u}_1 = -\frac{35}{32}\mathbf{e}_{\{2\}} - \frac{175\mathbf{e}_{\{3\}}}{32}$, and $\mathbf{u}_2 = \frac{64\mathbf{e}_{\{1\}}}{35} + \frac{4\mathbf{e}_{\{2\}}}{7} + \frac{4\mathbf{e}_{\{3\}}}{35}$.

Solution 8.2:

$$z_1 = \begin{pmatrix} 1 & 0 & 0 & 0 & 1 & 0 & 0 & 0 \\ 0 & 1 & 0 & 0 & 0 & 1 & 0 & 0 \\ 0 & 0 & 1 & 0 & 0 & 0 & 1 & 0 \\ 0 & 0 & 0 & 1 & 0 & 0 & 0 & 1 \\ -1 & 0 & 0 & 0 & -1 & 0 & 0 & 0 \\ 0 & -1 & 0 & 0 & 0 & -1 & 0 & 0 \\ 0 & 0 & -1 & 0 & 0 & 0 & -1 & 0 \\ 0 & 0 & 0 & -1 & 0 & 0 & 0 & -1 \end{pmatrix}$$

$$z_2 = \begin{pmatrix} 1 & 0 & 1 & 0 & 0 & 0 & 0 & 0 \\ 0 & 1 & 0 & 1 & 0 & 0 & 0 & 0 \\ -1 & 0 & -1 & 0 & 0 & 0 & 0 & 0 \\ 0 & -1 & 0 & -1 & 0 & 0 & 0 & 0 \\ 0 & 0 & 0 & 0 & 1 & 0 & 1 & 0 \\ 0 & 0 & 0 & 0 & 0 & 1 & 0 & 1 \\ 0 & 0 & 0 & 0 & -1 & 0 & -1 & 0 \\ 0 & 0 & 0 & 0 & 0 & -1 & 0 & -1 \end{pmatrix}$$

$$z_{\{1,2\}} = \begin{pmatrix} 1 & 0 & 1 & 0 & 1 & 0 & 1 & 0 \\ 0 & 1 & 0 & 1 & 0 & 1 & 0 & 1 \\ -1 & 0 & -1 & 0 & -1 & 0 & -1 & 0 \\ 0 & -1 & 0 & -1 & 0 & -1 & 0 & -1 \\ -1 & 0 & -1 & 0 & -1 & 0 & -1 & 0 \\ 0 & -1 & 0 & -1 & 0 & -1 & 0 & -1 \\ 1 & 0 & 1 & 0 & 1 & 0 & 1 & 0 \\ 0 & 1 & 0 & 1 & 0 & 1 & 0 & 1 \end{pmatrix}.$$

Solution 8.3:

$$\mathbf{e}_{\{1\}} \mapsto \begin{pmatrix} 0 & 0 & 1 & 0 \\ 0 & 0 & 0 & 1 \\ -1 & 0 & 0 & 0 \\ 0 & -1 & 0 & 0 \end{pmatrix}$$

$$\mathbf{e}_{\{2\}} \mapsto \begin{pmatrix} 0 & 0 & 0 & 1 \\ 0 & 0 & -1 & 0 \\ 0 & 1 & 0 & 0 \\ -1 & 0 & 0 & 0 \end{pmatrix}$$

Solution 8.4: First, note that multiplicative identity, $\mathbf{e}_\varnothing$, is indexed by a set of size zero so that ${\mathcal{B}_{p,q}}^+$ contains the identity. Secondly, the inverse of any element $\mathbf{e}_I$ is indexed by the same subset so that ${\mathcal{B}_{p,q}}^+$ is closed with respect to inverses. Finally, the symmetric difference of two sets of even cardinality is also of even cardinality so that ${\mathcal{B}_{p,q}}^+$ is closed under multiplication. Thus, ${\mathcal{B}_{p,q}}^+$ is a subgroup of $\mathcal{B}_{p,q}$.

To see that ${\mathcal{B}_{p,q}}^+$ is a normal subgroup, let $\mathbf{e}_I \in \mathcal{B}_{p,q}$ be fixed and consider conjugation of elements of ${\mathcal{B}_{p,q}}^+$. That is, consider $\mathbf{e}_I{\mathcal{B}_{p,q}}^+{\mathbf{e}_I}^{-1}$. Choosing arbitrary $\mathbf{e}_J \in {\mathcal{B}_{p,q}}^+$, one finds

$$\begin{aligned} \mathbf{e}_I\,\mathbf{e}_J\,{\mathbf{e}_I}^{-1} &= \vartheta(I,I)\mathbf{e}_I\,\mathbf{e}_J\,\mathbf{e}_I = \vartheta(I,I)\mathbf{e}_I\vartheta(J,I)\mathbf{e}_{J\triangle I} \\ &= \vartheta(I,I)\vartheta(J,I)\vartheta(I,J\triangle I)\mathbf{e}_{I\triangle(J\triangle I)} \\ &= \vartheta(I,I)\vartheta(J,I)\vartheta(I,J\triangle I)\mathbf{e}_J \in {\mathcal{B}_{p,q}}^+. \end{aligned}$$

Hence, the result.

Solution 9.1: Let $u = \zeta_{\{1\}} + \zeta_{\{2\}}$, and let $v = \zeta_{\{1\}} + \zeta_{\{2\}} + \zeta_{\{3\}}$.

Solution 9.2: For $1 \leq k \leq n$,

$$\left(\sum_{\varnothing\neq J\in 2^{[n]}} \zeta_J\right)^k = \sum_{\substack{(J_1,\ldots,J_k) \\ \varnothing\neq J_i\in 2^{[n]},1\leq i\leq k}} \zeta_{J_1}\cdots\zeta_{J_k}.$$

By the commutative and nilpotent properties of the generators, this reduces to

$$\sum_{\substack{\{J_1,\ldots,J_k\}\subset 2^{[n]} \\ \{J_\ell\}\text{ pairwise disjoint}}} k!\,\zeta_{J_1}\cdots\zeta_{J_k}.$$

Taking the Berezin integral of this term further reduces to only those k-subsets of the power set whose union is $[n]$. Dividing by $k!$ cancels the summation over all permutations and yields the number of partitions of $[n]$ with k nonempty subsets.

Solution 9.3: By definition, $e^A = \sum_{k=0}^{\infty} \frac{A^k}{k!}$. Further, noting that the nilpotent property of the generators $\{\zeta_i\}$ implies $A^k = 0$ for all $k > n$,

$$\int e^A \, d\zeta_n \cdots d\zeta_1 = \int \left(\sum_{k=0}^{n} \frac{A^k}{k!} \right) d\zeta_n \cdots d\zeta_1 = \sum_{k=0}^{n} \int \left(\frac{A^k}{k!} \right) d\zeta_n \cdots d\zeta_1.$$

By Proposition 9.2.2,

$$\int \frac{A^k}{k!} \, d\zeta_n \cdots d\zeta_1 = \left\{ {n \atop k} \right\},$$

so that summing over $k = 1, 2, \ldots, n$ gives the total number of partitions of $[n]$, which is the nth Bell number.

Solution 9.4: $1 + 2\zeta_{\{1\}} - 5\zeta_{\{2\}}$.

Solution 9.5: $4 - 3\zeta_{\{1\}} + 2\zeta_{\{2\}} - 2\zeta_{\{1,2\}}$.

Solution 9.6: $4 - 2\zeta_{\{2\}} - 3\zeta_{\{3\}} - 2\zeta_{\{1,2\}} - \zeta_{\{1,3\}} - \zeta_{\{2,3\}} + 3\zeta_{\{1,2,3\}}$.

Solution 9.7: $4 + 5\zeta_{\{1\}} - 2\zeta_{\{2\}} + 3\zeta_{\{3\}} - 4\zeta_{\{1,2\}} + \zeta_{\{1,3\}} + 5\zeta_{\{2,3\}}$.

Solution 9.8: $3 - \zeta_{\{1\}} - 3\zeta_{\{2\}} + \zeta_{\{3\}} - \zeta_{\{1,2\}} + 4\zeta_{\{1,3\}} + 3\zeta_{\{2,3\}} - \zeta_{\{1,2,3\}}$.

Solution 9.9: $3 + 4\zeta_{\{1\}} - \zeta_{\{2\}} - 2\zeta_{\{3\}} - 2\zeta_{\{1,2\}} + \zeta_{\{1,3\}} + \zeta_{\{2,3\}}$.

Solution 9.10: $a^{1/4} + \frac{b\zeta_{\{1,2\}}}{4a^{3/4}} + \frac{c\zeta_{\{1,4\}}}{4a^{3/4}} + \frac{d\zeta_{\{2,3,4\}}}{4a^{3/4}}$.

Solution 10.1:

$$\varphi_1(u_1) = 75\zeta_{\{1,2\}} + 60\zeta_{\{2,3\}} + 280\zeta_{\{1,2,3\}} - 75\zeta_{\{3\}} - 10.$$

Solution 10.2:

$$\varphi_2(u_2) = 10\zeta_{\{1,2\}} - \zeta_{\{1,2,3\}} - \zeta_{\{1\}} + \zeta_{\{2\}}.$$

Solution 10.3:

$$\varphi_3(u_3) = -3200\zeta_{\{2,3\}} - 2400\zeta_{\{1,2,3\}} - 244.$$

Solution 10.4:

$$\varphi_4(u_4) = 15\zeta_{\{1,3\}} + 11\zeta_{\{2,3\}} - 20\zeta_{\{2,4\}} + 18\zeta_{\{1,2,3\}} + 10\zeta_{\{1,2,4\}} - 28\zeta_{\{2,3,4\}} \\ -35\zeta_{\{1,2,3,4\}} + 15\zeta_{\{2\}} + 5\zeta_{\{3\}}.$$

Solution 10.5:

$$\varphi_5(u_5) = -4\zeta_{\{1,4\}} - 4\zeta_{\{2,3\}} - 4\zeta_{\{2,4\}} + 4\zeta_{\{3,4\}} + 40\zeta_{\{1,2,3\}} + 40\zeta_{\{1,2,4\}} \\ -52\zeta_{\{1,3,4\}} - 20\zeta_{\{2,3,4\}} + 190\zeta_{\{1,2,3,4\}} + 16\zeta_{\{1\}}.$$

Solution 10.6: The zeon discriminant Δ is invertible and $\Re\Delta > 0$, so there are two solutions:

$$u_1 = -1 - 3\zeta_{\{2\}} + \frac{2\zeta_{\{3\}}}{3} - \frac{20}{3}\zeta_{\{1,2\}} + \frac{2}{3}\zeta_{\{1,3\}} - \frac{10}{3}\zeta_{\{2,3\}} - 27\zeta_{\{1,2,3\}}$$
$$u_2 = -\frac{1}{3}5\zeta_{\{1\}} + \frac{\zeta_{\{2\}}}{3} - \frac{2\zeta_{\{3\}}}{3} + \frac{23}{9}\zeta_{\{1,2\}} - 4\zeta_{\{1,3\}} + \frac{139}{9}\zeta_{\{1,2,3\}}.$$

Solution 10.7: The zeon discriminant is $\Delta = 64 - 132\zeta_{\{1\}} + 44\zeta_{\{2\}} - 100\zeta_{\{1,2\}}$, so there are two solutions:

$$u_1 = 2 + \frac{7\zeta_{\{1\}}}{8} + \frac{7\zeta_{\{2\}}}{8} - \frac{409}{256}\zeta_{\{1,2\}}$$
$$u_2 = \frac{2}{3} - \frac{1}{72}7\zeta_{\{1\}} + \frac{\zeta_{\{2\}}}{72} + \frac{35\zeta_{\{1,2\}}}{6912}.$$

Solution 10.9:

$$\phi_1(u) = 1 + \frac{3}{5}\zeta_{\{1,2\}} + u\left(-\frac{3}{25}\zeta_{\{1,2\}} - \frac{\zeta_{\{1\}}}{25} - \frac{1}{5}\right)$$
$$\phi_2(u) = -\frac{3}{5}\zeta_{\{1,2\}} + u\left(\frac{3}{25}\zeta_{\{1,2\}} + \frac{\zeta_{\{1\}}}{25} + \frac{1}{5}\right)$$

$$\gamma(u) = 6 - \frac{1}{5}11\zeta_{\{1,2\}} + u\left(\frac{11}{25}\zeta_{\{1,2\}} + \frac{7\zeta_{\{1\}}}{25} - \frac{3}{5}\right).$$

Solution 11.1: Suppose $\lambda = \eta = \lim_{k\to\infty} \xi_k$, apply the definition and show that for any $\varepsilon > 0$, there exists $K \in \mathbb{N}$ such that $k \geq K \Rightarrow \|\lambda - \eta\| < \varepsilon$.

Solution 11.3: Hint: Show that if (ξ_k) is a zeon Cauchy sequence, then for any multi-index I, the sequence $(\langle \xi_k, \zeta_I \rangle)$ of zeon inner products is a Cauchy sequence of real numbers.

Solution 11.4: Hint: Suppose $(\Re\xi_k)$ does not converge to $\Re\xi$ and establish the contrapositive.

Solution 11.7: Hint: Apply the bounded convergence theorem for real sequences to the inner product sequences $(\langle \xi_k, \zeta_I \rangle)$.

Solution 11.12: One example:

$$\psi_j = \begin{cases} j\zeta_{\{2\}} & j \equiv 0 \pmod 2, \\ j\zeta_{\{1\}} & j \equiv 1 \pmod 2 \end{cases}$$

$$\xi_j = \begin{cases} \zeta_{\{1\}} & j \equiv 0 \pmod 2, \\ \zeta_{\{2\}} & j \equiv 1 \pmod 2 \end{cases}$$

For each $j \in \mathbb{N}$, it follows that $\xi_j\psi_j = j\zeta_{\{1,2\}}$.

Solution 11.15:

i. $\|u\|_\infty = 3$
ii. $\|v\|_1 = 14$
iii. $\|w\|_2 = \sqrt{71}$
iv. $\|uv\|_\infty = 21$.

Solution 11.16:

$$\sum_{\ell=0}^{\infty} ru^\ell = 16 + 16\zeta_{\{1,2\}} - 44\zeta_{\{1,3\}} - 160\zeta_{\{2,3\}} - 264\zeta_{\{1,2,3\}} + 12\zeta_{\{1\}} + 64\zeta_{\{2\}} - 16\zeta_{\{3\}}.$$

Solution 11.18: For each $k \geq 1$, $\left(1 + \frac{1}{k}\zeta_{\{1\}}\right)^k = 1 + \zeta_{\{1\}}$. The constant sequence converges trivially.

Solution 11.19:

$$\sum_{k=0}^{\infty} \left(\frac{1}{2} + \zeta_{\{1\}}\right)^k = 2 + 4\zeta_{\{1\}}.$$

Solution 11.20:

$$\sum_{k=0}^{\infty} \left(\frac{1}{2} + \zeta_{\{1\}} + \zeta_{\{2\}}\right)^k = 2 + 4\zeta_{\{1\}} + 4\zeta_{\{2\}} + 16\zeta_{\{1,2\}}.$$

Solution 11.21:

$$\sum_{k=0}^{\infty}\left(\frac{1}{2}+\zeta_{\{1\}}+\zeta_{\{2\}}+\zeta_{\{3\}}\right)^k = 2+4\zeta_{\{1\}}+4\zeta_{\{2\}}+4\zeta_{\{3\}} + 16\zeta_{\{1,2\}}+16\zeta_{\{1,3\}}+16\zeta_{\{2,3\}} + 96\zeta_{\{1,2,3\}}.$$

Solution 11.22:

$$\sum_{k=0}^{\infty}\left(\frac{1}{2}+\zeta_{\{1\}}+\cdots+\zeta_{\{n\}}\right)^k = \sum_{j=0}^{n} j!2^{j+1}\sum_{|I|=j}\zeta_I.$$

Solution 12.1:

$$A^{-1}=\begin{pmatrix} \frac{1}{2}-\frac{\zeta_{\{2\}}}{4} & -\zeta_{\{2\}} \\ -\frac{1}{8}3\zeta_{\{2\}}-\zeta_{\{1,2\}}-\frac{1}{4} & \frac{3\zeta_{\{2\}}}{2}+\frac{1}{2} \end{pmatrix}$$

Solution 12.2:

$$A^{-1}=\begin{pmatrix} \frac{3}{16}-\frac{5\zeta_{\{2\}}}{256} & \frac{5}{16}-\frac{3\zeta_{\{2\}}}{256} \\ \frac{45\zeta_{\{2\}}}{256}+\frac{5}{16} & \frac{27\zeta_{\{2\}}}{256}+\frac{3}{16} \end{pmatrix}$$

Solution 12.3:

$$A^{-1}=\begin{pmatrix} \frac{2\zeta_{\{1\}}}{7}-\frac{2}{7} & \frac{\zeta_{\{1\}}}{7}+\frac{1}{7} & \frac{4}{7}-\frac{4\zeta_{\{1\}}}{7} \\ \frac{\zeta_{\{1\}}}{7}-\frac{3}{14} & \frac{\zeta_{\{1\}}}{7}-\frac{1}{7} & -\frac{1}{7}2\zeta_{\{1\}}-\frac{1}{14} \\ \frac{2\zeta_{\{1\}}}{7}-\frac{3}{7} & \frac{2\zeta_{\{1\}}}{7}-\frac{2}{7} & \frac{6}{7}-\frac{4\zeta_{\{1\}}}{7} \end{pmatrix}$$

Solution 12.4:

$$A^{-1}=\begin{pmatrix} -\frac{\zeta_{\{1\}}}{2}-\frac{5}{3}\zeta_{\{1,2,3\}}+\frac{1}{2} & -\frac{\zeta_{\{1\}}}{2}-\frac{25}{9}\zeta_{\{1,2,3\}}+\frac{5}{6} \\ -\frac{\zeta_{\{1\}}}{2}-\zeta_{\{1,2,3\}}+\frac{1}{2} & -\frac{\zeta_{\{1\}}}{2}-\frac{5}{3}\zeta_{\{1,2,3\}}+\frac{1}{2} \end{pmatrix}$$

Solution 12.5:

$$A_1{}^{\dagger}=\begin{pmatrix} 0 \\ -\zeta_{\{1\}}+2\zeta_{\{1,3\}}-\zeta_{\{2,3\}}-\zeta_{\{1,2,3\}} \\ \zeta_{\{1\}}+2\zeta_{\{1,3\}}+\zeta_{\{2,3\}}+2\zeta_{\{1,2,3\}} \end{pmatrix}$$

$$A_1{}^{\dagger}A_1=\left(3\zeta_{\{1\}}+2\zeta_{\{1,3\}}+3\zeta_{\{2,3\}}+17\zeta_{\{1,2,3\}}\right)$$

$$\|A_1\|_{zF}=\sqrt{\Re\left(\mathrm{tr}(A_1{}^{\dagger}A_1)^{\star}\right)}=\sqrt{17}.$$

$$A_2{}^\dagger = \begin{pmatrix} 2\zeta_{\{1\}} + 2\zeta_{\{3\}} + \zeta_{\{1,2\}} - \zeta_{\{1,3\}} & -\zeta_{\{1\}} - \zeta_{\{2,3\}} & 0 \end{pmatrix}$$

$$A_2{}^\dagger A_2 = \begin{pmatrix} -2\zeta_{\{1,2\}}+2\zeta_{\{1,3\}}-2\zeta_{\{2,3\}}+10\zeta_{\{1,2,3\}} & \zeta_{\{1,2\}}-\zeta_{\{1,3\}}-2\zeta_{\{1,2,3\}} & 0 \\ -2\zeta_{\{1,3\}}-2\zeta_{\{1,2,3\}} & 2\zeta_{\{1,2,3\}} & 0 \\ 0 & 0 & 0 \end{pmatrix}$$

$$\|A_2\|_{zF} = \sqrt{\Re\left(\mathrm{tr}(A_2{}^\dagger A_2)^\star\right)} = 2\sqrt{3}.$$

$$A_3{}^\dagger = \begin{pmatrix} 2\zeta_{\{1,3\}} - 1 & 2\zeta_{\{1,2,3\}} - \zeta_{\{3\}} \\ \zeta_{\{3\}} & -\zeta_{\{2\}} - \zeta_{\{1,2,3\}} + 2 \end{pmatrix}$$

$$A_3{}^\dagger A_3 = \begin{pmatrix} 6\zeta_{\{1,2,3\}}-2\zeta_{\{2\}} & 2\zeta_{\{1,2\}}-2\zeta_{\{2,3\}} \\ -\zeta_{\{3\}}+\zeta_{\{1,2\}}+4\zeta_{\{1,3\}}-2 & \zeta_{\{2\}}-2\zeta_{\{3\}}-2\zeta_{\{1,3\}}+11\zeta_{\{1,2,3\}}-2 \end{pmatrix}$$

$$\|A_3\|_{zF} = \sqrt{\Re\left(\mathrm{tr}(A_3{}^\dagger A_3)^\star\right)} = \sqrt{17}.$$

$$A_4{}^\dagger = \begin{pmatrix} \zeta_{\{1\}} - \zeta_{\{1,3\}} & 2\zeta_{\{1,2\}} + 2\zeta_{\{1,2,3\}} \end{pmatrix}$$

$$A_4{}^\dagger A_4 = \begin{pmatrix} 2\zeta_{\{1,2,3\}} - \zeta_{\{1,2\}} & 0 \\ 2\zeta_{\{1\}} & 4\zeta_{\{1,2\}} + 8\zeta_{\{1,2,3\}} \end{pmatrix}$$

$$\|A_4\|_{zF} = \sqrt{\Re\left(\mathrm{tr}(A_4{}^\dagger A_4)^\star\right)} = \sqrt{10}.$$

$$A_5{}^\dagger = \begin{pmatrix} \zeta_{\{1\}} & -\zeta_{\{1,2\}} \\ 2 & 2\zeta_{\{2\}} + \zeta_{\{3\}} + 2\zeta_{\{1,2\}} + 2\zeta_{\{1,3\}} \end{pmatrix}$$

$$A_5{}^\dagger A_5 = \begin{pmatrix} 5\zeta_{\{1,2,3\}} & 0 \\ 4\zeta_{\{2\}} + 4\zeta_{\{3\}} + 2\zeta_{\{1,2\}} + 3\zeta_{\{1,3\}} & 6\zeta_{\{2,3\}} + 14\zeta_{\{1,2,3\}} \end{pmatrix}$$

$$\|A_5\|_{zF} = \sqrt{\Re\left(\mathrm{tr}(A_5{}^\dagger A_5)^\star\right)} = \sqrt{19}.$$

Solution 12.6: A counterexample: Let

$$A = \begin{pmatrix} \zeta_{\{1\}} + \zeta_{\{2\}} + \zeta_{\{3\}} & \zeta_{\{1\}} + \zeta_{\{2\}} + \zeta_{\{3\}} \\ \zeta_{\{1\}} + \zeta_{\{2\}} + \zeta_{\{3\}} & \zeta_{\{1\}} + \zeta_{\{2\}} + \zeta_{\{3\}} \end{pmatrix},$$

and let $B = A$. Then, $\|A\|_{\mathrm{zF}} = \|B\|_{\mathrm{zF}} = 2\sqrt{3}$ so that

$$\|AB\|_{\mathrm{zF}} = \|A^2\|_{\mathrm{zF}} = 8\sqrt{3} < 12 = \|A\|_{\mathrm{zF}}\|B\|_{\mathrm{zF}}.$$

Solution 12.8: Applying the inequality

$$\|AB\|_{\mathrm{zF}} \leq 2^{n/2} c_1 \sqrt{r_1 c_2} \|A\|_{\mathrm{zF}} \|B\|_{\mathrm{zF}},$$

it follows that $\|AB\|_{\mathrm{zF}} \leq 2^{5/2} 2\sqrt{3} \|A\|_{\mathrm{zF}} \|B\|_{\mathrm{zF}}$. Hence, $k = 8\sqrt{6}$.

Solution 12.9: Applying the inequality

$$\|AB\|_{\mathrm{zF}} \leq 2^{n/2} c_1 \sqrt{r_1 c_2} \|A\|_{\mathrm{zF}} \|B\|_{\mathrm{zF}},$$

it follows that $\|AB\|_{\mathrm{zF}} \leq 2^{n/2} m^2 \|A\|_{\mathrm{zF}} \|B\|_{\mathrm{zF}}$.

Solution 13.1:

(a)

$$\begin{aligned} q &= \frac{1}{3} - \frac{1}{9} 11\zeta_{\{1\}} + \frac{7\zeta_{\{3\}}}{9} - \frac{4\zeta_{\{2\}}}{9} + \frac{46}{27}\zeta_{\{1,2\}} - \frac{82}{27}\zeta_{\{1,3\}} \\ &\quad - \frac{32}{27}\zeta_{\{2,3\}} + \frac{139}{27}\zeta_{\{1,2,3\}} \\ r &= 1 - 3\zeta_{\{1,2\}} + 7\zeta_{\{1,3\}} + 4\zeta_{\{2,3\}} - 6\zeta_{\{1,2,3\}}. \end{aligned}$$

(b)

$$\begin{aligned} q &= \frac{3}{2} - \frac{1}{2} 15\zeta_{\{1\}} - 3\zeta_{\{3\}} - \frac{\zeta_{\{2\}}}{2} + 10\zeta_{\{1,2\}} + 12\zeta_{\{1,3\}} \\ &\quad + \frac{5}{2}\zeta_{\{2,3\}} - \frac{59}{2}\zeta_{\{1,2,3\}} \\ r &= \frac{3}{2} - 3\zeta_{\{1,3\}} + \zeta_{\{2,3\}} - \zeta_{\{1,2,3\}}. \end{aligned}$$

(c)

$$\begin{aligned} q &= \frac{1}{2} - \frac{\zeta_{\{1\}}}{6} + \frac{\zeta_{\{2\}}}{3} + \zeta_{\{3\}} + \frac{1}{18}\zeta_{\{1,2\}} - \frac{1}{6}\zeta_{\{1,3\}} \\ &\quad + \frac{1}{9}\zeta_{\{2,3\}} + \frac{31}{54}\zeta_{\{1,2,3\}} \\ r &= \frac{3}{2} + \frac{5}{3}\zeta_{\{1,2\}} - \frac{1}{3}\zeta_{\{1,3\}} - \frac{17}{3}\zeta_{\{2,3\}} + \frac{41}{9}\zeta_{\{1,2,3\}}. \end{aligned}$$

Solution 13.2: The factorization is $u = \prod_{j=1}^{6} w_j$, where $w_1 = 1 + 10\zeta_{\{1\}}$, $w_2 = 1 + -2\zeta_{\{2\}}$, $w_3 = 1 + 71\zeta_{\{1,2\}}$, $w_4 = 1 + 21\zeta_{\{1,3\}}$, $w_5 = 1 - 3\zeta_{\{2,3\}}$, and $w_6 = 1 + 92\zeta_{\{1,2,3\}}$.

Solution 13.5:

i. $\sin u = \sin(2) - 3\sec^2(2)\zeta_{\{1\}} + 3\sec^2(2)\zeta_{\{3\}} - 3\cos(2)\zeta_{\{1\}} + 3\cos(2)\zeta_{\{3\}} + 9\sin(2)\zeta_{\{1,3\}} - 2\cos(2)\zeta_{\{1,2\}} + 6\sin(2)\zeta_{\{1,2,3\}}$,

ii. $\cos u = \cos(2) + 3\sin(2)\zeta_{\{1\}} - 3\sin(2)\zeta_{\{3\}} + 2\sin(2)\zeta_{\{1,2\}} + 9\cos(2)\zeta_{\{1,3\}} + 6\cos(2)\zeta_{\{1,2,3\}}$,

iii. $\tan u = \tan(2) - 2\sec^2(2)\zeta_{\{1,2\}} - 18\tan(2)\sec^2(2)\zeta_{\{1,3\}} - 12\tan(2)\sec^2(2)\zeta_{\{1,2,3\}}$.

Solution 13.8: Setting $u = z + \sqrt{z^2 - 1}$, one verifies that

$$\begin{aligned}\cosh\left(z + \sqrt{z^2-1}\right) &= \frac{\exp(u) + \exp(-u)}{2} \\ &= \frac{1}{2}\left(z + \sqrt{z^2-1} + \left(z + \sqrt{z^2-1}\right)^{-1}\right) \\ &= \frac{1}{2}\left(z + \sqrt{z^2-1} + z - \sqrt{z^2+1}\right) \\ &= z\end{aligned}$$

so that $\cosh^{-1} z = \log(z + \sqrt{z^2-1})$.

Solution 13.9: Setting $u = \frac{1}{2}\log\left(\frac{1+z}{1-z}\right)$, it follows that

$$\begin{aligned}\tanh\left(\frac{1}{2}\log\left(\frac{1+z}{1-z}\right)\right) &= \frac{\exp(u) - \exp(-u)}{\exp(u) + \exp(-u)} \\ &= \frac{\left(\frac{1+z}{1-z}\right)^{1/2} - \left(\frac{1-z}{1+z}\right)^{1/2}}{\left(\frac{1+z}{1-z}\right)^{1/2} + \left(\frac{1-z}{1+z}\right)^{1/2}} \\ &= \frac{\left(\frac{1+z}{1-z}\right) + \left(\frac{1-z}{1+z}\right) - 2}{\left(\frac{1+z}{1-z}\right) - \left(\frac{1-z}{1+z}\right)}\end{aligned}$$

$$\begin{aligned}&= \frac{(1+z)^2+(1-z)^2-2(1-z^2)}{(1+z)^2-(1-z)^2}\\&= \frac{4z^2}{4z}\\&= z.\end{aligned}$$

Solution 14.3: Since $\varphi'(u) = 3u^2$, the critical points of φ are elements $u \in \mathcal{C}\ell_n{}^{\text{nil}}$ such that $3u^2 = 0$. Hence, any element that is nilpotent of index 2 is a solution.

Solution 14.4: The proof follows naturally from the Proposition 14.3.4 and Lemma 14.3.6; i.e.,

$$\psi'(t) = \varphi'(t)\frac{1}{\gamma(t)} + \varphi(t)\left(\frac{1}{\gamma(t)}\right)' = \frac{\gamma(t)\varphi'(t) - \gamma'(t)\varphi(t)}{(\gamma(t))^2}.$$

Solution 14.5:

$$\begin{aligned}(\varphi \circ g)'(c) &= \sum_{I \in 2^{[n]}} (f_I \circ g)'(c)\zeta_I\\&= \sum_{I \in 2^{[n]}} f_I{}'(g(c))g'(c)\zeta_I\\&= g'(c) \sum_{I \in 2^{[n]}} f_I{}'(g(c))\zeta_I\\&= g'(c)\varphi'(g(c)).\end{aligned}$$

Solution 15.1: Begin by observing that for path $\mathbf{u} = u_1 \cdots u_k$ on vertices $U \subset V$, $\omega_{\mathbf{u}}$ is given by the following:

$$\omega_{\mathbf{u}} = \zeta_U x_{u_1} x_{u_2} \cdots x_{u_k}.$$

Solution 15.2: Show that $\dim(\Omega_n) = \sum_{j=0}^{n} j!$.

Solution 15.4: The element ξ represents an enumeration of all paths of length 8 from vertex v_1 to vertex v_5 requiring 3 steps in the first frame, 1 step in the second frame, and 4 steps in the fourth frame.

Solution 17.1:

$$\begin{aligned}(\overline{a \wedge b}) \rightarrow c &\equiv (a \wedge b) \vee c\\&\equiv (a \vee c) \wedge (b \vee c).\end{aligned}$$

Solution 17.4: Note that the result is trivial for $|\mathfrak{c}| = 1$. Assuming the result holds for $|\mathfrak{c}| = k-1$, extend to a clause $a \vee \mathfrak{c}$ containing k literals:

$$\begin{aligned}\Psi((a \vee \mathfrak{c}) \wedge (a \vee \mathfrak{c})) &= \Psi(a \vee \mathfrak{c})\Psi(a \vee \mathfrak{c}) \\ &= (\Psi(a) + \Psi(\mathfrak{c}) - \Psi(a)\Psi(\mathfrak{c}))^2 \\ &= \Psi(a)^2 + \Psi(\mathfrak{c})^2 - \Psi(a)^2\Psi(\mathfrak{c})^2 \\ &\quad +2\left(\Psi(a)\Psi(\mathfrak{c}) - \Psi(a)\Psi(\mathfrak{c}) - \Psi(a)\Psi(\mathfrak{c})\right) \\ &= \Psi(a) + \Psi(\mathfrak{c}) - \Psi(a)\Psi(\mathfrak{c}) \\ &= \Psi(a \vee \mathfrak{c}).\end{aligned}$$

The result now holds by induction on the number of literals appearing in $\mathfrak{c}$.

Solution 17.5: The equivalence $\overline{x_i} \vee \overline{x_j} \equiv \overline{x_i \wedge x_j}$ is preserved by

$$\begin{aligned}\Psi(\overline{x_i} \vee \overline{x_j}) &= \widetilde{\varepsilon_{\{i\}}} + \widetilde{\varepsilon_{\{j\}}} - \widetilde{\varepsilon_{\{i,j\}}} \\ &= (1 - \varepsilon_{\{i\}}) + (1 - \varepsilon_{\{j\}}) - (1 - \varepsilon_{\{i\}})(1 - \varepsilon_{\{j\}}) \\ &= 1 - \varepsilon_{\{i\}} + 1 - \varepsilon_{\{j\}} - \left(1 - \varepsilon_{\{i\}} - \varepsilon_{\{j\}} + \varepsilon_{\{i,j\}}\right) \\ &= 1 - \varepsilon_{\{i,j\}} \\ &= \Psi(\overline{x_i \wedge x_j}).\end{aligned}$$

The equivalence $\overline{x_i \vee x_j} \equiv \overline{x_i} \wedge \overline{x_j}$ is preserved by

$$\begin{aligned}\overline{x_i \vee x_j} &\mapsto 1 - (\varepsilon_{\{i\}} + \varepsilon_{\{j\}} - \varepsilon_{\{i,j\}}) \\ \overline{x_i} \wedge \overline{x_j} &\mapsto \widetilde{\varepsilon_{\{i\}}}\widetilde{\varepsilon_{\{j\}}}.\end{aligned}$$

It follows naturally that

$$\begin{aligned}\Psi(\overline{x_i \vee x_j}) &= 1 - (\varepsilon_{\{i\}} + \varepsilon_{\{j\}} - \varepsilon_{\{i,j\}}) \\ &= 1 - \varepsilon_{\{i\}} - \varepsilon_{\{j\}} + \varepsilon_{\{i,j\}} \\ &= (1 - \varepsilon_{\{i\}})(1 - \varepsilon_{\{j\}}) \\ &= \widetilde{\varepsilon_{\{i\}}}\widetilde{\varepsilon_{\{j\}}} \\ &= \widetilde{\varepsilon_{\{i,j\}}} \\ &= \Psi(\overline{x_i} \wedge \overline{x_j}).\end{aligned}$$

Solution 17.9: Every $\mathfrak{J}$-class of $\mathcal{J}_n$ is classified, and then the regular classes are identified. From each regular $\mathfrak{J}$-class one idempotent element is chosen and the maximal subgroup at $\varepsilon_\varnothing$ is computed.

Each element is in its own $\mathfrak{J}$-class with no equivalent idempotent elements, giving $|\mathcal{J}_n| = 2^n$ unique idempotents. The maximal subgroups are found to be $G_{\varepsilon_\varnothing} = \{\varepsilon_\varnothing\}$ and $G_{\varepsilon_I} = \{\varepsilon_I\}$ for arbitrary non-trivial idempotent ε_I.

Enumerating the idempotent elements $\{f_1, \dots, f_{2^n}\}$ and letting k_i be the number of conjugacy classes in G_{f_i}, the number of irreducible representations is thus

$$\sum_{i=1}^{2^n} k_i = \sum_{i=1}^{2^n} 1 = 2^n.$$

Solution 17.10: Let $G = (V, E)$ by the graph of a k-SAT formula $\mathcal{F}$ having m clauses and n literals. To show

$$\Phi_{G'}{}^k = k! \sum_{\substack{\{I \subseteq V, |I|=m\} \\ I \text{ independent in} G'}} \phi(I)\zeta_{\mathcal{N}(I)},$$

consider the multinomial expansion

$$\begin{aligned} \Phi_{G'}{}^m &= \left(\sum_{v\in V} \psi(v) = \sum_{v\in V} \lambda(v)\zeta_{\mathcal{N}(v)}\right)^m \\ &= \sum_{t_1+\cdots+t_n=m} \binom{m}{t_1, t_2, \dots, t_n} \prod_{i=1}^n (\lambda(v_i))^{t_i} \zeta_{\mathcal{N}(v_i)}{}^{t_i}. \end{aligned}$$

By the null-square property of zeon generators, the only surviving terms of the sum correspond to n-tuples $(t_1, \dots, t_n) \in \{0, 1\}^n$. Further, any product

$$\zeta_{\mathcal{N}(v_i)}\zeta_{\mathcal{N}(v_j)} = \begin{cases} \zeta_{\mathcal{N}(\{v_i, v_j\})} & \{v_i, v_j\} \notin E, \\ 0 & \{v_i, v_j\} \in E. \end{cases}$$

Hence, the only nonzero terms of the product $\prod_{i=1}^n (\lambda(v_i))^{t_i} \zeta_{\mathcal{N}(v_i)}{}^{t_i}$ are of the form

$$\prod_{i=1}^n (\lambda(v_i))^{t_i} \zeta_{\mathcal{N}(v_i)}{}^{t_i} = \prod_{i\in I} (\lambda(v_i))\zeta_{\mathcal{N}(I)}$$

where I is an independent set of m vertices in G. Setting $\phi(I) = \prod_{i\in I} \lambda(v_i)$ and recalling the idempotent properties of the labeling λ completes the proof.

Solution 18.1: Let $\mathbf{x}, \mathbf{y} \in V$ be arbitrary, and decompose $\mathbf{x}, \mathbf{y}$ into components lying in $\mathcal{E}_{\lambda_1}$ and $\mathcal{E}_{\lambda_2}$. Then

$$\begin{aligned}
\langle \mathbf{x}, \Phi_{\mathrm{u}}(\mathbf{y})\rangle_Q &= \left\langle (\mathbf{x}_1 + \mathbf{x}_2)\widetilde{\Phi_{\mathrm{u}}(\mathbf{y})}\right\rangle_0 \\
&= \left\langle (\mathbf{x}_1 + \mathbf{x}_2)\left((-1)^{\sharp \mathrm{u}-1}\mathbf{y}_1 + (-1)^{\sharp \mathrm{u}}\mathbf{y}_2\right)\right\rangle_0 \\
&= \left\langle \mathbf{x}_1((-1)^{\sharp \mathrm{u}-1}\mathbf{y}_1 + (-1)^{\sharp \mathrm{u}}\mathbf{y}_2) + \mathbf{x}_2((-1)^{\sharp \mathrm{u}-1}\mathbf{y}_1 \right. \\
&\quad \left. + (-1)^{\sharp \mathrm{u}}\mathbf{y}_2)\right\rangle_0 \\
&= \left\langle \left((-1)^{\sharp \mathrm{u}-1}\mathbf{x}_1 + (-1)^{\sharp \mathrm{u}}\mathbf{x}_2\right)(\mathbf{y}_1 + \mathbf{y}_2)\right\rangle_0 \\
&= \left\langle \mathrm{u}(\mathbf{x}_1 + \mathbf{x}_2)\mathrm{u}^{-1}(\mathbf{y}_1 + \mathbf{y}_2)\right\rangle_0 \\
&= \langle \Phi_{\mathrm{u}}(\mathbf{x}), \mathbf{y}\rangle_Q. \qquad (20.1)
\end{aligned}$$

Solution 19.1: Let $\mathbf{v}_0 = (1, \ldots, 1)$ and observe that $L(\mathbf{v}_0) = \mathbf{0}$.

Solution 19.5: Let n be any odd integer, let $\kappa = 2\binom{n}{(n-1)/2}$, and let A be the adjacency matrix of the middle-levels subgraph $\mathcal{M}_n$ of the hypercube $\mathcal{Q}_n$. Suppose $A \twoheadrightarrow \Phi \in \mathcal{L}(\mathfrak{F}_\kappa)$ and $A \twoheadrightarrow \Psi \in \mathcal{L}(\mathfrak{Z}_\kappa)$. Then,

$$\mathrm{tr}(\sigma\Phi \odot \Psi^\star) \neq 0.$$

Bibliography

[1] R. Abłamowicz, G. Sobczyk, Eds., *Lectures on Clifford (Geometric) Algebras and Applications*, Birkhäuser, Boston, 2003.

[2] M. Albert, A. Frieze, B. Reed. Multicolored Hamilton cycles, *Electronic J. Combin.* **2** (1995), #R10.

[3] R. Abłamowicz, Computation of Non-Comutative Gröbner Bases in Grassmann and Clifford Algebras, *Advances in Applied Clifford Algebras*, **20** (2010), 447-476.

[4] N.M. Atakishiyev, K.B. Wolf, Fractional Fourier-Kravchuk transform, *J. Opt. Soc. Amer. A*, **147** (1997), 1467-1477.

[5] N.I. Akhiezer, I.M. Glazman, *Theory of Linear Operators in Hilbert Space*, Dover, New York, 1993.

[6] G. Aragón-González, J. L. Aragón, M. A. Rodríguez-Andrade, L. Verde-Star, Reflections, Rotations, and Pythagorean Numbers, *Advances in Applied Clifford Algebras*, **19** (2008), 1-14.

[7] J. Babu, S. Chandran, D. Rajendraprasad. Heterochromatic paths in edge colored graphs without small cycles and heterochromatic-triangle-free graphs, *European Journal of Combinatorics*, **48** (2015), 110-126. `http://dx.doi.org/10.1016/j.ejc.2015.02.014`

[8] T. Batard, M. Berthier, C. Saint-Jean, Clifford-Fourier Transform for Color Image Processing, in *Geometric Algebra Computing*, E. Bayro-Corrochano, G. Scheuermann, Eds., Springer London, pp. 135-162, 2010. `http://dx.doi.org/10.1007/978-1-84996-108-0_8`

[9] W.E. Bayliss, *Geometric Algebra Workbook*, `http://web4.uwindsor.ca/users/b/baylis/main.nsf` [Accessed April 3, 2019.]

[10] J. Ben Slimane, R. Schott, Y-Q. Song, G.S. Staples, E. Tsiontsiou, Operator Calculus Algorithms for Multi-Constrained Paths, *Int. J. Math. Comput. Sci.* **10** (2015), 69-104. `http://ijmcs.future-in-tech.net/10.1/R-Jamila.pdf`

[11] F.A. Berezin, *Introduction to Superanalysis*, D. Reidel Publishing Co., Dordrecht, 1987.

[12] A. Best, M. Kliegl, S. Mead-Gluchacki, C. Tamon, Mixing of quantum walks on generalized hypercubes, *International Journal of Quantum Information*, **6** (2008), 1135-1148.

[13] A. Biere, A. Biere, M. Heule, H. van Maaren, T. Walsh. Handbook of Satisfiability: Volume 185 Frontiers in Artificial Intelligence and Applications, IOS Press Amsterdam, The Netherlands, The Netherlands, 2009. ISBN:9781586039295.

[14] G. Boole, *An Investigation of the Laws of Thought on Which are Founded the Mathematical Theories of Logic and Probabilities*, McMillan, 1854, Reprinted with corrections, Dover Publications, New York, NY, 1958. (Reissued by Cambridge University Press, 2009; ISBN 978-1-108-00153-3.)

[15] H.J. Broersma, X.L. Li, G. Woeginger, S.G. Zhang, Paths and cycles in colored graphs, *Australasian J. Combin.*, **31** (2005), 297-309.

[16] M. Budinich, P. Budinich, A spinorial formulation of the maximum clique problem of a graph, *J. Math. Phys.*, **47** (2006), 043502.

[17] M. Budinich. The Boolean satisfiability problem in Clifford algebra, *Theoretical Computer Science* (2019). In press. `https://doi.org/10.1016/j.tcs.2019.03.027`.

[18] E. Cartan, *The Theory of Spinors*, Hermann, Paris, 1966.

[19] C. Cassiday, G.S. Staples, On representations of semigroups having hypercube-like Cayley graphs, *Clifford Analysis, Clifford Algebras and Their Applications*, **4** (2015), 111-130.

[20] S. Chaiken, A combinatorial proof of the all minors matrix tree theorem, *SIAM J. Alg. Disc. Meth.*, **3** (1982), 319-329.

[21] H. Chen, X. Li. Long heterochromatic paths in edge-colored graphs, *Electronic Journal of Combinatorics*, **12** (2005), #R33.

[22] C. Chevalley, *The Construction and Study of Certain Important Algebras*, Math. Soc. of Japan, Tokyo, 1955.

[23] A.H. Clifford, G.B. Preston, *The algebraic theory of semigroups, Vol. I*, Mathematical Surveys, No. 7. American Mathematical Society, Providence, R.I., 1961.

[24] Coddens, G., Spinor approach to the rotation and reflection groups, *Eur. J. Phys.*, **23** (2002), 549-564.

[25] D. Cohen, *Basic Techniques of Combinatorial Theory*, Wiley, 1978.

[26] S.A. Cook. The complexity of theorem-proving procedures, *Proceedings of the 3rd Annual ACM Symposium on Theory of Computing*, 151-158. `doi:10.1145/800157.805047`

[27] H. Cruz-Sánchez, G.S. Staples, R. Schott, Y-Q. Song, Operator calculus approach to minimal paths: Precomputed routing in a store-and-forward satellite constellation, *Proceedings of IEEE Globecom 2012, Anaheim, USA, December 3-7*, 3438-3443.

[28] A. Davis, G.S. Staples. Zeon and idem-Clifford formulations of Boolean satisfiability, *Adv. Appl. Clifford Alg.* (2019) 29:60. `http://dx.doi.org/10.1007/s00006-019-0978-8`.

[29] P. Delsarte, Bounds for restricted codes, by linear programming, *Philips Res. Reports*, **27** (1972), 272-289.

[30] P. Delsarte, Four fundamental parameters of a code and their combinatorial significance, *Info. & Control*, **23** (1973), 407-438.

[31] P. Delsarte, An algebraic approach to the association schemes of coding

theory, *Philips Research Reports Supplements*, No. 10, N.V. Philips' Gloeilampenfabrieken, Eindhoven, Netherlands, 1973.

[32] Dollar, L.M. & Staples, G.S. Zeon Roots, *Adv. Appl. Clifford Algebras*, **27** (2017), 1133-1145. `http://dx.doi.org/10.1007/s00006-016-0732-4`

[33] L. Dorst, D. Fontijne, *Efficient Algorithms for Factorization and Join of Blades*, in Geometric Algebra Computing, E. Bayro-Corrochano, G. Scheuermann, Eds., Springer, London, 2010, pp. 457-476.

[34] C.F. Dunkl, A Krawtchouk polynomial addition theorem and wreath products of symmetric groups, *Indiana Univ. Math. J.*, **25** (1976), 335-358.

[35] C.F. Dunkl, D.F. Ramirez, Krawtchouk polynomials and the symmetrization of hypergraphs, *SIAM J. Math. Anal.*, **5** (1974), 351-366.

[36] P. Erdös, Zs. Tuza. Rainbow Hamiltonian paths and canonically colored subgraphs in infinite complete graphs, *Mathematica Pannonica*, **1** (1990), 5-13.

[37] P. Erdös, Zs. Tuza. Rainbow subgraphs in edge-colorings of complete graphs, *Ann. Discrete Math.*, **55** (1993), 81-88.

[38] P. Feinsilver, R. Fitzgerald, The spectrum of symmetric Kravchuk matrices, *Linear Algebra and Its Applications*, **235** (1996), 121-139.

[39] P. Feinsilver, U. Franz, R. Schott, Duality and multiplicative processes on quantum groups, *J. Th. Prob.*, **10** (1997), 795-818.

[40] P. Feinsilver, J. Kocik, Krawtchouk polynomials and Krawtchouk matrices, *Recent Advances in Applied Probability*, R. Baeza-Yates, J. Glaz, H. Gzyl, J. Hüsler, J.L. Palacios, (Eds.), Springer, 2005, pp. 115-141.

[41] P. Feinsilver, J. Kocik, Krawtchouk matrices from classical and quantum random walks, *Contemporary Mathematics*, **287** (2001), 83-96.

[42] P. Feinsilver, R. Schott, On Krawtchouk transforms, *Intelligent Computer Mathematics*, Proc. 10th Intl. Conf. AISC 2010, LNAI 6167, 64-75.

[43] P. Feinsilver. Zeon algebra, Fock space, and Markov chains, *Commun. Stoch. Anal.*, **2** 263-275, (2008).

[44] P. Flajolet, R. Schott, Non-overlapping partitions, continued fractions, Bessel functions, and a divergent series, *Europ. J. Combinatorics*, **11** (1990), 421-432.

[45] D. Fontijne, *Efficient Implementation of Geometric Algebra*, Ph.D. thesis, University of Amsterdam, 2007.

[46] J.B. Fraleigh, *A First Course in Abstract Algebra, 7th Ed.*, Addison Wesley, Boston, 2003.

[47] S.H. Friedberg, A.J. Insel, L.E. Spence, *Linear Algebra 4th Ed.*, Prentice-Hall, Upper Saddle River, 2003.

[48] A. Friedman, *Foundations of Modern Analysis*, Dover, New York, 1982.

[49] A.M. Frieze, B.A. Reed. Polychromatic Hamilton cycles, *Discrete Math.*, **118** (1993), 69-74.

[50] A. M. Frydryszak. Nilpotent quantum mechanics, qubits, and flavors of entanglement, arXiv:0810.3016v1 [quant-ph].

[51] I.P. Goulden, D.M. Jackson. The enumeration of directed closed Euler trails and directed Hamiltonian circuits by Lagrangian methods, *Europ. J. Combinatorics*, **2** (1981), 131-135.

[52] S. Gudder, *Quantum Probability*, Academic Press, Boston, 1988.
[53] A. Gyáfás. Vertex coverings by monochromatic paths and cycles, *Journal of Graph Theory*, **7** (1983), 131-135.
[54] E. Haake, G.S. Staples, Zeros of zeon polynomials and the zeon quadratic formula, *Advances in Applied Clifford Algebras*, (2019) 29:21.
[55] G. Harris, G.S. Staples, Spinorial formulations of graph problems, *Advances in Applied Clifford Algebras*, **22** (2012), 59-77.
[56] Y. Hashimoto, A. Hora, N. Obata. *Central limit theorems for large graphs: Method of quantum decomposition*, J. Math. Phys. **44** (2003), 71-88.
[57] J. Helmstetter, Factorization of Lipschitzian elements, *Advances in Applied Clifford Algebras*, **24** (2014), 675-712.
[58] Z. Izhakian, J. Rhodes, B. Steinberg, Representation Theory of Finite Semigroups Over Semirings, *J. Alg.*, **336** 139-157, (2011).
[59] R.M. Karp. Reducibility among combinatorial problems, in *Complexity of Computer Computations: Proc. of a Symp. on the Complexity of Computer Computations*, R.E. Miller and J.W. Thatcher, Eds., The IBM Research Symposia Series, New York, NY: Plenum Press, 1972, pp. 85-103.
[60] G. Kirchhoff. Über die Auflösung der Gleichungen, auf welche man bei der untersuchung der linearen verteilung galvanischer Ströme geführt wird, *Ann. Phys. Chem.*, **72** (1847), 497-508.
[61] T. Koornwinder, Krawtchouk polynomials, a unification of two different group theoretic interpretations, *SIAM J. Math. Anal.*, **13** (1982), 1011-1023.
[62] V.I. Levenstein, Krawtchouk polynomials and universal bounds for codes and design in Hamming spaces, *IEEE Transactions on Information Theory*, **41** (1995), 1303-1321.
[63] K. Li, Rapidly Mixing Random Walks on Hypercubes with Application to Dynamic Tree Evolution, *Parallel and Distributed Processing Symposium, 2005. Proceedings. 19th IEEE International.* 4-8 April, 2005. `doi:10.1109/IPDPS.2005.372`
[64] X. Li, S. Zhang, H. Broersma, Paths and cycles in colored graphs, *Electronic Notes in Discrete Mathematics*, **8** (2001), 128-132.
[65] T. Lindell, G.S. Staples, Norm inequalities in zeon algebras, *Advances in Applied Clifford Algebras*, (2019) 29:13.
[66] C.J. Liu. Enumeration of Hamiltonian cycles and paths in a graph, *Proc. Amer. Math. Soc.*, **111** (1991), 289-296.
[67] P. Lounesto, *Clifford Algebras and Spinors*, Cambridge University Press, Cambridge, 2001.
[68] P. Lounesto, E. Latvamaa, Conformal Transformations and Clifford Algebras, *Proc. Am. Math. Soc.*, **79** (1980), 533-538.
[69] F. J. MacWilliams, N. J. A. Sloane, *Theory of Error-Correcting Codes*, North-Holland, 1977.
[70] P.A. Meyer, *Quantum Probability for Probabilists*, Lecture Notes in Mathematics 1538, Springer-Verlag, Berlin,1995.
[71] C. Moore, A. Russell, Quantum walks on the hypercube, *Randomization and Approximation Techniques in Computer Science*, Lecture Notes in

Computer Science **2483**, Springer, 2002.
[72] T. Mü, Proof of the middle levels conjecture, *Proceedings of the London Mathematical Society*, **112** (2016), 677-713.
[73] B. Nefzi, R. Schott, Y.Q. Song, G.S. Staples, E. Tsionsiou. An operator calculus approach for multi-constrained routing in wireless sensor networks. *Proceedings of ACM MobiHoc, Hangzhou, CHINA, June 22-25, 2015*, ACM New York, NY, USA, 2015, pp. 367-376.
[74] A.F. Neto. Higher order derivatives of trigonometric functions, Stirling numbers of the second kind, and zeon algebra, *Journal of Integer Sequences*, **17** (2014), Article 14.9.3.
[75] A.F. Neto. Carlitz's identity for the Bernoulli numbers and zeon algebra, *J. Integer Sequences*, **18** (2015), Article 15.5.6.
[76] A. F. Neto. P.H.R. dos Anjos, Zeon algebra and combinatorial identities, *SIAM Review*, **56** (2014), 353-370.
[77] K.R. Parthasarathy, *An Introduction to Quantum Stochastic Calculus*, Birkhäuser Verlag, Basel, 1992.
[78] C. Perwass, *Geometric Algebra with Applications in Engineering*, Springer-Verlag, Berlin, 2009.
[79] I. Porteous, Lecture 2: Mathematical structure of Clifford algebras. *Lectures on Clifford (Geometric) Algebras and Applications*, R. Abłamowicz, G. Sobczyk, Eds., Birkhäuser, Boston, 2003.
[80] I. Porteous, *Clifford Algebras and the Classical Groups*, Cambridge Studies in Advanced Mathematics: 50, Cambridge University Press, Cambridge, 1995.
[81] H. Raynaud. Sur le circuit hamiltonien hi-colore dans les graphes orientes. *Periodica Math. Hung.*, **3** (1973), 289-297.
[82] J. Rhodes, Y. Zalcstein, Elementary Representation and Character Theory of Finite Semigroups and Its Application, *Monoids and semigroups with applications (Berkeley, CA, 1989)*, pp. 334-367. World Sci. Publ., River Edge, NJ, 1991.
[83] G.-C. Rota, T. Wallstrom, Stochastic integrals: a combinatorial approach, *Ann. Prob.* **25**, 1257-1283, (1997).
[84] J.J. Rotman, *A First Course in Abstract Algebra with Applications*, Prentice Hall, Upper Saddle River, 2006.
[85] R. Schott, G.S. Staples, Nilpotent adjacency matrices, random graphs, and quantum random variables, *J. Phys. A: Math. Theor.*, **41** 155205, (2008).
[86] R. Schott, G.S. Staples. Random walks in Clifford algebras of arbitrary signature as walks on directed hypercubes, *Markov Processes and Related Fields*, **14** (2008), 515-542.
[87] R. Schott, G.S. Staples. Zeons, lattices of partitions, and free probability, *Comm. Stoch. Anal.*, **4** (2010), 311-334.
[88] R. Schott, G.S. Staples. Nilpotent adjacency matrices and random graphs, *Ars Combinatoria*, **98** (2011), 225-239.
[89] R. Schott, G.S. Staples. Connected components and evolution of random graphs: an algebraic approach, *J. Alg. Comb.*, **35** (2012), 141-156. `http://dx.doi.org/10.1007/s10801-011-0297-1`

[90] R. Schott, G.S. Staples. Partitions and Clifford algebras, *Eur. J. Comb.* **29** (2008) 1133-1138.

[91] R. Schott, G.S. Staples, *Operator Calculus on Graphs (Theory and Applications in Computer Science)*, Imperial College Press, London, 2012. ISBN 978-1-84816-876-3.

[92] A. Sengupta, *Representing Finite Groups: A Semisimple Approach*, Springer, 2012. ISBN 9781461412304.

[93] J.P. Serre, *Linear Representations of Finite Groups*, Graduate Texts in Mathematics, Springer-Verlag, New York, 1977. ISBN 0-387-90190-6.

[94] D. Shale, W.F. Stinespring, States of the Clifford algebra, *Annals of Math*, **80** (1964), 365-381.

[95] R.P. Stanley, (1999), *Enumerative Combinatorics. Vol. 2*, Cambridge Studies in Advanced Mathematics, **62**, Cambridge University Press, 1999. ISBN 978-0-521-56069-6.

[96] G.S. Staples. Clifford-algebraic random walks on the hypercube, *Advances in Applied Clifford Algebras*, **15** (2005), 213-232.

[97] G.S. Staples. Graph-theoretic approach to stochastic integrals with Clifford algebras, *J. Theor. Prob.*, **20** 257-274, (2007).

[98] G.S. Staples, A new adjacency matrix for finite graphs, *Advances in Applied Clifford Algebras*, **18** (2008), 979-991.

[99] G.S. Staples, Kravchuk polynomials and induced/reduced operators on Clifford algebras, *Complex Analysis and Operator Theory*, **9** (2015), 445-478. `http://dx.doi.org/10.1007/s11785-014-0377-z`.

[100] G.S. Staples, Hamiltonian cycle enumeration via fermion-zeon convolution, International Journal of Theoretical Physics, **56** (2017), 3923-3934. `http://dx.doi.org/10.1007/s10773-017-3381-z`.

[101] G.S. Staples. Differential calculus of zeon functions, *Advances in Applied Clifford Algebras*, (2019) 29:25.

[102] G.S. Staples. CliffMath: Clifford algebra computations in Mathematica, 2008-2018. `http://www.siue.edu/~sstaple/index_files/research.html`. [Accessed Feb. 13, 2019.]

[103] G.S. Staples, T. Stellhorn. Zeons, orthozeons, and graph colorings, *Advances in Applied Clifford Algebras*, **27** (2017), 1825-1845. `http://dx.doi.org/10.1007/s00006-016-0732-4`

[104] G.S. Staples, A. Weygandt. Elementary functions and factorizations of zeons, *Advances in Applied Clifford Algebras*, (2018) 28:12.

[105] G.S. Staples, D. Wylie, Clifford algebra decompositions of conformal orthogonal group elements, *Clifford Analysis, Clifford Algebras and Their Applications*, **4** (2015), 223-240.

[106] G. Szëgo, *Orthogonal Polynomials*, Amer. Math. Soc., Providence, 1955.

[107] A. Tucker, *Applied Combinatorics*, Fifth Edition, John Wiley & Sons, Hoboken, 2007.

[108] Y. Vizel, G. Weissenbacher, S. Malik. Boolean satisfiability solvers and their applications in model checking, *Proc. of IEEE*, **103** (2015), 2021-2035. `https://doi.org/10.1109/JPROC.2015.2455034`

[109] S. Warner, *Modern Algebra*, Dover Publications, 1990.

[110] E.W. Weisstein, "Stirling Number of the Second Kind." From MathWorld-A Wolfram Web Resource. `http://mathworld.wolfram.com/StirlingNumberoftheSecondKind.html` [Accessed April 3, 2019.]

[111] D. West, Revolving Door (Middle Levels) Conjecture: Open Problems-Graph Theory and Combinatorics. `http://www.math.uiuc.edu/~west/openp/revolving.html`[Accessed April 3, 2019.]

[112] D. West, *Introduction to Graph Theory, Second Ed.*, Prentice Hall, Upper Saddle River, 2001.

[113] R. Wilson, *Introduction to Graph Theory 3rd Ed.*, Longman Group Limited, 1985.

Index

www.ingramcontent.com/pod-product-compliance
Lightning Source LLC
Chambersburg PA
CBHW050537190326
41458CB00007B/1811
* 9 7 8 9 8 1 1 2 0 2 5 7 5 *